Hurricane Risk

Volume 2

Series Editor
Jennifer M. Collins, School of Geosciences, NES 107, University of South Florida, Tampa, FL, USA

Hurricanes are one of nature's most destructive agents and widespread interest surrounds the possibility that future hurricanes will increase in both frequency and intensity, causing devastation to vulnerable communities effecting their way of life and livelihoods. The Risk Prediction Initiative has organized a unique series of conferences bringing a new focus to the subject of impacts and risks from hurricanes, and how they may be changing under evolving climate regimes. The goals are to advance the understanding of risk, rather than solely the meteorology of storms and the observation and modelling of the climate.

Jennifer M. Collins • James M. Done
Editors

Hurricane Risk in a Changing Climate

A Moody's
Analytics
Company

Editors
Jennifer M. Collins
School of Geosciences
University of South Florida
Tampa, FL, USA

James M. Done
National Center for Atmospheric Research
Boulder, CO, USA

ISSN 2662-3064　　　　　　　ISSN 2662-3072　(electronic)
Hurricane Risk
ISBN 978-3-031-08567-3　　　ISBN 978-3-031-08568-0　(eBook)
https://doi.org/10.1007/978-3-031-08568-0

© The Editor(s) (if applicable) and The Author(s), under exclusive license to Springer Nature Switzerland AG 2022
8 chapters are licensed under the terms of the Creative Commons Attribution 4.0 International License (http://creativecommons.org/licenses/by/4.0/). For further details see licence information in the chapters.
This work is subject to copyright. All rights are solely and exclusively licensed by the Publisher, whether the whole or part of the material is concerned, specifically the rights of translation, reprinting, reuse of illustrations, recitation, broadcasting, reproduction on microfilms or in any other physical way, and transmission or information storage and retrieval, electronic adaptation, computer software, or by similar or dissimilar methodology now known or hereafter developed.
The use of general descriptive names, registered names, trademarks, service marks, etc. in this publication does not imply, even in the absence of a specific statement, that such names are exempt from the relevant protective laws and regulations and therefore free for general use.
The publisher, the authors, and the editors are safe to assume that the advice and information in this book are believed to be true and accurate at the date of publication. Neither the publisher nor the authors or the editors give a warranty, expressed or implied, with respect to the material contained herein or for any errors or omissions that may have been made. The publisher remains neutral with regard to jurisdictional claims in published maps and institutional affiliations.

This Springer imprint is published by the registered company Springer Nature Switzerland AG
The registered company address is: Gewerbestrasse 11, 6330 Cham, Switzerland

Foreword

Tempests in a Greenhouse: Assessing Coastal Risk in a Changing Climate

Among the greatest threats posed by climate change is the risk to hundreds of millions of coastal inhabitants from both rising seas and more intense hurricanes. This heightened societal exposure has profound implications for people and communities, various stakeholders and businesses, and the stability and resilience of our built and natural environment. In this important edited volume by Jennifer M. Collins and James Done, key experts in the field provide a blueprint for assessing and managing the coming risk through an assessment of the collective evidence from models and observations, and insights from a variety of case studies and approaches, with a range of perspectives that reflects current uncertainties about projections and impacts.

It's worth taking stock of what we already know about climate change and coastal risk. Even in the absence of changing tropical cyclone characteristics, for example, sea level rise means worse storm surges. The 14-foot storm surge at Battery Park from 2012 Superstorm Sandy, for example, contained the better part of a foot of global sea level rise. That meant 25 more square miles of flooding and billions of dollars in additional damage.

Tropical cyclones, moreover, are reaching higher peak intensities, driven by the additional energy that comes from increased evaporation from warmer oceans, and the latent heating of the atmosphere that comes when that rising moisture condenses into clouds and rainfall. That means more direct wind damage from landfalling storms, but, all else being equal, greater intensities come with bigger storm surges, and more coastal flooding.

The additional moisture in the storms leads to greater inland rainfall rates and flooding. That means more catastrophic "compound flooding" events where storm surge backs up estuaries as flooding rains continue to fall. The two worst flooding events in US history—the 2017 landfall of Hurricane Harvey in Texas and the 2018 landfall of Hurricane Florence in the Carolinas—happened within the past 5 years. That's not a coincidence.

We're now dealing with more dangerous and damaging storms. Some might call these natural disasters, but how "natural" are they, really? The scientific consensus is that we wouldn't be seeing these stronger, wetter storms if not for the human factor of carbon pollution and a warming climate. We should probably be calling these increasingly dangerous storms "unnatural disasters."

When it comes to policy and planning, there's also much we know. We know enough, for example, to justify substantial mitigation, that is, taking actions to ramp down carbon emissions and stop the continued warming of the planet that is fueling the growing coastal threat. And we know enough to justify adaptation—taking measures to instill greater coastal resilience in the face of both the heightened threat that already exists and the additional threat we are committed to in the form of continued sea level rise.

But there are many gaps in our knowledge and understanding that remain in both the scientific and policy realm. Will we see more landfalling hurricanes or not? If so, where? Are tropical islands at particular risk? Think Puerto Rico and the deadly landfall of category 5 Maria in 2017.

Are there better ways to define hurricane risk that would aid emergency planners? For example, are there metrics of damage we can define that incorporate compound impacts of wind and flood damage? How are damages and insured losses likely to change in the future? There are uncertainties in the climate model projections and complications that arise from the differences in how flood and wind damage is insured that create nontrivial challenges here. How can "counterfactual" analysis (consideration of what damages might have resulted if the details of a particular event had been different in various ways) be used by insurers to better assess risk? And last but not least, how do socioeconomic and demographic factors influence exposure and risk and how we go about reducing it?

The value of this book is its effort to identify the gaps, fill them in where possible, and, perhaps most importantly, provide a rigorous, scientifically based framework for continued updating and refining of our methods for assessing and reducing coastal risk from tropical cyclones. It is a must-have resource for scientists, insurers, reinsurers, real estate managers, city planners, emergency planners, policymakers, and anyone connected in one way or another to matters involving coastal risk.

Department of Meteorology and Atmospheric Science, Penn State University, PA, USA	Michael E. Mann

Preface

This book provides a source reference for both risk managers and climate scientists for topics on the interface between tropical cyclones, climate, and risk. Critical questions of hurricane risk and climate change are laid out in the first two chapters. Chapter 1 presents an essential and foundational perspective on the role of the insurance sector in determining societal resilience to hurricanes in a changing climate. It outlines the requirements for incorporating climate change into risk management strategies, and the opportunities this presents for collaborative work between the scientific community and the private sector. Chapter 2 then asks us to think more deeply about what is meant by risk. It reminds us of the fundamental principles, and adds novel interpretations of the characteristics of risk. These principals provide a foundation from which to understand the assessment and management of risk.

Hurricane risk is typically focused on the subset of storms that make landfall. Chapter 3 is the first in a sequence of chapters exploring the hazard and its impact in exposed coastal regions. It explores the controls on the inland behavior of tropical cyclones and their regional variations, focusing on the duration of damaging winds and their inland extent. Chapter 4 then utilizes a combination of extreme value theory and non-parametric regression methods to assess the impact of different climate modes on the occurrences of landfalling hurricanes along the US coastline. By breaking the shoreline into small segments, this chapter also reveals the spatial variation of landfall frequency during different climate conditions, which is critical to mitigation planning by local agencies. Chapter 5 draws attention to landfall events with extreme compound impacts (joint wind and precipitation). The estimated return period of these events could benefit the emergency planning not only from the intensity of the storm but a more direct measure of potential risks. Chapters 6 and 7 revolve around the threat from tropical cyclones on small tropical islands. Tropical islands are particularly vulnerable due to the exposure of frequent tropical cyclone occurrences and the lack of resources to mitigate risks. The destructive wind risk of small islands on a global scale is investigated both spatially and temporarily in Chap. 6, while Chap. 7 proposes a simple, open-source model for assessing tropical cyclone wind risk in Bermuda. Besides being free and easily accessible, less computational power and data input required by the model could serve as a quick assessment tool for jurisdictions with limited resource availability.

Chapter 8 combines climate change science, hazard risk models, and loss models to explore future losses. How the relative contributions to loss from wind and water may change because damage from flood is not typically covered in the residential market is a key concern for private insurers. This study addresses the concern by considering how, under an extreme climate scenario, climate change may impact hurricane frequency and damage. Chapters 9 and 10 explore the value of the counterfactual approach for risk assessment. Downward counterfactual analysis quantitatively estimates how our observed history could have been worse. Insurers can increase risk awareness, stress-test risk management frameworks, and inform decision-making by operationalizing downward counterfactuals. Chapter 9 is particularly relevant considering climate change and provides a methodology for insurers to operationalize downward counterfactuals using tropical cyclone catastrophe models. Considering three recent major hurricanes that were near misses for Miami, the authors reveal downward counterfactuals that produce insured losses many times greater than what transpired. The potential value of objective counterfactual analysis, as methods evolve, is highlighted again in Chap. 10 as the authors show, through the use of counterfactual analysis, the magnitude of volatility and spatial variability in hurricane landfalls in single cities and wider regions along the US coastline.

The final four chapters focus on policy-relevant understanding of hurricane vulnerability, risk-related decision-making, and/or building design. Chapter 11 sheds light on multiple methods that have been used to assess and quantify TC vulnerability with a focus on open-source methods. The chapter also discusses structural, economic, and social (or demographic) vulnerability approaches. Chapter 12 explores factors that influence the decision to obtain flood or wind insurance coverage (behavioral, personal, and socio-demographic) to examine how decision-making processes differ when the threat of a hurricane is looming, compared to a situation when no hurricane is forecasted in the near future. This chapter also demonstrates the complexity of property insurance in the USA and shows the difference in the determinants of insurance coverage purchase across various policyholder types. Chapter 13 examines the role of social networks on preparedness and evacuation decision-making, with a focus on residents in public housing. Studies such as this can assist public housing authority leaders and those in local emergency management to consider policies and practices to promote the use of strong social connections in disaster planning and evacuation decision-making. The final chapter, Chap. 14, considers designing safe, affordable hurricane-resistant housing and discusses how we can adapt our dwellings to our climate crisis so that people of all means can live safely and securely in the years ahead.

Tampa, FL, USA Jennifer M. Collins
Boulder, CO, USA James Done
Tampa, FL, USA Yi-Jie Zhu

Acknowledgments

The authors would like to thank the expert reviewers for their time and careful review of the chapters. The following reviewers reviewed one of the chapters (most anonymously at the time of review): Terri Adams, Howard University; Steven G. Bowen, Aon; David N. Bresch, ETH Zurich; Jeffrey Czajkowski, National Association of Insurance Commissioners; Meri Davlasheridze, Texas A&M University at Galveston; Kelsey Ellis, University of Tennessee; Stanley Goldenberg, NOAA-Atlantic Oceanographic and Meteorological Laboratory-Hurricane Research Division; Mona Hemmati, Columbia University; Kelly Hereid, Liberty Mutual; Greg Holland, NCAR; Jan Kleinn, Kleinn Risk Management/WSL Institute for Snow and Avalanche Research SLF; Yolanda C. Lin, University of New Mexico; Thomas Mortlock, Macquarie University/Aon; Siddharth Narayan, East Carolina University; Paul Northrop, University College London; Tom Philp, Maximum Information; Christa L. Remington, University of South Florida; Dail Rowe, RenaissanceRe Risk Sciences; Geoffrey Saville, Willis Towers Watson; Daniel Smith, James Cook University; Suz Tolwinski-Ward, AIR Worldwide; and Gordon Woo, Risk Management Solutions (RMS). In addition, we thank additional reviewers who remain anonymous. The authors are grateful for the assistance of Yi-Jie Zhu, University of South Florida (USF) with this book, particularly with communication, some logistical aspects, and co-writing the preface. We also appreciate the suggestions given by Danya Awshah. We would like to acknowledge the following people who have co-organized the Symposium on Hurricane Risk in a Changing Climate with Jennifer M. Collins: Jan Kleinn, Kleinn Risk Management/WSL Institute for Snow and Avalanche Research SLF; Angela Burnett, Government of The Virgin Islands; Kerry Emanuel, MIT; Greg Holland, NCAR; Adam Sobel, Columbia University; Yi-Jie Zhu, University of South Florida; Mark Guishard, Bermuda Weather Service; and Jim Neumann, Industrial Economics for their assistance on the earlier conference plans prior to COVID-19. This Symposium on Hurricane Risk in a Changing Climate was the inspiration for this book, and we likewise acknowledge our symposium sponsors: RMS, MS Amlin, the American Association of Geographers, USF, and the School of Geosciences (USF). We are also indebted to Dr. Michael Mann for writing the foreword. His leadership in climate change science and its communication provides a valuable perspective from which to set the scene for this book. Finally, the authors deeply appreciate the productive collaboration with the professionals at Springer, particularly Margaret Deignan, Joseph Daniel, and the copy-editing team.

Contents

1. Hurricane Risk Management Strategies for Insurers in a Changing Climate 1
   Kelly A. Hereid

2. Characteristics of Risk 25
   Jan Kleinn, Dörte Aller, and Matthias Oplatka

3. The Response of Hurricane Inland Penetration to the Nearshore Translation Speed 43
   Yi-Jie Zhu and Jennifer M. Collins

4. Estimating North Atlantic Hurricane Landfall Counts and Intensities in a Non-stationary Climate 57
   Jo Kaczmarska, Hugo C. Winter, Florian Arfeuille, and Enrica Bellone

5. Analysis of the Future Change in Frequency of Tropical Cyclone-Related Impacts Due to Compound Extreme Events 87
   Patrick A. Harr, Antoni Jordi, and Luke Madaus

6. Current and Future Tropical Cyclone Wind Risk in the Small Island Developing States 121
   Nadia Bloemendaal and E. E. Koks

7. Development of a Simple, Open-Source Hurricane Wind Risk Model for Bermuda with a Sensitivity Test on Decadal Variability 143
   Pinelopi Loizou, Mark Guishard, Kevin Mayall, Pier Luigi Vidale, Kevin I. Hodges, and Silke Dierer

8   Climate Change Impacts to Hurricane-Induced Wind
    and Storm Surge Losses for Three Major Metropolitan
    Regions in the U.S. 161
    Peter J. Sousounis, Roger Grenier, Jonathan Schneyer,
    and Dan Raizman

9   Downward Counterfactual Analysis in Insurance
    Tropical Cyclone Models: A Miami Case Study 207
    Cameron J. Rye and Jessica A. Boyd

10 Identifying Limitations when Deriving Probabilistic
    Views of North Atlantic Hurricane Hazard
    from Counterfactual Ensemble NWP Re-forecasts 233
    Tom J. Philp, Adrian J. Champion, Kevin I. Hodges, Catherine Pigott,
    Andrew MacFarlane, George Wragg, and Steve Zhao

11 Estimating Tropical Cyclone Vulnerability: A Review
    of Different Open-Source Approaches 255
    Katy M. Wilson, Jane W. Baldwin, and Rachel M. Young

12 Assessing the Drivers of Intrinsically Complex Hurricane
    Insurance Purchases: Lessons Learned from Survey
    Data in Florida 283
    Juan Zhang, Jeffrey Czajkowski, W. J. Wouter Botzen,
    Peter J. Robinson, and Max Tesselaar

13 Exploring the Role of Social Networks in Hurricane
    Preparedness Planning: A Study of Public Housing Residents 323
    Robin Ersing, Beverly Ward, and Jennifer M. Collins

14 Geohome: Resilient Housing for Climate Hazard Mitigation 341
    George Elvin

Contributors

Dörte Aller Aller Risk Management, Zurich, Switzerland

Florian Arfeuille Model Development, Risk Management Solutions Ltd., London, UK

Jane W. Baldwin University of California Irvine, Irvine, CA, USA
Lamont-Doherty Earth Observatory, Columbia University, Palisades, NY, USA

Enrica Bellone Model Development, Risk Management Solutions Ltd., London, UK

Nadia Bloemendaal Institute for Environmental Studies, Vrije Universiteit, Amsterdam, The Netherlands
Allianz SE, Munich, Germany

W. J. Wouter Botzen Department of Environmental Economics, Institute for Environmental Studies, VU University Amsterdam, Amsterdam, The Netherlands

Jessica A. Boyd MS Amlin, The Leadenhall Building, London, UK

Adrian J. Champion University of Exeter, Exeter, UK

Jennifer M. Collins School of Geosciences, University of South Florida, Tampa, FL, USA

Jeffrey Czajkowski Center for Insurance Policy and Research, National Association of Insurance Commissioners, Kansas City, MO, USA

Silke Dierer Axis Capital, Zurich, Switzerland

George Elvin College of Design, North Carolina State University, Raleigh, NC, USA

Robin Ersing University of South Florida, School of Public Affairs, Tampa, FL, USA

Roger Grenier AIR Worldwide, Boston, MA, USA

Mark Guishard Bermuda Weather Service, Bermuda Airport Authority, St. George's, Bermuda

Patrick A. Harr Jupiter Intelligence, San Mateo, CA, USA

Kelly A. Hereid Liberty Mutual Insurance, Corporate Enterprise Risk Management, Oakland, CA, USA

Kevin I. Hodges National Centre for Atmospheric Science, Department of Meteorology, University of Reading, Reading, UK

Antoni Jordi Jupiter Intelligence, San Mateo, CA, USA

Jo Kaczmarska Model Development, Risk Management Solutions Ltd., London, UK

Jan Kleinn Kleinn Risk Management, Zurich, Switzerland
WSL Institute for Snow and Avalanche Research SLF, Davos, Switzerland

Elco E. Koks Institute for Environmental Studies, Vrije Universiteit, Amsterdam, The Netherlands
Oxford Programme for Sustainable Infrastructure Systems, Environmental Change Institute, University of Oxford, Oxford, UK

Pinelopi Loizou Department of Meteorology, University of Reading, Reading, UK

Andrew MacFarlane AXA XL, Hamilton, Bermuda

Luke Madaus Jupiter Intelligence, San Mateo, CA, USA

Kevin Mayall Locus Ltd, Hamilton, Bermuda

Matthias Oplatka Office of Waste, Water, Energy and Air (AWEL) of the Canton of Zurich, Zürich, Switzerland

Tom J. Philp Maximum Information, London, UK

Catherine Pigott AXA XL, London, UK

Dan Raizman AIR Worldwide, Boston, MA, USA

Peter J. Robinson Department of Environmental Economics, Institute for Environmental Studies, VU University Amsterdam, Amsterdam, The Netherlands

Cameron J. Rye MS Amlin, The Leadenhall Building, London, UK

Jonathan Schneyer Willis Re, Boston, MA, USA

Peter J. Sousounis AIR Worldwide, Boston, MA, USA

Max Tesselaar Department of Environmental Economics, Institute for Environmental Studies, VU University Amsterdam, Amsterdam, The Netherlands

Pier Luigi Vidale National Centre for Atmospheric Science, Department of Meteorology, University of Reading, Reading, UK

Beverly Ward BG Ward and Associates, LLC, Tampa, FL, USA

Katy M. Wilson Graduate School of Arts and Sciences, Columbia University, New York, NY, USA

Hugo C. Winter Model Development, Risk Management Solutions Ltd., London, UK

George Wragg AXA XL, London, UK

Rachel M. Young School of Public and International Affairs, Princeton University, Princeton, NJ, USA

Juan Zhang College of Business, Eastern Kentucky University, Richmond, KY, USA

Steve Zhao AXA XL, London, UK

Yi-Jie Zhu School of Geosciences, University of South Florida, Tampa, FL, USA

Chapter 1
Hurricane Risk Management Strategies for Insurers in a Changing Climate

Kelly A. Hereid

Abstract The insurance sector is among the earliest private sector industries to see substantial losses associated with climate change, and the risk signals embedded in insurance premiums could be a key driver for how climate adaptation decisions are made by individuals and communities. Therefore, risk assessments made in the insurance sector are critical for determining how the built environment responds to and rebuilds from natural disasters, and how society prices in the changes in disaster risk associated with climate change. Catastrophe models are the key tool the insurance community can use to bridge the gap between climate models and the extreme events that are the primary way in which society experiences climate change. The following discussion provides concrete suggestions for extending established catastrophe risk management practices to include emerging climate risks. Incorporating climate change into hurricane risk management requires risk managers to 1. Explore catastrophe and climate model sensitivity to inputs in the built environment, 2. Divide hurricane risk by subperil so risk management strategies can be tailored to variable levels of uncertainty, and 3. Translate climate impacts into metrics that tie directly to decision-making and business outcomes. Each of these steps also highlights opportunities for collaboration between the insurance sector and scientific community. The general strategy of assessing sensitivity to model inputs, tailoring risk management strategies to the level of uncertainty in the hazard, and producing outputs that are useful for end users is broadly applicable for climate data services across a wide variety of sectors.

Keywords Insurance · Climate · Catastrophe Modeling · Hurricane · Risk management

K. A. Hereid (✉)
Liberty Mutual Insurance, Corporate Enterprise Risk Management, Oakland, CA, USA
e-mail: Kelly.Hereid@libertymutual.com

© The Author(s), under exclusive license to Springer Nature Switzerland AG 2022
J. M. Collins, J. M. Done (eds.), *Hurricane Risk in a Changing Climate*, Hurricane Risk 2, https://doi.org/10.1007/978-3-031-08568-0_1

1.1 Catastrophe Models: A Key Tool for Linking Climate Change to Risk

Insurance industry disaster risk is generally managed through tools known as catastrophe models. Catastrophe models were widely adopted in the aftermath of Hurricane Andrew making landfall in Florida in 1992, and further solidified following the Northridge earthquake in California in 1994. Because the most extreme events are not frequent enough to apply standard actuarial techniques with a basis in statistical theory of large sample sizes, these models offer a powerful tool to fill out gaps in historical coverage for rare and extreme events. Historical records of disasters are heterogeneous in time and space, and even lengthy historical records are typically insufficient to characterize events in the 100–250 year return period range commonly assessed in the insurance industry. Even where historical records are available, they might have impacted a building stock that was materially different from what exists today, with many fewer buildings in areas at high risk, paired with older building codes that would not reflect a current view of damage potential. Catastrophe models allow insurers to assess the risk of plausible events that may not be in the historical record and capture the effects of changes in exposed building stock over time.

Catastrophe models vary somewhat in form, but generally come with the same four basic components. Catastrophe models take in exposure as an input, in the form of a database representing the locations that an insurer covers in its policies, describing their location, occupancy (such as single-family residential, or a commercial office building), construction type (wood frame, steel), age, and other details of how the structure was built. The physical environment is modeled through a hazard module, which is an event set of thousands of years of hypothetical events, including tropical cyclones, earthquakes, floods, wildfires, etc. The hazard and exposure feed into a vulnerability module, which builds a relationship between the level of hazard and type of building – for example, the percent damage that a single-family wood frame home will experience at a given wind speed in a hurricane. Vulnerability typically reflects a combination of engineering expertise and tuning to historical losses. Lastly, the amount of damage produced by the vulnerability component is fed into a financial module, which applies insurance policy terms like deductibles and limits.

Catastrophe models do not attempt to capture the same types of events as a typical climate model. While climate models are best at producing average projected changes in large scale patterns over long time periods, catastrophe models are designed to capture low frequency/high severity hazards in the tail of a probability distribution in single year increments at high spatial resolution. To capture extreme tails of the hazard distribution, catastrophe models blend physical modeling with statistical sampling of hazard parameters derived from the historical record, such as number of storms, genesis locations, or storm trajectory by latitude/longitude (Hall and Jewson 2007). Monte Carlo simulation of storm characteristics is computationally cheap compared with the high resolution physical climate models needed to represent extreme events like hurricanes, enabling catastrophe models to produce tens or hundreds of thousands of years of simulated disasters to realistically reflect rare but high impact events.

Although insurers have long been broadly aware of climate risk from an enterprise risk and regulatory standpoint, the catastrophe modeling tools that are most commonly used in daily decision-making and that feed into risk appetite reflect a snapshot of risk at a particular time. Since catastrophe models are drawn from a statistical distribution built from historical data, they are inherently designed for short-term climate impacts. At present, they do not explicitly incorporate most forms of forward-looking climate risk.

The statistical distributions of events used to build a catastrophe model become less and less representative of potential risk with climate projections that are further in the future, where events may happen that are outside the distribution of historical experience. However, climate model skill in attributing climate change impacts to extreme events tends to increase for most hazards further in the future, as external forcing continues to increase relative to internal natural variability. Therefore, leveraging these complementary tools in combination can help to highlight how climate change could impact extreme events, and how changes in extreme events might play out in societal impacts and economic losses.

Although catastrophe models were originally designed for the insurance sector, there are academic models that capture at least some components of the insurance industry catastrophe models. Hazard models are the most common, such as the Columbia HAZard Model (Lee et al. 2018) and Synthetic Tropical cyclOne geneRation Model (STORM) (Bloemendaal et al. 2020). However, the hazard and vulnerability components of a catastrophe model that are used to produce a loss are not fully independent. With two unknown variables for any given loss (hazard and vulnerability), the solution to produce that loss is not unique. Two models could produce the exact same loss for an event, with one arriving at that estimate through a higher assumption of hazard but a lower assumption of vulnerability, and the other producing a lower amount of hazard but higher vulnerability. Therefore, loss-producing catastrophe models need to be carefully ground-truthed using observational data on wind and storm surge footprints, as well as engineering studies that link particular wind speeds to a proportion of damage.

Tying the hazard and vulnerability modules together has historically limited the development of academic or public sector catastrophe models, as private sector companies have access to insurance claims data that are generally not publicly accessible, as well as combined scientific and engineering expertise that is difficult to replicate in the academic space. Ongoing work to develop more open source or publicly available catastrophe models could help to spread the risk insights that are accessible in the insurance industry to a broader set of end users in the public and private sectors. For example, catastrophe models only capture direct physical damage to the insurable built environment – they are not yet able to capture broader societal impacts such as transition risk, detailed supply chain impacts, or climate-driven migration or social upheaval, all valuable potential contributions from participants beyond the insurance industry.

For the purposes of this discussion, the risk analysis methodology will focus on North Atlantic hurricane risk, although much of what follows is applicable to other tropical cyclone basins as well. The hurricane hazards highlighted are not intended to be comprehensive, but rather representative of how hazards may interact with the built environment.

1.2 Three Recommendations for Risk Modelers

The following sections will break down three key considerations to advance meaningful and actionable climate risk analytics within the framework of insurance risk management.

1. Exposure: Explore sensitivity of climate-driven losses to inputs from the built environment.
2. Hazard: Quantify scientific confidence and impact by subperil.
3. Translating climate data for decision-makers.

Each of these considerations highlights opportunities for leveraging catastrophe models to capture economic and societal impacts of climate change on hurricane risk. However, each recommendation also has space for further improvement with better tools and advances in the basic science. Therefore, each section is followed by a "Research Opportunities" subheading, showing key data or analytical gaps that are an opportunity to improve quantification of the societal impacts of hurricanes in a changing climate.

1.2.1 Recommendation 1 – Exposure: Explore Sensitivity of Climate-Driven Losses to Inputs from the Built Environment

Climate hazards do not occur in a vacuum – they can only become disasters when they intersect with an impacted community. Catastrophe models are particularly adept at capturing the impact of climate hazards on the built environment, which may be subject to changes in risk that do not scale linearly with the change in hazard. However, this means that assessing climate impacts requires stringent data collection processes on exposure inputs, or the resulting modeled impacts will not be meaningful.

1.2.1.1 Modeling Water Risk in the Built Environment

Several of the most immediately changing hurricane subperils, such as storm surge and inland rainfall, are making hurricane damage wetter. This poses a particular challenge for risk modelers, since water perils require a much finer degree of detail on exposed properties to produce meaningful views of risk compared with more developed hazards like wind and earthquake. Wind risk, for example, is relatively forgiving of minor errors in modeling the location of a building, as the average risk from wind does not vary particularly dramatically at small spatial scales along the coast. Flood risk, however, may vary widely even within a single building parcel, and minor errors in location can easily cause properties to be modeled in the middle of retention basins or roadside ditches, leading to dramatic overestimates in risk. For this reason, modeling "climate risk" often requires substantial investments in improved data about exposed building locations in a book of business, long before examining any changes in physical hazard. Advances in geocoding and capturing important building characteristics like basements and building elevation can dramatically improve the realism of modeling hurricane-driven water risks like surge and inland rainfall.

Furthermore, at the scale of metropolitan areas, hurricane flood risk may not be distributed equitably. Redlining was a historical discriminatory practice that deemed communities populated by racial and ethnic minorities, and particularly Black communities, to be risky investments. Many communities that experienced redlining are disproportionately located in areas of higher flood hazard (Katz 2021). Redlining systematically reduced property values as well, since it constrained access to mortgages and related financing, limiting the tax base for investments in flood defenses like stormwater drainage infrastructure. After a disaster has occurred, homeowners are more likely to receive post-disaster funding to rebuild, both through federal and insurance assistance - this homeownership bias exacerbates unequal disaster impacts in communities with a higher proportion of renters (Fussell 2015). Climate-related water risks compound this historical inequality, potentially resulting in "climate gentrification" where only the wealthy have the resources to harden their homes, lobby for defensive infrastructure, and rebuild after major disasters. Investments in capturing socioeconomic data for use in physical risk models (Tedesco et al. 2021) can help to predict where such impacts are likely, highlighting communities that would benefit from additional resources for protective measures and climate adaptation.

1.2.1.2 Spatial Distribution of Hazard vs. Built Environment

Even in the absence of changes in event frequency, risk and losses may change if different areas are exposed to hurricane risk in the future. A prime example of where this may play out is through poleward migration. There is at least some suggestion that migration of storms toward higher latitudes may be occurring in the historical record in some basins (Kossin et al. 2014), which is most likely to be occurring beyond an extent explainable by natural variability in the North Pacific basin (Knutson et al. 2019). This pattern may be driven by a change in tropical cyclone genesis location (Daloz and Camargo 2018), which could theoretically lead to declines in risk at very low latitudes. However, mid-latitude metropolitan areas may see increases in risk whether poleward migration comes from migrating genesis locations or an expansion in the parts of the ocean with sea surface temperatures warm enough to support tropical cyclones.

Moving stronger storms over regions whose building codes were designed based on historical wind return periods can increase losses through higher building vulnerability, independent of any changes in storm frequency. Building codes and enforcement are not standardized and vary widely from state to state, but in general, more northerly states are less likely to design for high tropical cyclone wind risk. Even absent these building code considerations, exposing mid-latitude metropolitan areas like New York and Boston to hurricane risk, when these events have been historically rare, may substantially change the population that could be exposed to hurricane risk in the future.

1.2.1.3 Critical Thresholds

Even where hazards are smoothly varying over time, the intersection between hazard and the built environment may lead to sharp discontinuities in risk. Hurricane Sandy highlighted the cascading impacts that ensue when a hazard intersects with critical pieces of infrastructure, as when Sandy's storm surge entered the New York City subway system and took the storm's impact from damaging to catastrophic. Every metropolitan area has a point when damage becomes massively disruptive – whether it is public transit like New York, a critical piece of the power grid (also inundated in New York during Sandy, causing the blackout in lower Manhattan), or the point when a downtown center is inundated. For coastal cities in particular, storm surge sitting on top of sea level rise brings many pieces of critical infrastructure closer to the point of failure. Without explicit defensive measures, cities may find that infrastructure designed to be defended against a 1 in 100 or 1 in 250 year return period storm surge may actually be impacted by the 1 in 50 or 1 in 30 year storm surge in the future. Different pieces of infrastructure may have different levels of risk tolerance – for example, a municipality is likely to be more comfortable with a road being occasionally inundated compared with a nuclear plant. Mapping and quantifying the risk to critical facilities is a key component of preparing communities for future extreme events.

1.2.1.4 Exposure Research Opportunities

1. Publicly available exposure data

The insurance sector has access to a rich pool of data about the built environment, both through individual portfolios of business that they write, as well as "industry exposure" datasets produced by catastrophe modeling companies. However, this data has some notable limitations. Most importantly, it is expensive to build and maintain, so it is generally proprietary and expensive to license. Detailed data about the built environment is necessary for quantifying costs of increased hazard, but the granularity of public datasets is generally not sufficient to capture high gradient hazards like flood, and it lacks an ability to distinguish critical infrastructure. Investments in public data available to the research community would fill a critical gap in quantifying economic impacts of climate change.

Defining economic costs of extreme hazards like hurricanes is particularly critical in a policy environment where decisions about mitigation and adaptation pathways are at least partially based on the expected economic costs of climate change. Extreme hazards are a key driver of economic loss, but they are minimally captured in the economic models widely used in the financial sector to model transition risk, which could lead to underestimation of the long-term costs of climate change on society.

2. Tools to capture detailed exposure characteristics

Hurricane water subperils, as well as other high-resolution hazards like inland flood and wildfire, require highly detailed information about the built environment to produce meaningful model results. Investments in remote sensing and machine learning to better identify details like building footprints and first floor elevations are critical to meaningfully incorporate emerging climate hazards into the risk models used throughout the insurance sector. This is an active area of private sector investment, but it is also an area where academic researchers could contribute meaningfully to managing climate risk.

3. Economic effects of physical risk on exposure values and investments

Forward-looking views of physical risk require projections of future exposures, along with the usual forecasts of physical hazards. However, physical hazards may have impacts on property values, which can complicate and materially affect the expected cost of extreme events like hurricanes. Coastal flood is a key example of the complexity of assessing changes in future economic risks from hurricanes, with knock-on implications for both catastrophe and transition risk modeling.

Coastal flood risk is not only composed of storm surge from hurricanes. For a given location, sea level rise manifests first through changes in coastal flood event frequencies, beginning with extreme hurricane surge events, and eventually moving through routine high tide flooding to complete inundation. Sunny day or "nuisance" flooding, high-frequency flooding driven by high tides combined with sea level rise, does not fall within the traditional purview of hurricane catastrophe models, although it can be exacerbated by bypassing hurricanes. However, the mechanism of stacking high tide events on top of current and future sea level rise scenarios works in much the same way as storm surge in a catastrophe model. While often treated separately, storm surge and other types of coastal flood actually represent a continuum of coastal hazard, which may intersect in ways that could drive larger-scale climate risks.

While individual losses are small compared with storm surge, high frequency coastal flooding events may have systematic risk potential due to their impact on property values. Structures that experience routine flooding may be harder to sell, and ultimately lose value or grow in value more slowly than their lower-risk neighbors (Keys and Mulder 2020), as damages accumulate and insurance premiums rise from repeated claims. Since property values determine the amount of property taxes collected, this may limit a critical potential funding source for future coastal flood defenses. Coastal states that lack state income tax, like Florida and Texas, have fewer obvious local funding sources if coastal property tax bases were to be impacted by a local decline in property values.

Property value risk is an area where insurers may be impacted on both sides of their balance sheet – in physical hazards driven by their portfolios of insured risks, as well as economic hazards within their investments. Municipal bonds, investments in the mortgage market, and other investments tied to a clear physical location could potentially see areas of clash, where a large hurricane could hypothetically trigger losses both from direct physical damage, as well as longer-term slowing or reversal of investment returns concentrated in the same location. While investments are chosen with a certain level of risk appetite for an economic shock, climate-driven financial risk has a different timeline for recovery than traditional economic shocks – unlike a recession, sea level is not expected to retreat globally for centuries or more, so some assets may not be recoverable. However, with a few exceptions like those noted above, investments typically have limited information about the physical locations of their facilities, a critical data gap for incorporating hazard assessments for extreme events like hurricanes into an investment portfolio.

Despite its importance for societal impacts of hurricanes, the interaction between nuisance flooding and storm surges, particularly regarding socioeconomic risk, is comparatively understudied. For example, is a major storm surge event more likely to cause widespread mortgage defaults if the community has already been stressed by nuisance floods depressing property values? The interaction between high frequency and low frequency coastal flood events is also currently not generally incorporated into catastrophe modeling, suggesting a potential avenue for collaboration between the climate modeling, catastrophe modeling, and economic communities.

1.2.2 Recommendation 2 – Hazard: Quantify Scientific Confidence and Impact by Subperil

For a complex hazard like hurricanes, the question, "how does climate change affect hurricane risk?" is so broad as to be virtually meaningless. Within the framework of climate modeling, hurricanes are comparatively small scale and extreme events, so the science remains in flux for some aspects of the hazard. However, some characteristics of hurricanes have seen clear, immediate impacts in the current climate. Rather than attempting to roll hazards with wildly variable levels of uncertainty together, it is useful to break apart hurricanes into their associated subperils and apply different risk management strategies based on their corresponding scientific confidence and materiality of impact.

High-confidence perils, like increases in storm surge inundation, are explicitly included in catastrophe models in the current climate, and could be added in near-term (5–10 year) time steps that are relevant for underwriting, pricing, and portfolio planning. For lower-confidence perils, like hurricane frequency, an exploratory approach that examines how much the hazard would have to change to affect the business (rather than prescribed time-based scenarios) is more useful for planning purposes, and allows the organization to assess when and if action would be needed.

1.2.2.1 Water Subperils

Storm Surge

Storm surge is the clearest and most immediate impact of climate change on tropical cyclone risk. While the question of whether the component of the surge caused by the storm itself is changing remains open (Grinsted et al. 2012), sea level rise provides a higher platform for every incoming storm surge (Knutson et al. 2020), increasing both depth and the inland extent of the surge inundation footprint. The impact of sea level rise is compounded by the locations of major concentrations of exposure, which are often cities built on deltas that subside naturally, and in some cases whose subsidence is enhanced by factors like groundwater withdrawals (Nicholls et al. 2021).

Catastrophe models are built using sea levels that are close to those of the present day, so this climate-related trend is explicitly included in modeled loss results. However, forward-looking sea level rise scenarios, particularly projecting out into the medium term where portfolio optimization decisions are made, are in comparatively early stages of development, and are not consistently included in model results from many of the largest catastrophe model vendors.

At the time scales that are most important for decision-makers in the insurance sector (up to 2050 at the latest), uncertainties in sea level rise even among a wide range of emissions scenarios are comparatively small, ranging from median [likely] global mean sea level rise by 2046–2065 of 0.24 m [0.17–0.32 m] in RCP 2.6 to 0.30 m [0.22–0.38 m] in RCP 8.5 (*Intergovernmental Panel on Climate Change* 2014). While projecting global mean sea level changes down to the local scale required for catastrophe modeling is more challenging, the emergence of gridded products like that produced by NOAA in the US (Sweet 2017) makes it possible to produce reasonably localized sea level projections. Capturing this regional level of detail is critical for interpreting changes in storm surge risk over time, as individual cities do not experience global mean sea level change. For example, portions of the US East Coast are experiencing sea level rise that is 50% or more above the global mean, predominantly but not exclusively driven by post-glacial isostatic rebound (Piecuch et al. 2018).

Catastrophe model results highlight the impact that climate change can have on tropical cyclone subperils, even within the historical record. A study led by Lloyd's in the aftermath of Hurricane Sandy found that the amount of sea level rise since

1950 had increased ground up losses by 30% in the New York metropolitan area (Maynard et al. 2014). Similarly, this modeling framework can be used to assign economic losses to the anthropogenic portion of sea level rise from historical storm surge events (Strauss et al. 2021). Increases in loss are not limited to the increase in storm surge depth; higher storm surges also lead to greater inland extents of storm surge footprints, with the greatest increase in extent on low, flat coastlines. Consequently, loss increases may not be linear, but rather may have step functions as surge footprints are able to reach major concentrations of buildings or critical exposure that would have stayed dry with a lower baseline sea level. With adjustments for sea level rise, catastrophe models could be valuable tools to highlight these critical discontinuities in loss potential.

Rainfall-Induced Flooding

Hurricane rainfall is generally expected to intensify with climate change. The Clausius-Clapeyron equation shows a ~ 7% increase in the moisture that the atmosphere could potentially hold with every 1 °C temperature increase. Therefore, a warmer atmosphere itself provides a potential mechanism to enhance hurricane rainfall. Rainfall increases could also hypothetically be enhanced by stalling or other dynamical factors, as was the case for Hurricane Harvey over Houston and Hurricane Florence in the Carolinas (van Oldenborgh et al. 2017; Emanuel 2017a; Wang et al. 2018). Rain damage may not necessarily correlate well with wind damage, as major inland rainfall events have also been driven by tropical storms (Tropical Storm Allison, 2001 in Houston) and unnamed but near-tropical circulations (Baton Rouge flooding, 2016). This effect is enhanced along parts of the Gulf Coast by the "brown ocean" effect, allowing storms to maintain their structure longer over land (Andersen and Shepherd 2014).

Rainfall impacts may also be enhanced by their intersection with the built environment. For example, Hurricane Harvey produced the greatest amount of rainfall for a US tropical cyclone in records going back to the 1880s, with seven rainfall stations breaking the prior record held by Hurricane Hiki in Hawaii in 1950 (Blake and Zelinsky 2018). This could be thought of as particularly unlucky, to have such a historic event happen right over Houston, the fifth-largest metropolitan area in the United States (United States Census Bureau 2021). Alternatively, it could be interpreted that the record US tropical cyclone rainfall event and dramatic flood impacts happened in Harvey specifically *because* it happened to occur over such a large metropolitan area.

A variety of methods have been proposed to enhance rainfall over metropolitan areas, including surface roughness provided by buildings (Zhang et al. 2018) and local aerosol pollution (Pan et al. 2020). Similarly, impervious surfaces like buildings, roads, and parking lots can enhance flood risk by limiting avenues for rainfall to drain into the ground, forcing it to behave as overland flow instead and compounding the impact of increased rainfall on flood behavior (Sebastian et al. 2019).

As this Houston example shows, cities are in the position of being uniquely exposed to enhanced hurricane rainfall risk. Increased rainfall inputs are expected from warmer air and sea surface temperatures, with potential further enhancements specifically in tropical systems. The presence of a city may itself enhance local precipitation. And impervious cover prevents rainwater from absorbing into the ground, leading to overland flow into rivers and other drainage networks. All of this is happening where populations exposed to hurricane risk are the most concentrated. This puts cities at an uncomfortable confluence of compounding climate and exposure-driven hurricane flood hazard, so they would greatly benefit from more sophisticated risk modeling.

Although historically not captured explicitly in catastrophe models in the US, hurricane rainfall is a subperil that gained particular interest in the aftermath of Hurricane Harvey, and is now more broadly incorporated into hazard modeling. However, given the complexity of a peril that is changing via climate-driven hazard impacts, land use impacts, and whose loss history reflects a blend of public sector (mostly residential flood losses, in the National Flood Insurance Program) and private sector (larger commercial flood losses) datasets, this is an emerging hazard in the catastrophe modeling community that will likely continue to evolve as climate impacts and attribution research solidifies.

Compound Flood

Many major coastal metropolitan areas are located along estuaries or deltas that are susceptible to both riverine flood and coastal flood risk. A tropical cyclone can drive both risks occurring at the same time, with storm surge running up waterways and preventing river waters from draining as quickly, paired with intense inland rainfall enhancing river flood at the same time. Therefore, although the return period of a flood height can be calculated independently for coastal and river floods, best practice would account for the strong correlation between these perils in coastal areas that are exposed to storm surges. Furthermore, compounding riverine and coastal flood events will be exacerbated by sea level rise (Moftakhari et al. 2017). The science of compound hazards is comparatively immature, and it presents an opportunity to quantify risk in areas where human exposure and impacts are particularly high. Capturing this correlated hazard footprint is an area that is under development at many catastrophe modeling vendors.

1.2.2.2 Wind Subperil

Wind Frequency

Changes in hurricane frequency and intensity due to climate change are a key question posed by regulatory bodies, which partially reflects the fact that this is a problem which catastrophe models are ideally suited to address. The hazard portion of a catastrophe model is a database of hypothetical events and their associated frequencies, which may be constant or vary by event, to produce a hazard distribution that matches the historical record. Manually adjusting the frequencies of these events, or resimulating event catalogs by resampling, is a comparatively straightforward exercise that allows for exploring a range of frequency and intensity change scenarios.

Unfortunately, frequency is one of the areas of hurricane science where immediate climate-related impacts are least clear. Even at a global scale, there is credible disagreement even on the direction of potential frequency changes, much less the magnitude of those changes (Knutson et al. 2020; Lee et al. 2020). In the insurance industry, tropical cyclone risk is not managed at the global scale, but rather based on landfall frequencies at the country scale. In the case of the US, landfall frequencies are further downscaled to a regional scale, with loss tolerances designed to maintain capital adequacy around specific regions like the US Gulf or Northeast. Projecting future changes in frequency of hurricane landfalls at the scale of a few states at sub-decadal time scales lies well beyond the current state of climate science.

Wind Intensity

Much like frequency changes, catastrophe models are suitable tools to assess the impacts in changes in intensity. Across a large stochastic catalog of hypothetical events, intensity changes can be alternatively interpreted as a change in frequency. For example, a 5% increase in the intensity of a 100mph hurricane could instead be thought of as an increase in the frequency of 105mph events, shifting the entire probability distribution of hurricane intensities towards stronger events.

Changes in intensity are less useful without an understanding of how that change is distributed across the spectrum of storms. Intensity changes are often reported as a mean change across all storms, either globally or within individual basins (Knutson et al. 2020). But an increase in intensity concentrated in low category storms would have a very different loss distribution than a similar change among stronger storms – building codes along the coast typically are designed to protect at least against weaker storms, but lower category storms are much more frequent than more intense ones. Catastrophe models could assess competing influences on loss with a change in intensity distribution, but a flat change across all storms may be less applicable to risk management.

Wind Risk Management Strategies

The limited forward-looking skill in hurricane frequency projections and comparative lack of detail in intensity projections pose a challenge to risk modelers. Even when the scientific community can make reasonable frequency projections for global tropical cyclone activity, that skill typically declines when extrapolated to a single basin. And importantly, basin activity does not cause property damage – landfall statistics are the ultimate measure of real wind risk to an insured portfolio. However, these limitations do not mean that insurers lack tools to manage future hurricane wind perils.

A key consideration for insurers involves assessing which parts of the historical record should be included when determining *current* frequency distributions. In the North Atlantic hurricane basin, much of this debate centers on the application of a correction to account for Atlantic multidecadal variability, a pattern which strongly influences hurricane activity via increased sea surface temperatures, decreased sea level pressure, and reduced vertical wind shear in the tropical Atlantic (Goldenberg et al. 2001), and which is correlated with greatly increased frequency of major hurricanes (Klotzbach et al. 2015), but may not actually represent a physical oscillation with any predictive skill (Waite et al. 2020; Mann et al. 2021). Catastrophe models have the option to represent hurricane frequencies associated with the positive phase of the AMO, such as by producing medium-term forecasts or conditioning frequencies on observational periods in the same AMO phase rather than the full historical record, but usage of these views would be complicated if the Atlantic multidecadal variability lacks forward-looking skill.

Accounting for the degree of historical human influence on hurricane frequency is controversial as well. Sea surface temperatures are unlikely to return to their preindustrial state any time in the near future, but does removing older storms limit the view of internal variability? In the Atlantic, the 1960s–1980s experienced a low number of hurricanes, but was this driven by anthropogenic aerosol pollution (Murakami et al. 2020) that is not expected to return? Even if reasonable projections of basin-scale event frequency can be made, can we disentangle the influence of sub-basin processes like wind shear patterns, which may materially modify landfalls (Kossin 2017) in ways that may not stay consistent with future climate change (Ting et al. 2019)? Quantifying the climate change impacts that have already occurred through the last century is key for understanding what baseline to use for current and future wind risk.

Although a consensus on future changes in overall frequency is lacking, modeling is more consistent in predictions for the most intense storms – generally producing an increase in the proportion of the most intense storms (Category 4–5 storms or Category 3–5, depending on the study), with an increase in absolute number in all but those models where an overwhelming decline in total storm number overwhelms the increasing proportion of stronger storms (Knutson et al. 2020). Furthermore, there is evidence that this trend is appearing even in recent historical records in the Atlantic (Kossin et al. 2020), which may reflect a blend of natural and climate variability, but places the change within the time frame that is most relevant for insurance analytics.

While major hurricanes make up a small fraction of total storms, they contribute disproportionately to damage and loss. Estimates that account for inflation, population, and growth of wealth arrive at roughly 80% of all US hurricane losses coming from storms of Category 3 or above, even though they only account for about a third of US landfalls (much of the remainder comes from surge-heavy storms like Hurricanes Sandy and Ike). Similarly, the most intense Category 4 and above storms, which produce ~10% of US landfalls, are responsible for roughly half of all US hurricane losses (Weinkle et al. 2018). Furthermore, as the US population continues to concentrate in major coastal cities that are exposed to wind risk, losses will be compounded irrespective of any underlying climate trends. For insurers concerned about capital adequacy and near-term climate impacts, it is clear that climate sensitivity testing should focus on changes in high category hurricanes, where the science is comparatively clearer, and where changes in frequency can drive the largest potential impacts in losses.

The impact of large losses from major hurricanes on the US property insurance market is clearly visible in the Rate on Line index maintained by the broker Guy Carpenter (Guy Carpenter 2020). After a substantial spike in market prices for US property risk following the extremely active 2004–05 Atlantic hurricane seasons, there was a long decline that coincided with the US major hurricane drought (Hall and Hereid 2015). Market prices did not consistently reverse their trajectory until after the US major hurricane drought was broken in the 2017 Atlantic hurricane season. These market fluctuations likely have a variety of causes, including a recency bias on the part of decision-makers. An increase in available capital competing for premium dollars, due to a lack of events requiring claims payments, was particularly important in the reinsurance and insurance-linked securities markets. There are disputes about how to define a major hurricane drought (Hart et al. 2016), and pressure may be a more consistent representation of a storm's damage potential than wind speed (Klotzbach et al. 2020), but however a major storm is defined, the comparative lull of the most intense US hurricane wind events clearly was associated with a long-term fluctuation in the US property insurance market. This substantial market impact was driven by a gap in intense hurricanes that likely occurred by chance, but changes in the background frequency of major hurricanes could change the odds that such a drought could recur in the future.

Changes in hurricane wind risk may threaten insurers from an operational and claims handling perspective as well. Rapid intensification is a particular threat, since it makes hurricane forecasting unusually difficult, is a key driver of major hurricanes (Lee et al. 2016), and may worsen with future climate change (Emanuel 2017b, Bhatia et al. 2019). From the perspective of a building, it doesn't matter how quickly a hurricane intensifies – damage is determined by the intensity of the local wind field. However, major insurers use track and intensity forecasts to position teams of claims adjusters in advance of a storm, and may bring in external contractors for repairs if the local labor market is unlikely to be sufficient to support post-event rebuilding. Lack of predictive skill in rapid intensification may cause claims teams to underestimate the needed response, slowing claims handling or recovery due to underestimated pre-deployment. If the events that cause the most economic damage are the least predictable, that should be reflected in how insurers prepare to respond immediately before and after a hurricane makes landfall.

1.2.2.3 Hazard Research Opportunities

1. Localizing water impacts

Risk managers are charged with managing water risks at the spatial scale of individual locations. This means that spatially-averaged climate impacts on water hazards, like global sea level rise on storm surges, are less useful than detailed spatial projections. Gridded sea level products are starting to fill that gap. However, adding storm surge on top should ideally not behave like a bathtub, but rather reflect the changing hydrodynamics of surge with a new baseline, to better capture inland surge inundation.

As noted in the rainfall section, there is also some evidence for important impacts on hazard related to the presence of the built environment, such as extent of impervious cover. Cities pose the greatest potential for risk accumulation in a single event, so capturing these local-scale impacts on hazard is important to correctly represent risk where the exposed population is the largest.

2. Deeper understanding of tail wind impacts

As noted above, the highest intensity storms drive the overwhelming majority of economic losses from hurricanes in the US. However, the most extreme storms are also the most difficult to represent in climate models, making this a critical research target to quantify the most societally impactful risk.

3. Thinking beyond frequency

A variety of emerging research topics on wind risk have been proposed for hurricanes, including changes in rapid intensification (Emanuel 2017b), inland decay rates (Li and Chakraborty 2020), and track changes (Shuai and Ralf 2021). Many of these risks would have material impacts on property and life and safety risk without touching the frequency or severity of overall hurricane counts. Splitting hurricane risk by subperil highlights these emerging hurricane risk management topics, and although confidence is currently low, further research could help to explore ways that hurricanes may produce societal impacts in a more sophisticated way than simply checking if there will be more or fewer hurricanes.

4. Explicitly modeling impacts of green infrastructure on hazard

Catastrophe models have opportunities for enhancement that could make them more effective partners in climate adaptation. For example, interest has grown in recent years in "green infrastructure" to build coastal flood defenses against storm surge that utilize ecosystem services from mangroves, marshes, oyster banks, and the like to minimize surge damage. There is some evidence that these ecosystem services can produce material reductions in storm surge damage and losses, at times with clearly quantified economic benefits defined in the scientific literature (Narayan et al. 2017; del Valle et al. 2020; Reguero et al. 2021), but there are opportunities to further explore cost-benefit tradeoffs between mobile coastal wetlands compared with hard infrastructure like seawalls, which catastrophe models could be well-suited to address.

Catastrophe models in their present state are designed to reflect a snapshot of risk at a single point in time, so leveraging the time-varying development of coastal flood risk presents an opportunity to blend the strengths of both climate and catastrophe models. By quantifying expected annual losses associated with storm surges under different defense assumptions, a catastrophe model coupled with a probability distribution of sea level rise out of a climate model could put an explicit price tag on the choice of an effective but inflexible piece of gray infrastructure like a seawall, compared with a coastal ecosystem that perhaps does not stop every surge, but can dynamically respond to changing sea levels. Without such tools, hard coastal defense projects like seawalls that make sense with current sea levels cannot be sufficiently compared with mobile ecosystems like marshes, which may migrate inland without human intervention assuming no inland obstructions.

Similarly, a common way that cities attempt to reduce the impacts of increased extreme rainfall (from hurricanes or otherwise) is via green stormwater management – building in green spaces, rain gardens, and other strategies to reduce impervious surfaces, absorb excess rainfall, and minimize flood peaks. Despite clear benefits to flood risk, with knock-on improvements in groundwater storage and other ecosystem services (Prudencio and Null 2018), catastrophe models generally lack the ability to explicitly give credit for the risk reduction produced by these activities, which could perversely disincentivize this key adaptation strategy, since local spending on risk reduction would not lead to a corresponding reduction in modeled loss impacts.

1.2.3 Recommendation 3 – Translating Climate Data for Decision-Makers

1.2.3.1 Time Horizon

There has been a longstanding disconnect between the timing required to make decisions in a business environment in comparison with the time frames that generally receive the most attention in the scientific literature. The scientific community often focuses on the ~2050–2100 time frame, where climate impacts are more likely to have emerged from natural variability. At this time scale, anthropogenic forcing is large relative to natural variability, making trends and impacts on extreme events clearer. Conversely, the insurance business wants to know what to expect for the upcoming year, to correspond with the length of a standard insurance contract, up to perhaps a decade in the future, for its underwriting and portfolio management strategy. While insurers may do qualitative horizon-scanning exercises beyond this time frame to consider broader product strategy and emerging risks, quantitative risk assessments beyond the portfolio steering time frame have limited practical applicability in the insurance decision-making framework. This time frame is heavily impacted by interannual and decadal scale variability, where predictive skill in climate models is low, so at first glance it is difficult to give clear guidance. Similarly, seasonal and sub-seasonal forecasting is available that relies heavily on climate features like El Niño/La Niña, but for contracts that are written to cover an entire year, the relative benefit to insurance companies of adjusting their risk appetite to match a forecast at the current level of forecast skill is marginal (Emanuel et al. 2012).

However, this view does not consider that the models used to manage risk are built using historical data. While recent events are included in regular catastrophe model updates, these models also include events from past decades that occurred in a climate that will not return within the next century. A key potential space to incorporate climate risk comes from understanding how much change in hazard has already occurred relative to a historical dataset that may span up to the last 50–100 years. Trends in extreme hazards like hurricanes represent a complex mix of internal natural variability, potentially short-term external forcings like aerosol pollution, and long-term anthropogenic climate change, so disentangling the influence of each for various hurricane subperils would help risk managers to account for trends in historical data while preserving as much of the distribution of natural variability as possible. For example, catastrophe models are not limited to only using storm surges within the last few years, even though sea levels have risen throughout the full historical record. By extracting the component of historical storm tides that came from wind only, the full distribution of historical surge events is preserved, and then added on top of current sea levels. This explicitly accounts for known climate trends up to present while still leveraging the full variability present in over a century of historical data. In this sense, attribution exercises that quantify how much climate change has impacted individual recent storms or trends in events is highly valuable within the insurance sector, since they can then be explicitly modeled while preserving a historical statistical distribution that better captures decadal-scale natural climate variability.

Selecting an appropriate time horizon for risk assessment has also emerged as a concern for climate risk disclosure in the regulatory space. Different parts of the public and private sector have different risk tolerances and planning horizons. Insurers tend to focus on immediate and short-term quantitative climate impacts, because their risk is driven by extreme events like hurricanes, more than changes in mean conditions that are better understood at longer time periods. Infrastructure planners may also need to think about extreme events, such as when designing a seawall for future storm surges, but the design lifetime of a structure is much longer than an annual insurance contract, which can be adjusted dynamically as new climate risks emerge. Even within infrastructure planning, a nuclear power plant likely has a lower tolerance for a failure in hazard defenses than a local street. A risk management framework that includes (1) critically assessing the time scale at which material climate risks emerge, (2) cumulative tolerance for negative outcomes, and (3) the ability to adjust as the climate changes is a useful thought exercise for determining climate risk tolerance both within and outside of the insurance sector.

Within insurance, the short-term, quantitative assessment of climate risk can be supplemented by longer term, qualitative analysis of emerging risks and mitigation that are poorly captured in current tools. For example, it is likely that building codes will continue to improve in the face of more destructive hurricanes, but predicting the timing and magnitude of these reductions in building vulnerability is not currently possible, lowering the skill of long-term projections in hurricane economic losses. These differing levels of skill at different time horizons must also be clearly articulated in the regulatory space, to avoid counterproductive or maladaptive decision-making driven by quantitative modeling at longer time horizons that exceed scientific knowledge.

1.2.3.2 Risk Metrics

Climate risk has often been treated as a type of emerging risk within the insurance sector, with a focus on exploratory scenarios that are designed to highlight potential impacts across different parts of the business (underwriting, investments, regulatory risk, etc.). While such scenarios are valuable in assessing broad-scale, interconnected risks at longer time scales, they have limited utility in active decision-making. A hypothetical scenario with a 50% increase in major hurricanes may appear alarming to a risk manager, but without a clear and short-term time frame for when such a change could be expected, it is difficult to justify making immediate adjustments to a portfolio of risks.

More valuable information from a risk management perspective comes when climate risk is translated to use the same risk metrics that are used in portfolio steering. This "normative approach" designs climate sensitivity tests around common business goals like profitability or maintaining sufficient capital (Rye et al. 2021). Rather than framing climate testing as an assessment of business impacts for a given amount of climate change, this approach allows risk managers to turn the question around, and instead run negative stress tests to determine what level of climate impact would be sufficient to affect the risk appetite of the business.

The insurance industry uses a wide range of metrics to plan for risk appetite, but some of the most common are tail risk metrics, expected losses, historical events, and extreme disaster scenarios (Rye et al. 2021). Tail metrics might include probable maximum loss (PML), a representative large loss at a low expected frequency such as a 100 year or 250 year return period, or tail value at risk (TVaR), the average of all losses above a given return period. In the insurance industry, tail metrics are often used to assess capital sufficiency, to ensure that enough reserves are held to pay claims after large events, and they are key drivers of an organization's risk tolerance. Expected loss, or average annual loss (AAL), is the amount of loss that occurs each year on average over a long period of time, which insurers must cover with premium revenue to maintain long-term profitability. Historical events and extreme disaster scenarios both represent known loss values, which can be used as standalone scenarios, or companies can use frequency to assess risk tolerance - for example, how often the company could sustain a Hurricane Andrew-level loss, or a Category 3 hurricane in New York City.

The value of a climate management approach driven by traditional risk metrics is that it drives climate risk into the language used by risk managers across the insurance industry. Climate science is no longer limited to the domain of scientific technical experts. As climate impacts move further into the broader society, the role of translation - tracking and understanding the rapidly evolving scientific literature, contextualizing the level of uncertainty, and turning climate hazards into decision-relevant metrics - will spread further into more climate-affected industries. Insurance offers an opportunity to set a clear example for how to turn science into decision-relevant data for practitioners throughout the economy.

1.2.3.3 Translating Impacts Research Opportunities

1. Hazard assessment at a wider range of time horizons

Decision-makers on the ground who will be directly impacted by climate change, be it insurance companies, city planners, commercial risk managers, or homeowners taking out a mortgage, do not have the luxury of waiting until 2100, when climate forcings are large enough to see clear signals emerge far beyond natural variability. Particularly for rare events like hurricanes, even large trends can take decades to emerge from historical records that feature substantial multidecadal variability, as has been the case in the literature looking at historical trends in major hurricanes (Vecchi et al. 2021).

Therefore, the end-users of hurricane climate information would greatly benefit from added research activity in the messy present. This is not a call for seasonal or decadal forecasting, per se, but rather a more robust quantification of how much change seen to date is driven by anthropogenic forcing, compared with how much can be confidently attributed to natural variability.

2. Engagement with end-users on time horizons and metrics

This discussion highlights key time horizons and metrics that are widely used in the insurance sector, but other climate data end-users are likely to have different needs. Infrastructure or city planners may want to look 50–75 years in the future, and a mortgage holder may want to understand their cumulative flood risk over the next 30 years. The work of engaging with climate data services end-users is slow and has struggled to gain priority from an incentives standpoint (funding, advancing toward tenure, etc.). Enhanced interaction with social scientists may help improve communication and co-development of climate analytics between the academic, public, and private sectors (Findlater et al. 2021).

1.3 Conclusion

In this article, I have presented three recommendations for how to leverage the capabilities of catastrophe risk management for dealing with climate change impacts on hurricane risk:

1. Risk modelers need to bring more data about the built environment to the table to meaningfully model emerging climate risks, particularly for water subperils like storm surge and inland rainfall. The substantial gap in capturing the built environment in the climate modeling space shines a bright light on one of the most important contributions that catastrophe models could make to quantify future societal impacts of climate-driven hurricane risk.
2. The unbundling of hurricane hazards into subperils that can be assessed for relative confidence in their intersection with broader climate trends offers a model that could be widely applied to other emerging climate hazards, like flood and wildfire risk. Where climate hazard confidence is higher, such as sea level rise, the near-term impacts on hurricane subperils like storm surge are clear enough that further dissecting of uncertainty will not fundamentally alter what is now a risk management and policy problem (Sobel 2021). However, lower confidence but high severity impacts like changes in hurricane frequency can be explored through normative risk assessment, and prioritized by their downstream societal and economic impacts.
3. Finally, incorporating climate science into catastrophe modeling is a clear case study highlighting the value of "climate translators" in the private sector. Calls for climate risk disclosure have leapfrogged the abilities of climate models (Fiedler et al. 2021), but that does not mean that climate models have nothing to offer the risk management community. Instead, emerging climate risks in hurricane risk management will require both climate scientists and catastrophe modelers to better understand the strengths and limitations of each other's data and toolkits. Translating climate science into actionable information for end-users is critical across the public and private sectors to help society adapt to the inevitable climate challenges to come.

The entire economy will have to grapple with the ongoing impacts of changes in extreme events like hurricanes. Insurance risk managers can leverage their decades of experience in working with models designed to quantify society's most extreme natural hazards, which will bring valuable leadership to the private sector response to climate change. The resources and expertise in the catastrophe risk management community are key partners to drive greater societal resilience in the face of climate change.

Acknowledgments Many thanks to Suz Tolwinski-Ward and Steve Bowen, whose insightful and constructive reviews have saved you from having to read the first version of this paper.

My thanks also to Matthew Shelton, who gave me time and space and support to write during this stupid pandemic.

Bibliography

Andersen TK, Shepherd JM (2014) A global spatiotemporal analysis of inland tropical cyclone maintenance or intensification. Int J Climatol 34:391–402. https://doi.org/10.1002/joc.3693

Bhatia KT, Vecchi GA, Knutson TR, Murakami H, Kossin J, Dixon KW, Whitlock CE (2019) Recent increases in tropical cyclone intensification rates. Nat Commun 10:635. https://doi.org/10.1038/s41467-019-08471-z

Blake E, Zelinsky D (2018) National Hurricane Center tropical cyclone report: Hurricane Harvey (AL092017)

Bloemendaal N, Haigh ID, de Moel H, Muis S, Haarsma RJ, Aerts JCJH (2020) Generation of a global synthetic tropical cyclone hazard dataset using STORM. Sci Data 7:40. https://doi.org/10.1038/s41597-020-0381-2

Daloz AS, Camargo SJ (2018) Is the poleward migration of tropical cyclone maximum intensity associated with a poleward migration of tropical cyclone genesis? Clim Dyn 50:705–715. https://doi.org/10.1007/s00382-017-3636-7

del Valle A, Eriksson M, Ishizawa OA, Miranda JJ (2020) Mangroves protect coastal economic activity from hurricanes. Proc Natl Acad Sci USA 117:265. https://doi.org/10.1073/pnas.1911617116

Emanuel K (2017a) Assessing the present and future probability of hurricane Harvey's rainfall. Proc Natl Acad Sci USA 114:12681. https://doi.org/10.1073/pnas.1716222114

Emanuel K (2017b) Will global warming make hurricane forecasting more difficult? Bull Am Meteorol Soc 98:495–501. https://doi.org/10.1175/BAMS-D-16-0134.1

Emanuel K, Fondriest F, Kossin J (2012) Potential economic value of seasonal hurricane forecasts. Weather Clim Soc 4:110–117. https://doi.org/10.1175/wcas-d-11-00017.1

Fiedler T, Pitman AJ, Mackenzie K, Wood N, Jakob C, Perkins-Kirkpatrick SE (2021) Business risk and the emergence of climate analytics. Nat Clim Chang 11:87–94. https://doi.org/10.1038/s41558-020-00984-6

Findlater K, Webber S, Kandlikar M, Donner S (2021) Climate services promise better decisions but mainly focus on better data. Nat Clim Chang 11:731–737

Fussell E (2015) The long-term recovery of New Orleans' population after hurricane Katrina. Am Behav Sci 59:1231–1245. https://doi.org/10.1177/0002764215591181

Goldenberg SB, Landsea CW, Mestas-Nunez AM, Gray WM (2001) The recent increase in Atlantic hurricane activity: causes and implications. Science 293:474–479. https://doi.org/10.1126/science.1060040

Grinsted A, Moore JC, Jevrejeva S (2012) Homogeneous record of Atlantic hurricane surge threat since 1923. Proc Natl Acad Sci USA 109:19601–19605. https://doi.org/10.1073/pnas.1209542109

Guy Carpenter (2020) Chart: U.S. Property Catastrophe Rate-On-Line (ROL) Index – 1990 to 2020. In: Chart: U.S. Property Catastrophe Rate-On-Line (ROL) Index – 1990 to 2020. https://www.gccapitalideas.com/2020/09/15/chart-u-s-property-catastrophe-rate-on-line-rol-index-1990-to-2020/. Accessed 25 May 2021

Hall T, Hereid K (2015) The frequency and duration of U.S. hurricane droughts. Geophys Res Lett 42:3482–3485. https://doi.org/10.1002/2015gl063652

Hall TM, Jewson S (2007) Statistical modelling of North Atlantic tropical cyclone tracks. Null 59:486–498. https://doi.org/10.1111/j.1600-0870.2007.00240.x

Hart RE, Chavas DR, Guishard MP (2016) The arbitrary definition of the current Atlantic major hurricane landfall drought. Bull Am Meteorol Soc 97:713–722. https://doi.org/10.1175/bams-d-15-00185.1

Intergovernmental Panel on Climate Change (2014) Climate change 2013: the physical science basis: working group I contribution to the fifth assessment report of the intergovernmental panel on climate change. Cambridge University Press

Katz L (2021) A racist past, a flooded future: formerly redlined areas have $107 billion worth of homes facing high flood risk—25% more than non-redlined areas. Redfin

Keys B, Mulder P (2020) Neglected no more: housing markets, mortgage lending, and sea level rise. National Bureau of Economic Research

Klotzbach P, Gray W, Fogarty C (2015) Active Atlantic hurricane era at its end? Nat Geosci 8:737–738. https://doi.org/10.1038/ngeo2529

Klotzbach PJ, Bell MM, Bowen SG, Gibney EJ, Knapp KR, Schreck CJ (2020) Surface pressure a more skillful predictor of normalized hurricane damage than maximum sustained wind. Bull Am Meteorol Soc 101:E830–E846. https://doi.org/10.1175/bams-d-19-0062.1

Knutson T, Camargo SJ, Chan JCL, Emanuel K, Ho C-H, Kossin J, Mohapatra M, Satoh M, Sugi M, Walsh K, Wu L (2019) Tropical cyclones and climate change assessment: part I: detection and attribution. Bull Am Meteorol Soc 100:1987–2007. https://doi.org/10.1175/bams-d-18-0189.1

Knutson T, Camargo SJ, Chan JCL, Emanuel K, Ho C-H, Kossin J, Mohapatra M, Satoh M, Sugi M, Walsh K, Wu L (2020) Tropical cyclones and climate change assessment: part II: projected response to anthropogenic warming. Bull Am Meteorol Soc 101:E303–E322. https://doi.org/10.1175/bams-d-18-0194.1

Kossin JP (2017) Hurricane intensification along United States coast suppressed during active hurricane periods. Nature 541:390–393. https://doi.org/10.1038/nature20783

Kossin JP, Emanuel KA, Vecchi GA (2014) The poleward migration of the location of tropical cyclone maximum intensity. Nature 509:349–352. https://doi.org/10.1038/nature13278

Kossin JP, Knapp KR, Olander TL, Velden CS (2020) Global increase in major tropical cyclone exceedance probability over the past four decades. Proc Natl Acad Sci USA 117:11975–11980. https://doi.org/10.1073/pnas.1920849117

Lee C-Y, Tippett MK, Sobel AH, Camargo SJ (2016) Rapid intensification and the bimodal distribution of tropical cyclone intensity. Nat Commun 7:10625. https://doi.org/10.1038/ncomms10625

Lee C-Y, Tippett MK, Sobel AH, Camargo SJ (2018) An environmentally forced tropical cyclone hazard model. J Adv Model Earth Syst 10:223–241. https://doi.org/10.1002/2017MS001186

Lee C-Y, Camargo SJ, Sobel AH, Tippett MK (2020) Statistical–dynamical downscaling projections of tropical cyclone activity in a warming climate: two diverging genesis scenarios. J Clim 33:4815–4834. https://doi.org/10.1175/JCLI-D-19-0452.1

Li L, Chakraborty P (2020) Slower decay of landfalling hurricanes in a warming world. Nature 587:230–234. https://doi.org/10.1038/s41586-020-2867-7

Mann ME, Steinman BA, Brouillette DJ, Miller SK (2021) Multidecadal climate oscillations during the past millennium driven by volcanic forcing. Science 371:1014–1019. https://doi.org/10.1126/science.abc5810

Maynard T, Beecroft N, Gonzalez S, Restell L, Toumi R (2014) Catastrophe modelling and climate change

Moftakhari HR, Salvadori G, AghaKouchak A, Sanders BF, Matthew RA (2017) Compounding effects of sea level rise and fluvial flooding. Proc Natl Acad Sci USA 114:9785–9790. https://doi.org/10.1073/pnas.1620325114

Murakami H, Delworth TL, Cooke WF, Zhao M, Xiang B, Hsu P-C (2020) Detected climatic change in global distribution of tropical cyclones. Proc Natl Acad Sci USA 117:10706–10714. https://doi.org/10.1073/pnas.1922500117

Narayan S, Beck MW, Wilson P, Thomas CJ, Guerrero A, Shepard CC, Reguero BG, Franco G, Ingram JC, Trespalacios D (2017) The value of coastal wetlands for flood damage reduction in the Northeastern USA. Sci Rep 7:9463. https://doi.org/10.1038/s41598-017-09269-z

Nicholls RJ, Lincke D, Hinkel J, Brown S, Vafeidis AT, Meyssignac B, Hanson SE, Merkens J-L, Fang J (2021) A global analysis of subsidence, relative sea-level change and coastal flood exposure. Nat Clim Chang 1–5. https://doi.org/10.1038/s41558-021-00993-z

Pan B, Wang Y, Logan T, Hsieh J, Jiang JH, Li Y, Zhang R (2020) Determinant role of aerosols from industrial sources in hurricane Harvey's catastrophe. Geophys Res Lett 47. https://doi.org/10.1029/2020gl090014

Piecuch CG, Huybers P, Hay CC, Kemp AC, Little CM, Mitrovica JX, Ponte RM, Tingley MP (2018) Origin of spatial variation in US East Coast sea-level trends during 1900–2017. Nature 564:400–404. https://doi.org/10.1038/s41586-018-0787-6

Prudencio L, Null SE (2018) Stormwater management and ecosystem services: a review. Environ Res Lett 13:033002. https://doi.org/10.1088/1748-9326/aaa81a

Reguero BG, Storlazzi CD, Gibbs AE, Shope JB, Cole AD, Cumming KA, Beck MW (2021) The value of US coral reefs for flood risk reduction. Nat Sustain 4:688–698. https://doi.org/10.1038/s41893-021-00706-6

Rye CJ, Boyd JA, Mitchell A (2021) Normative approach to risk management for insurers. Nat Clim Chang. https://doi.org/10.1038/s41558-021-01031-8

Sebastian A, Gori A, Blessing RB, van der Wiel K, Bass B (2019) Disentangling the impacts of human and environmental change on catchment response during Hurricane Harvey. Environ Res Lett 14:124023. https://doi.org/10.1088/1748-9326/ab5234

Shuai W, Ralf T (2021) Recent migration of tropical cyclones toward coasts. Science 371:514–517. https://doi.org/10.1126/science.abb9038

Sobel AH (2021) Usable climate science is adaptation science. Clim Chang 166. https://doi.org/10.1007/s10584-021-03108-x

Strauss BH, Orton PM, Bittermann K, Buchanan MK, Gilford DM, Kopp RE, Kulp S, Massey C, de Moel H, Vinogradov S (2021) Economic damages from Hurricane Sandy attributable to sea level rise caused by anthropogenic climate change. Nat Commun 12:2720. https://doi.org/10.1038/s41467-021-22838-1

Sweet WV (2017) Global and regional sea level rise scenarios for the United States

Tedesco M, Hultquist CG, de Sherbinin A (2021) A new dataset integrating public socioeconomic, physical risk, and housing data for climate justice metrics: a test-case study in Miami. Environ Justice. https://doi.org/10.1089/env.2021.0059

Ting M, Kossin JP, Camargo SJ, Li C (2019) Past and future hurricane intensity change along the U.S. East Coast. Sci Rep 9(7795). https://doi.org/10.1038/s41598-019-44252-w

United States Census Bureau (2021) Metropolitan and Micropolitan statistical areas population totals and components of change: 2010–2019

van Oldenborgh GJ, van der Wiel K, Sebastian A, Singh R, Arrighi J, Otto F, Haustein K, Li S, Vecchi G, Cullen H (2017) Attribution of extreme rainfall from Hurricane Harvey, august 2017. Environ Res Lett 12:124009. https://doi.org/10.1088/1748-9326/aa9ef2

Vecchi G, Landsea C, Zhang W, Villarini G, Knutson T (2021) Changes in Atlantic major hurricane frequency since the late-19th century. Nat Commun 12

Waite AJ, Klavans JM, Clement AC, Murphy LN, Liebetrau V, Eisenhauer A, Weger RJ, Swart PK (2020) Observational and model evidence for an important role for volcanic forcing driving Atlantic multidecadal variability over the last 600 years. Geophys Res Lett 47:e2020GL089428. https://doi.org/10.1029/2020GL089428

Wang S-YS, Zhao L, Yoon J-H, Klotzbach P, Gillies RR (2018) Quantitative attribution of climate effects on Hurricane Harvey's extreme rainfall in Texas. Environ Res Lett 13:054014. https://doi.org/10.1088/1748-9326/aabb85

Weinkle J, Landsea C, Collins D, Musulin R, Crompton RP, Klotzbach PJ, Pielke R (2018) Normalized hurricane damage in the continental United States 1900–2017. Nat Sustain 1:808–813. https://doi.org/10.1038/s41893-018-0165-2

Zhang W, Villarini G, Vecchi GA, Smith JA (2018) Urbanization exacerbated the rainfall and flooding caused by hurricane Harvey in Houston. Nature 563:384–388. https://doi.org/10.1038/s41586-018-0676-z

Chapter 2
Characteristics of Risk

Jan Kleinn, Dörte Aller, and Matthias Oplatka

Abstract Risk, as opposed to the physical hazard, always involves consequences like damages or losses. Risk is a concept which is difficult to grasp, partly because it cannot be directly measured. Furthermore, risk cannot be summarized in only a single number. Nevertheless, risk has to be characterized for a number of applications, ranging from risk assessment to risk communication, stakeholder involvement, and discussions on the level of acceptable risk.

The authors are providing an approach to characterize risk. These risk characteristics are the result of practical applications in natural hazard risk management. With these characteristics, risk can be used as a basis for decisions regarding the level of acceptable risk or for the appraisal of risk reduction measures, including the characterization of uncertainties in current or future risk.

Keywords Risk · Loss Frequency Curve · Expected annual loss · Return periods · Exceedance probabilities

2.1 Introduction and Motivation

Risk assessments and discussions on risk are instrumental in many natural hazard risk management applications. The authors have conducted a number of natural hazard mitigation projects, which included risk dialogue with stakeholders. In these

---

J. Kleinn (✉)
Kleinn Risk Management, Zürich, Switzerland

WSL Institute for Snow and Avalanche Research SLF, Davos, Switzerland
e-mail: kleinn@kleinn-risk.ch

D. Aller
Aller Risk Management, Zürich, Switzerland
e-mail: aller@aller-risk.ch

M. Oplatka
Office of Waste, Water, Energy and Air (AWEL) of the Canton of Zurich, Zürich, Switzerland
e-mail: matthias.oplatka@bd.zh.ch

© The Author(s) 2022
J. M. Collins, J. M. Done (eds.), *Hurricane Risk in a Changing Climate*,
Hurricane Risk 2, https://doi.org/10.1007/978-3-031-08568-0_2

projects, the authors have realized how important it is to characterize risks. This chapter is meant to be a write-up of risk characteristics, which have proven to be helpful in risk dialogue processes such as the approach used in the "Guide to an Accepted Risk" (Aller et al., in preparation). These characteristics are applicable to any kind of risk for which the consequences can be quantified or ranked. The consequences don't have the be converted to monetary values for the risk to be characterized.

The characteristics of risk proposed in this chapter are fostering the dialogue with stakeholders and helping in making risk analyses societally relevant. It can be seen as the basis for transdisciplinary frameworks as those proposed by Fischer et al. (2021) and Pohl et al. (2017).

2.1.1 Accepted Risk and Risk Dialogue

Accepted risk is the amount of risk a stakeholder is willing to accept. Risk dialogue and the determination of the accepted risk for all stakeholders in a participatory process are important ingredients in finding sustainable measures of risk reduction. If stakeholders are involved in an early stage of the risk analysis and risk management process, chances are better that the risks are understood, and mitigation measures are being supported by the entire community. Stakeholders involved in such risk dialogue processes can be people or entities directly or indirectly affected by the risks or by potential mitigation measures, policy makers, or risk takers (e.g., insurance companies or public entities).

Some of the stakeholders are mainly impacted by how frequently they are affected by a damage or loss, no matter how small it might be. Other stakeholders are more affected by the consequences of possible maximum impacts. Some stakeholders might be more interested in the probability of damages reaching or exceeding a certain threshold or the damage at certain probabilities.

Furthermore, stakeholders might be interested in different time horizons for their risk. Some are interested in the risk within the next months to years, some rather within the next decades.

The risk analysis in combination with the accepted risks of different stakeholders are an important basis to determine whether mitigation measures are needed to reduce the risks to an accepted level. For the operator of a street, the acceptable risk depends on the frequency of the street being flooded and the consequences of the flooding. It might be acceptable that the street is flooded once every other year as long as the street only has to be cleaned. For the operation of a regional medical emergency service, the level of accepted risk for the access road being flooded can depend on the availability of a redundant emergency service in the vicinity. If there is another emergency service in the vicinity, it is acceptable that the access road is being flooded as long as the interruption of the access is of short duration. If the emergency service is the only one in a larger region, a flooded access road is not acceptable.

Risk assessments help in the development of optimum combinations of mitigation measures to reduce risks to an acceptable level. They also serve as a basis for cost-benefit-analyses of potential measures (or combinations thereof) and therefore help in the process of prioritizing different mitigation measures. The cost-benefit-analyses used for natural hazard mitigation are similar to those used for climate adaptation, as described in Bresch and Aznar-Siguan (2021).

To be able to compare different risks, these risks have to be characterized using the same methods of risk analysis. Such a consistent risk characterization also helps in getting an integrated view of risks, not only concentrating on individual sectors.

2.2 Risk – An Introduction

The IPCC (2020) makes a clear distinction between hazard and risk. In this chapter, the term "risk" always involves some kind of consequence and does not only describe the physical hazard.

Risk is frequently defined as being the probability of losses or damages. In some economic applications, risk is defined as the future uncertainty of the deviation from an expected outcome. This is in line with the ISO 310000 standard on risk management, where risk is defined, independent of positive or negative impact, as the effect of uncertainty on objectives (International Standards Organization (ISO) 2018). In IPCC (2012), risk is defined as the product of probability and consequence. IPCC (2020) specifically defines risk as "the potential for adverse consequences".

In the natural hazard context, risk is often considered as being the product of probability and consequence or the severity of the consequence. Consequence in this case is the consequence of an impact, e.g., a physical damage, a monetary loss, or an interruption of services. The consequence is usually a combination of exposure and vulnerability for a particular hazard intensity. Risk can therefore be considered a combination of hazard, exposure, and vulnerability, similar to the definition of weather and climate risks used by the IPCC (2012). The concept used by the IPCC (2012) can therefore be applied to any kind of disaster risk, not only to weather and climate events (Fig. 2.1).

The notion of risk being the product of probability and consequence does not necessarily mean that it is a single probability multiplied with a single consequence. Such a single product of probability and consequence can be a reasonable approach for risks which only have a single probability, and in case of the event, a single consequence. Such risks are usually of technical nature with a failure probability and an associated consequence. This could e.g., be a power line, which is either fully functional or not functional at all. In the case of natural hazards, the entire range of consequences and their probabilities usually has to be considered. From frequent to rare events, from small to large consequences (Fig. 2.2). Frequently occurring weak storms only cause minor damage. Increasing intensity of storms leads to increasing damage. When using a single probability and consequence or a very limited number of values of probabilities and consequences to determine the risk, the users have to be aware that the risk might not be properly characterized. It is therefore important to

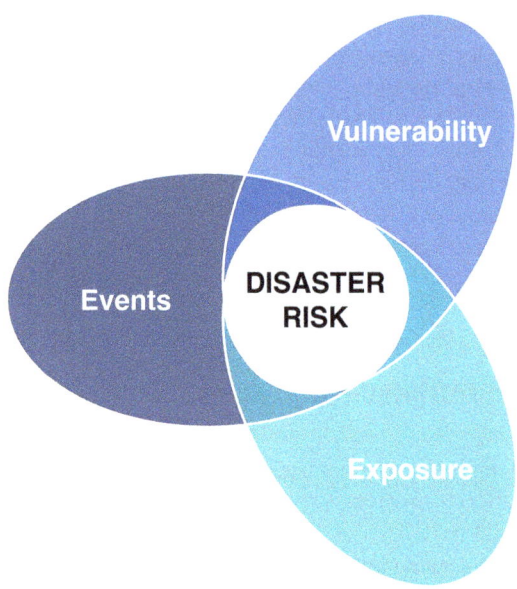

Fig. 2.1 Illustration of the concept of disaster risk, based on the illustration by IPCC (2012)

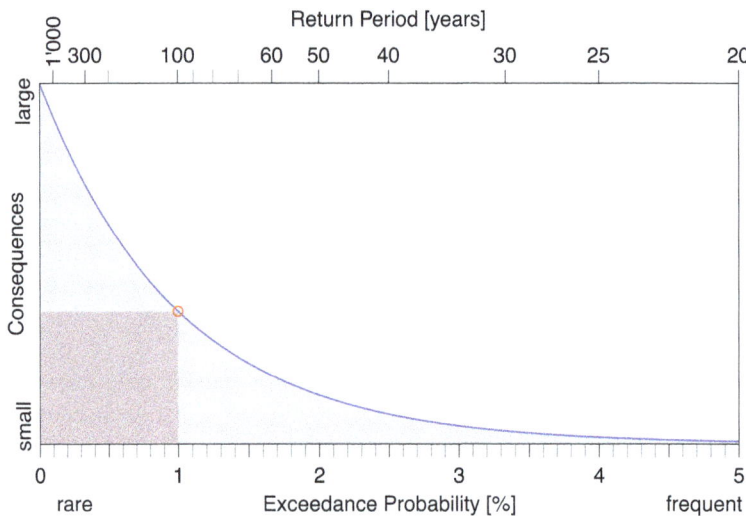

Fig. 2.2 Illustration of risk defined by a single point of probability and consequences or losses (red point and red surface) compared to using an entire loss frequency curve (blue curve and blue surface)

estimate the entire range of probabilities and consequences. This constitutes a damage curve, also called a loss curve, loss frequency curve, or damage exceedance probability curve (see e.g., Mitchell-Wallace et al. 2017 or Aznar-Siguan and Bresch

2019). The term loss frequency curve will be used in the remaining text. The loss frequency curve is explained in more detail in the following section.

2.2.1 Return Periods

Return periods are frequently used in the natural hazard context, especially in communicating with stakeholders. Return periods are usually used to characterize a particular event: "The wind speed of a certain storm at this location had a return period of 1 in 100 years". This is also called a "100-year event". Return periods are the inverse of exceedance probability and hence also linked to non-exceedance probabilities.

$$Return\ Period = \frac{1}{Exceedance\ Probability} = \frac{1}{1 - Non\text{-}Exceedance\ Prob}.$$

These probabilities and the corresponding return periods usually denote yearly probabilities in natural hazard applications. This needs to be made clear when communicating return periods to a non-technical audience.

The unit of years gives the impression that return periods are easier to understand by stakeholders. Return periods are another way of describing yearly exceedance or non-exceedance probabilities. Especially for longer return periods, the probability is commonly underestimated as it is believed that such an event can only happen once within this time-period. A 100-year event is not comparable to a 100-year-old person. This is frequently assumed by lay people in practical applications and therefore, they believe they're not affected.

It is often easier to communicate the probability of an event over a longer period of time rather than using return periods. The probability for a selected time period (Nyrs) given a return period (RetPer) can be easily calculated via the probability of the event not occurring:

$$P = 1 - \left(1 - \frac{1}{RetPer}\right)^{Nyrs}$$

The resulting probabilities for several return periods and time periods (e.g., lifetimes of buildings) are shown in Fig. 2.3.

If probabilities are considered over a longer period of time (e.g., the average lifetime of a building or infrastructure) of 50 years, even a 100-year return period event has a probability of occurrence of about 40% and a 300-year return period event of about 16%. This thought concept is commonly used in earthquake engineering, where the 475-year and 975-year return periods are used as they represent 10% and 5% of probability in 50 years (see Fig. 2.3). In communicating return periods to stakeholders, it is often favorable to convert these to probabilities over a

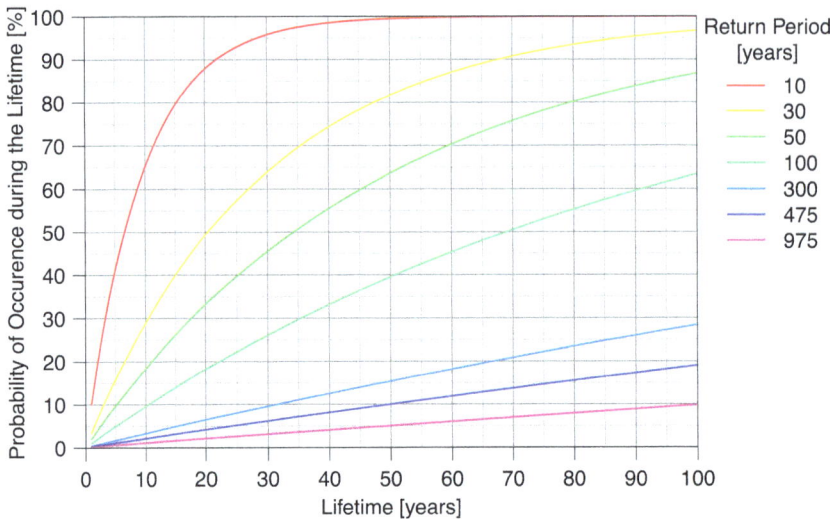

Fig. 2.3 Probability of different return period events over different lifetimes or time periods.

longer period of time. Otherwise, the probabilities of higher return periods are easily underestimated.

For the conversion to probabilities over longer time periods, the hazard probability is assumed to be constant over the entire time period being considered. Current estimates of return periods are usually based on past and current probabilities. Applying these probabilities to longer time periods might give the impression that they will stay constant in the future. This is usually not the case. Especially for hazards driven by meteorological events, probabilities are assumed to change with climate change.

2.2.2 Loss Frequency Curve

Comparing different risks is much easier if entire loss frequency curves are available for the risks to be compared. The loss frequency curve needs to represent losses or damages for a whole range of return periods or probabilities. It doesn't matter, whether the loss frequency curve is defined and shown by exceedance probabilities or non-exceedance probabilities (see Fig. 2.4). The term of exceedance and non-exceedance probability highlights that it is the probability for a certain value (of loss or damage in this case) to be either exceeded or not reached, not the probability of exactly the value.

Loss frequency curves based on exceedance probabilities are commonly used in insurance applications with the associated risk metrics. When communicating to diverse stakeholders, loss frequency curves based on non-exceedance probabilities

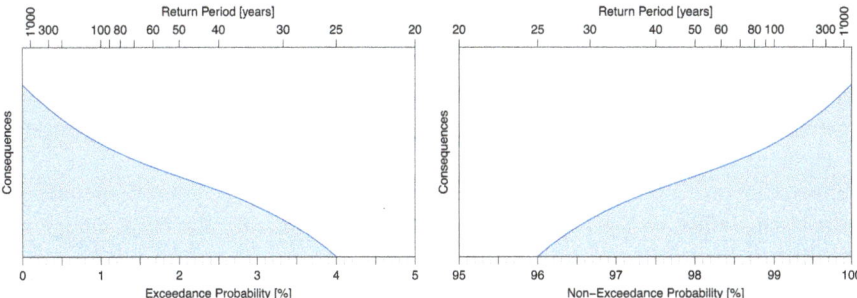

Fig. 2.4 Illustration of loss frequency curves with exceedance probabilities (left) and non-exceedance probabilities (right) as x-axis. Return periods corresponding to exceedance and non-exceedance probabilities are added at the top of the panels

have revealed to be easier to communicate. Practical applications in Switzerland have shown that stakeholders are using return periods as they are commonly used in hazard mapping. Increasing return periods along the x-axis, together with increasing consequences, are therefore favorable.

Loss frequency curves do not necessarily need to have a monetary scale for the consequences. The consequences can also be provided in terms of indices or any other metrics, e.g., people affected. Other examples might be the interruption of certain services (e.g., fresh water supply) or the intangible value of cultural heritage sites. The detail needed in determining loss frequency curves depends on the impact of having more or less detail on the use of the curve and/or the decision to be taken based on the risk assessment.

Note that loss frequency curves represent a snapshot in time of the risk. Risk is changing over time for various reasons. These changes will be covered in a later section of this chapter.

2.3 Characteristics of Risk

The question is: how can risk be characterized? Several characteristics of loss frequency curves are important:

- The shape of the loss frequency curve
- The probability of first loss
- The surface underneath the loss frequency curve (i.e., the integral): the expected annual loss
- Loss values at reference return periods, e.g., the 30-, 100-, 300-, and 1'000-year return periods, which are commonly used in hazard mapping in Switzerland.

Each of these characteristics will be treated in the following sections.

2.3.1 Shape of the Loss Frequency Curve

The shape of the loss frequency curve is providing information, whether the risk is dominated by frequently occurring events with small consequences or by rare events with large consequences (Fig. 2.5). The shape of the loss frequency curve is furthermore providing information whether there are any jumps in the consequences which can influence the (acceptable) risk, and the possible choice of mitigation measures.

If the loss frequency curve is estimated based on a limited number of points of consequence and frequency, the possible impact of the shape of the curve between these points (e.g., possible jumps) on the acceptable risk and the possible choice of mitigation measures needs to be estimated. In case of a big impact, the shape of the curve needs to be analyzed in more detail.

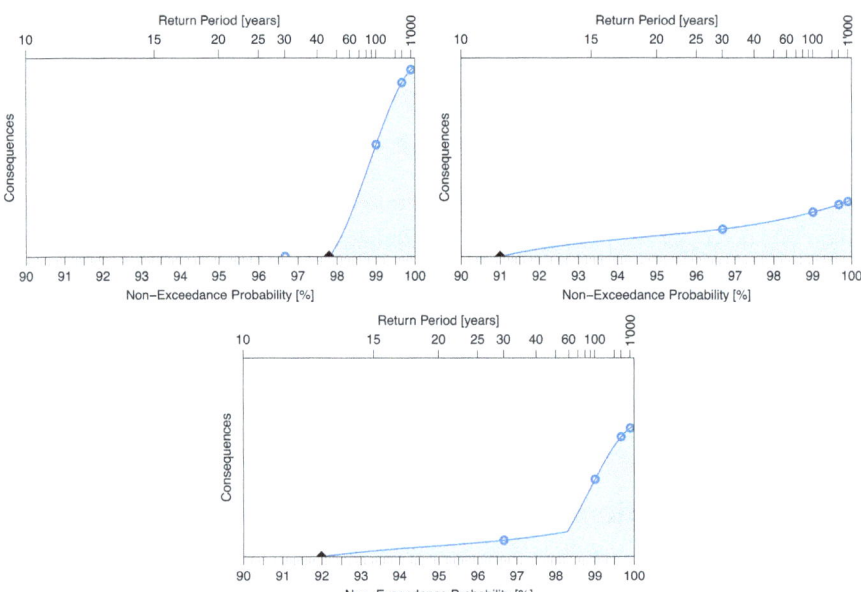

Fig. 2.5 Illustration of several characteristics of risk: Shape of the curve is denoted by lines. Loss levels at reference return periods (30-, 100-, 300-, and 1'000-year) are highlighted by blue points and probabilities of first loss by black diamonds. The surface underneath the curve is illustrated by light color, it is identical for all three examples. The top-left panel could be an example for a metropolitan area with rare natural hazard events which cause large losses, the top-right panel a region with very low population density experiencing regular natural hazard events and a very limited maximum loss. The bottom panel could be a case, where a certain degree of protection is provided up to a certain intensity and therefore probability and losses increasing rapidly thereafter

2.3.2 Probability of First Loss

Another characteristic is the probability at which the first consequences occur. This is also providing information regarding up to which probability no consequences occur.

The probability of first damages can have an impact on the (accepted) risk. Furthermore, the dealing with risks can be different for hazards with frequent damages compared to hazards with rare damages. Storms (frequent) and earthquakes (rare) are hazards with such a difference in probability of first loss (Fig. 2.5).

2.3.3 Surface Underneath Loss Frequency Curve: Expected Annual Loss

The surface underneath the loss frequency curve (i.e., the integral of the curve) is equal to the expected annual loss or also called average annual loss. The expected annual loss is an important metric in the insurance industry for defining the appropriate risk premium needed to be compensated for taking a risk. It is also an important metric for cost-benefit analyses of mitigation measures. The difference between expected annual loss with and without measure allows for the determination of the benefit of the measure, which can then be compared to the expected cost of the measure.

For the surface to be visually equal to the expected annual loss, the consequences or losses have to be shown in a linear scale and the probabilities have to be shown in either exceedance or non-exceedance probabilities. Other scales than linear consequences and (non-)exceedance probabilities are used for loss frequency curves, e.g., using a logarithmic scale of return periods or a logarithmic scale of consequences. These other scales might be more suitable for certain applications to highlight specific differences of two risks, but they do not allow a direct visual comparison of the expected annual loss of different loss frequency curves (see Fig. 2.6). The human eye is weighting by surface in figures. This might lead to misinterpretations if scales other than linear consequences and (non-)exceedance probabilities are used. The calculated number of the expected annual loss is always comparable though.

2.3.4 Consequences at Reference Return Periods

It is common to communicate consequences or losses at specific reference return periods, i.e., using specific points from the loss frequency curve only. In the insurance and reinsurance industry, the 100-year and 250-year losses are common metrics for risk management. The 100-year return period has also been used for the

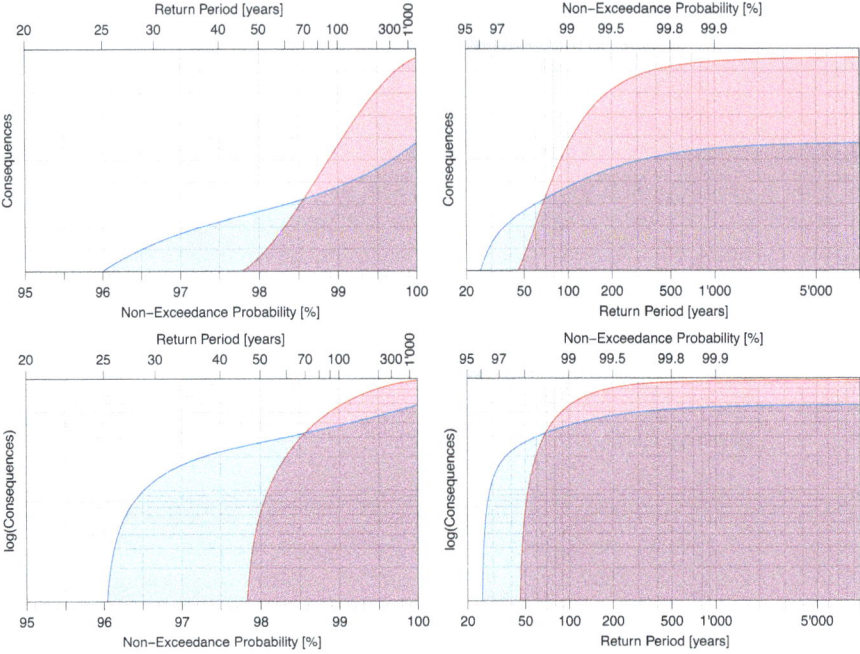

Fig. 2.6 Illustration of two loss frequency curves with identical expected annual loss using different scales for probability and consequences. The top-left panel is using linear axes for consequences and non-exceedance probabilities. The top-right panel is using a logarithmic scale of return periods and suggesting the annual expected loss of the red risk to be much bigger than of the blue risk. The bottom-left panel is using a logarithmic scale for consequences and hence suggesting the blue risk being much larger than the red risk. The scales of the bottom-right panel are logarithmic return periods and logarithmic consequences. The surfaces of the two risks look very similar again. The expected annual loss can only be visually compared in the top-left panel, where axes of linear consequences and linear non-exceedance probabilities are used and hence the surfaces are correctly depicted

"1-in-100 initiative" (UN 2014). In other cases, consequences are only available for specific return periods and the loss frequency curve has to be constructed from these data points. In Swiss hazard mapping, 30-, 100-, 300-, and 1'000-year hazard is mapped and hence consequences are usually available for these return periods.

2.3.5 Relationship Between Risk Characteristics

The four characteristics of risk are somewhat linked to each other, which is highlighted in this section.

Risks can have a similar expected annual loss while having very different probability of first loss or shape of the loss frequency curve. Figure 2.5 is illustrating three loss frequency curves with identical expected annual loss but with different probabilities of first loss and different shapes. By only using the expected annual loss, these risks could not be distinguished.

The probability of first loss itself can have a large impact on the expected annual loss, as shown in the following example: Changing the probability of a first loss from 1-in-25 years to 1-in-20 years or 1-in-30 years can have a significant impact on the expected annual loss while the loss frequency curve is remaining the same for return periods larger than 1-in-50 years (Fig. 2.7). Even if the probability of first loss is not precisely known, a reasonable or rough estimate should be used for the risk analysis.

If only a limited number of values are available for consequences and their probabilities, the ranges in-between these values have to be estimated. The lowest expected annual loss would result from consequences staying constant until the next value of non-exceedance probability is reached. The highest expected annual loss results from consequences jumping to the next level right beyond each value of non-exceedance probabilities (see Fig. 2.8). Reality usually lies somewhere in-between those two extremes. If the final decision is not impacted by this range of possibilities, no further investigation is needed. Otherwise, more detailed analyses or at least estimates are needed to define the loss frequency curve between the limited number of values being available.

If there is not enough data available to define the entire loss frequency curve with high accuracy, it is relevant to check whether the accuracy is adequate for the application and the decisions to be taken based on this risk analysis. Making a few reasonable assumptions is sometimes providing sufficient accuracy. The probability

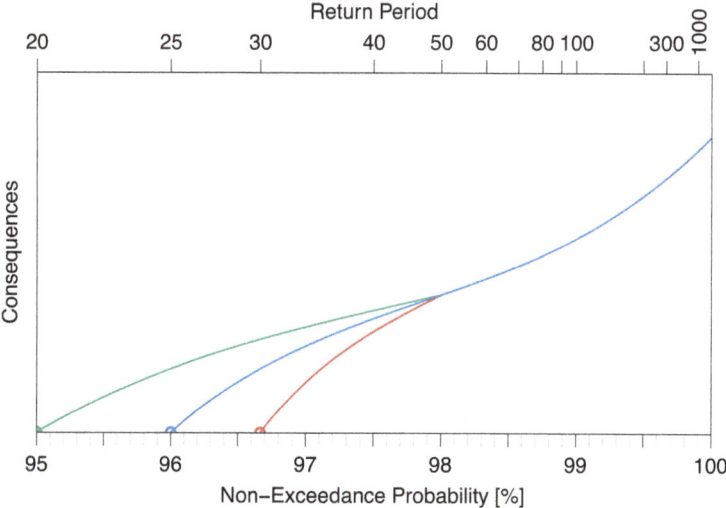

Fig. 2.7 Example of loss frequency curves being identical beyond 50 years of return period with different probabilities of first loss. The difference in expected annual loss is -10% when changing the probability of first loss from 1-in-25 years (blue) to 1-in-30 years (red) and $+14\%$ for 1-in-25 years to 1-in-20 years (green). Probability of first loss is highlighted by points

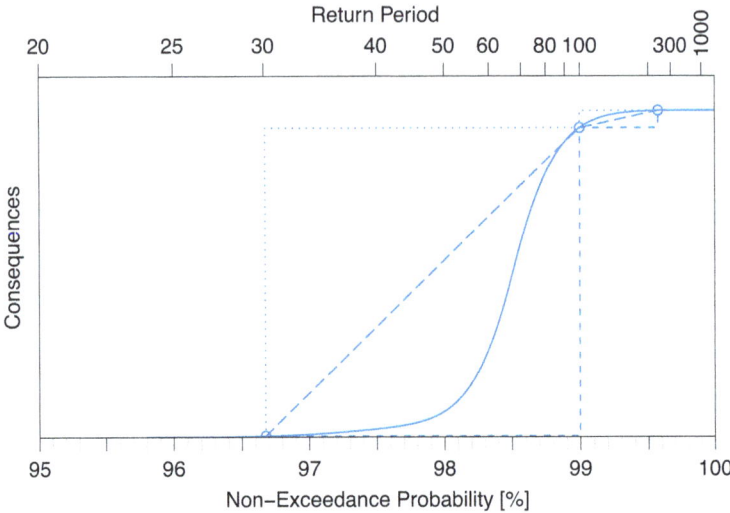

Fig. 2.8 Illustration of the uncertainty when estimating a loss frequency curve based on a limited number of values (points). The short-dashed line highlights the curve providing the lowest possible expected annual loss based on the limited number of values, the dotted line the curve providing the highest possible expected annual loss. The long-dashed line denotes an approximation by linear interpolation. We usually don't know, what the "real" loss frequency curve (solid line) looks like

of a first loss for flooding e.g., can be approximated by using the probability of water reaching a property or building.

A common method for deriving the entire loss frequency curve is the use of probabilistic modelling, so that a large range of probabilities and consequences can be covered (e.g., Aznar-Siguan and Bresch 2019; Golnaraghi et al. 2018; or Mitchell-Wallace et al. 2017).

For comprehensive risk management, it is important to know the characteristics of risk presented above. These characteristics are important for developing optimum risk mitigation measures and support risk communication and risk dialogue. Loss frequency curves can be different depending on the stakeholder, the object, and its function. These differences become apparent in risk dialogue with stakeholders.

2.4 Uncertainties

For risk management purposes, it is important to know whether uncertainties can have an impact on the decisions, which are based on the risk analysis. Depending on the application, some of the uncertainties can be more important than others. Not all uncertainties in risk analyses can be assessed or quantified. Depending on the application and the impacts of the uncertainties, a pessimistic rough estimate can be sufficient. As long as the decision is not affected by the range of loss frequency curves provided by the uncertainties, no effort is needed to reduce these uncertainties. If flood damage cannot increase with increasing flood levels, or if the flood level is still far away from causing (further) damage, higher precision is not needed.

Uncertainties in risk analyses are a combination of uncertainties in all three components of risk (Fig. 2.1): hazard, exposure, and vulnerability.

- Hazard uncertainties are usually a result of limited hazard data or changes in hazard over time. More frequent events are usually estimated with less uncertainty than rare and extreme events.
- Exposure data are usually not comprehensively publicly available. Number of persons, values of buildings and content, infrastructure services, etc. therefore need to be estimated, which leads to uncertainties.
- In general, there is a large variability of how structures, their functions, and their users react to hazard intensities. For larger portfolios, vulnerabilities are commonly derived from past damage events, which are associated with a number of uncertainties.

Uncertainties can also be divided into uncertainties in consequences and uncertainties in frequency. Uncertainties in consequences are a combination of uncertainties in exposure, vulnerability, and hazard intensity. These uncertainties are due to uncertainties in the estimated values of consequences, which are usually larger for less frequent events (Fig. 2.9). Uncertainties in the hazard probabilities are usually also smaller for frequent events than for rare events (Fig. 2.10).

Starting with pessimistic and rough estimates is sufficiently precise in most cases to see whether this rough first risk analysis is adequate for the application. Such a risk analysis can be based on pessimistic assumptions for the probability of first loss and the loss amount at certain return periods. If no action (e.g., protective measures) is needed with this pessimistic approach, no effort needs to be put into achieving a more precise risk analysis. If action needs to be taken dependent on whether the upper or the lower bound of uncertainty is chosen, a sensitivity analysis of the rough risk assessment is advisable. Further investigations to reduce the uncertainties are only needed if the decision can change within the uncertainty range. It is always recommended to find out what the largest drivers of the different risk characteristics are.

2.5 Future Changes

Similar to uncertainties, future changes affecting risk analyses can also be the result from changes in each of the three components of risk:

- Changes in hazard driven by meteorological events are mainly expected from climate change. These might be changes in frequency and intensity as well as extent of natural hazard events. The changes are most likely non-linear and can

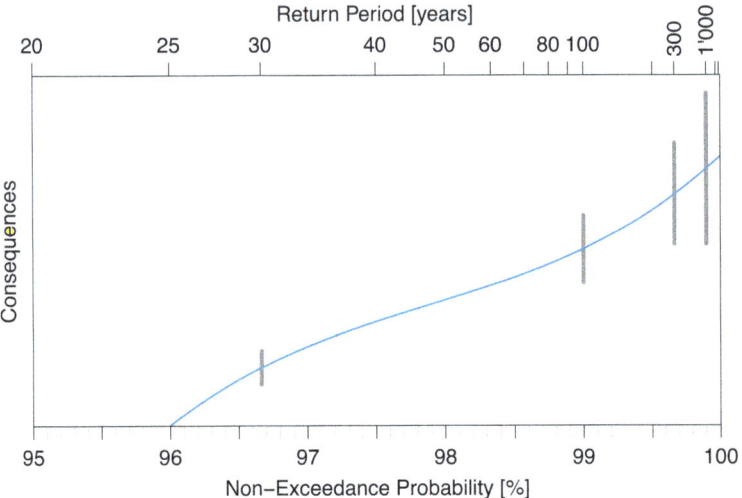

Fig. 2.9 Illustration of uncertainties in the consequences, which are usually smaller for more frequent events. Vertical grey lines denote the uncertainties in the estimates of consequences for a set of given probabilities. Further uncertainties arise between the given probabilities, highlighted in Fig. 2.8

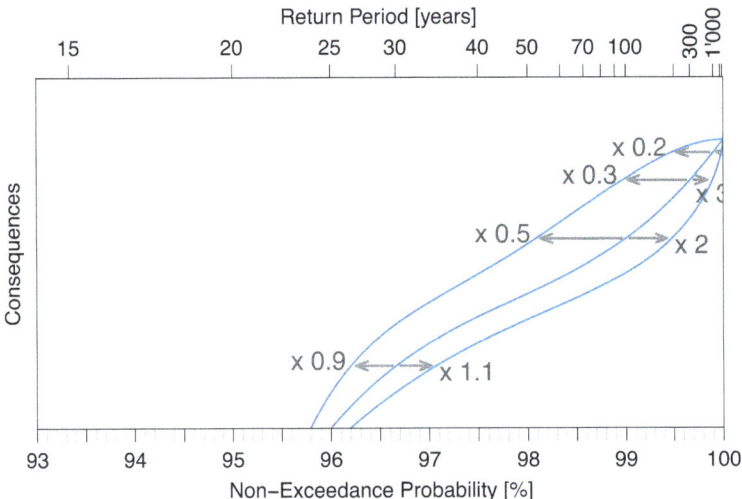

Fig. 2.10 Illustration of uncertainties in non-exceedance probabilities. This uncertainty is based on the assumption, that non-exceedance probability estimates for frequent events have less uncertainty (+/− 10% for the 1-in-30-year event) than those for rare events (factor of 5 for the 1-in-1'000-year event). Large uncertainties for rare events have less of an impact on the overall risk than smaller uncertainties for frequent events

usually not be summarized by adding a factor to the current hazard. Additional compound and cascading effects or hazards have to be expected in the future.
- Changes in exposure are expected from land use changes, building activities, or changes in building stock, which is itself related to changes in population density. Exposure changes can also depend on changes in building values, i.e., when buildings are replaced by more valuable ones.
- Vulnerabilities also change over time as land use, building stock, building practices, and building utilization change over time.

Changes in the view of any of these three components can also be a result of increased scientific knowledge, such as more data, new methods, or new insights into the processes being available.

Similar to the assessment of uncertainties, it is important to find out in a first step whether the potential changes will influence the final decision. Only if this is the case, the potential changes need to be investigated further. Such an investigation should include an analysis of which of the different potential changes has the largest impact on the risk.

To illustrate the importance of including changes in frequent events, a simple approach is used to estimate the impact on risk of potential changes in hazard frequencies. Return periods are halved and doubled in a loss frequency curve (Fig. 2.11). This simple approach shows that the impact of the changes from medium and frequent events is much bigger than the changes of the extreme events. The frequent smaller events are therefore as important as the rare extreme events when considering changes in the overall risk.

2.6 Conclusions

Risk cannot be summarized in a single number only. A loss frequency curve is needed to characterize the risk. It can be useful to differentiate the risk of different stakeholders or objects and hence use several loss frequency curves. The loss frequency curves can be characterized by several metrics like the shape of the loss frequency curve, the probability of a first loss, the expected annual loss, and the loss at reference return periods. Information about possible jumps in the loss frequency curve or the probability of exceeding certain losses can further help characterize the risk. Using a single one of these metrics is usually not enough to characterize the risk for the given application. For most applications, it is advisable to take several of these characteristics into account. The expected annual loss is needed for cost-benefit-analysis. For some of the stakeholders, the probability of first loss or the loss at a certain return period is important. The importance of the different characteristics depends on the individual application the risk assessment is used for. The accuracy needed in the risk assessment is also dependent on the application. It is therefore always helpful to have the final application of the risk assessment in mind from the beginning. Furthermore, when reviewing or interpreting an existing risk

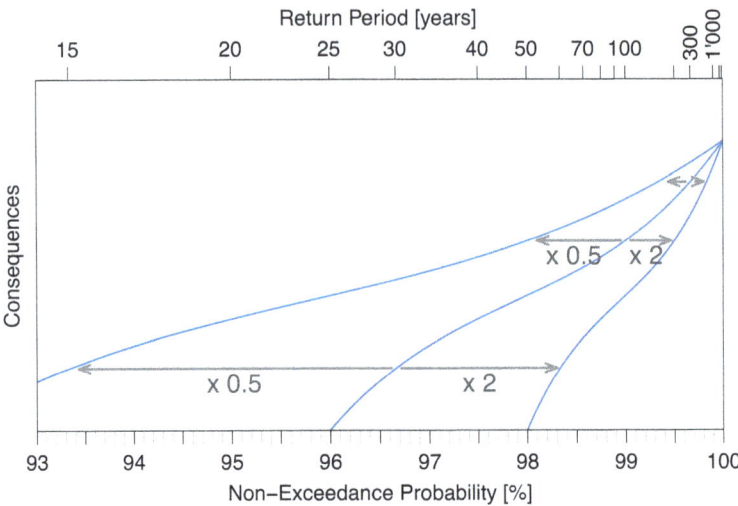

Fig. 2.11 Illustration of a loss frequency curve when return periods are halved and doubled as assumption for future changes. This illustration is highlighting the large impact of medium and frequent events on some of the risk characteristics like the probability of first loss and the expected annual loss compared to the changes from extreme events

analysis, it is useful to know for which application this risk analysis has been performed to avoid misinterpretations.

With the impact of uncertainties and potential changes in mind, it is possible to estimate how detailed potential changes like climate change or changes to population density and building stock need to be taken into account in risk assessment. A detailed inclusion of potential changes is not needed if their impact on the overall risk assessment is negligible. Uncertainty estimates can also help in determining which parameters have the largest influence on the risk metrics and therefore are most important to be investigated further.

Questions about risk frequently appear in risk dialogue and discussions on accepted risk. These questions can usually be answered by loss frequency curves, even if the loss frequency curves have large uncertainties. Not all of these uncertainties have an influence on the final decision and therefore not all of these uncertainties need to be quantified in detail or reduced.

Communicating risk is a big challenge, and the concept of return periods is sometimes problematic as it leads to the underestimation of probabilities. Using annual probabilities is favorable. Using loss frequency curves based on non-exceedance probabilities compared to exceedance probabilities (Fig. 2.4) makes communication with stakeholders easier.

Characterizing risk helps in raising awareness about risks and hence serves as a basis to find an optimum integral combination of measures for risk management (PLANAT 2018) out of all measures being possible. Characterizing risk is therefore a basis for sustainable risk management.

Acknowledgments Most of the content in this publication is based on work developed within the project "Common risk assessment of natural hazards" within the framework "Pilot program Adaptation to climate change" by the Swiss Federal authorities.[1] The authors would like to acknowledge the Office of Waste, Water, Energy and Air (AWEL) of the Canton of Zurich and the Federal Office for Civil Protection (FOCP) for their financial support of this project.

We thank the two reviewers for their valuable comments, which helped improve the clarity of this chapter.

References

Aller D, Kleinn J, Oplatka M, Wyser G (In preparation) Guide to an accepted risk. Report on the project C.06 "Common risk assessment of natural hazards" within the framework "Pilot program Adaptation to climate change" by the Swiss Federal authorities

Aznar-Siguan G, Bresch DN (2019) CLIMADA v1: a global weather and climate risk assessment platform. Geosci Model Dev 12:3085–3097. https://doi.org/10.5194/gmd-12-3085-2019

Bresch DN, Aznar-Siguan G (2021) CLIMADA v1.4.1: towards a globally consistent adaptation options appraisal tool. Geosci Model Dev 14:351–363. https://doi.org/10.5194/gmd-14-351-2021

Fischer LJ, Wernli H, Bresch DN (2021) Widening the common space to reduce the gap between climate science and decision-making in industry. Climate Services 23:100237. https://doi.org/10.1016/j.cliser.2021.100237

Golnaraghi M, Nunn P, Muir- Wood R, Guin J, Whitaker D, Slingo J, Asrar G, Branagan I, Lemcke G, Souch C, Jean M, Allmann A, Jahn M, Bresch DN, Khalil P, Beck M (2018) Managing physical climate risk: leveraging innovations in catastrophe risk modelling: research brief. The Geneva Association—International Association for the Study of Insurance Economics

International Standards Organization (ISO) (2018) Risk Managment: ISO 31000:2018

IPCC (2012) Summary for policymakers. In: Field CB, Barros V, Stocker TF, Qin D, Dokken DJ, Ebi KL, Mastrandrea MD, Mach KJ, Plattner GK, Allen SK, Tignor M, Midgley PM (eds) Managing the risks of extreme events and disasters to advance climate change adaptation. A special report of working groups I and II of the Intergovernmental panel on climate change. Cambridge University Press, Cambridge/New York, pp 1–19

IPCC (2020) The concept of risk in the IPCC Sixth Assessment Report: a summary of cross-Working Group discussions. Guidance for IPCC authors

Mitchell-Wallace K, Jones M, Hillier J, Foote M (eds) (2017) Natural catastrophe risk management and modelling: a practitioner's guide. Wiley-Blackwell, 536p. ISBN: 978-1-118-90604-0

PLANAT (2018) Management of risks from natural hazards. Strategy 2018. National Platform for Natural Hazards PLANAT, Bern

Pohl C, Pius K, Stauffacher M (2017) Ten reflective steps for rendering research societally relevant. GAIA Ecol Perspect Sci Soc 26:43–51

UN (2014) Integrating risks into the financial system: the 1-in-100 initiative, action statement. Climate Summit 2014, UN Headquarters

---

[1] More information on the project in German, French and Italian can be found on the project webpage: https://www.nccs.admin.ch/nccs/de/home/massnahmen/pak/projektephase2/pilotprojekte-zur-anpassung-an-den-klimawandel%2D%2Dcluster%2D%2Dmanagem/c-06-gemeinsame-risikobetrachtung-von-naturgefahren-.html

Open Access This chapter is licensed under the terms of the Creative Commons Attribution 4.0 International License (http://creativecommons.org/licenses/by/4.0/), which permits use, sharing, adaptation, distribution and reproduction in any medium or format, as long as you give appropriate credit to the original author(s) and the source, provide a link to the Creative Commons license and indicate if changes were made.

The images or other third party material in this chapter are included in the chapter's Creative Commons license, unless indicated otherwise in a credit line to the material. If material is not included in the chapter's Creative Commons license and your intended use is not permitted by statutory regulation or exceeds the permitted use, you will need to obtain permission directly from the copyright holder.

Chapter 3
The Response of Hurricane Inland Penetration to the Nearshore Translation Speed

Yi-Jie Zhu and Jennifer M. Collins

Abstract The destructive wind force from inland moving hurricanes may penetrate a considerable distance from the coast. The hurricane inland intensity decay and the translation speed are critical components for estimating the damage potential for inland regions. By employing the modified version of the decay period and the decay distance, this study investigates the relationship between hurricane inland decay and nearshore translation speed. We find that the variation between the decay period and decay distance can be well explained by the change of translation speed within 24 hours (h) before and after landfall. For cases that were still tropical storms within 24 h of landfall, the decay distance is found to be an exponential function of the translation speed during the first 24 h after landfall. The duration of inland destructive wind force as reflected by the decay period shows strong spatial variations along the U.S. coast. The outcome of this study may facilitate hurricane inland wind modeling as well as mitigation planning.

Keywords Hurricane · Landfall wind decay · Translation speed

3.1 Introduction

The dissipation of landfalling hurricanes has recently been brought into the spotlight with the detection of potentially slower over land intensity decay in a warmer climate (Li and Chakraborty 2020). Meanwhile, the translation speed, particularly around the time of landfall, has also received significant attention from the public since extremely slow-moving Hurricane Harvey (2017) and Hurricane Florence (2018) caused tremendous amounts of inland flooding. The destruction of hurricanes that propagate far inland or stall over a location is usually underestimated and can cause an unexpected loss of life and economic loss (Lorsolo and Uhlhorn 2020).

Y.-J. Zhu (✉) · J. M. Collins
School of Geosciences, University of South Florida, Tampa, FL, USA
e-mail: yijiezhu@usf.edu

The intensity decay of U.S. landfalling hurricanes is conventionally described by the exponential decay function as proposed by Kaplan and DeMaria (1995). The spatial distribution of the inland wind profile and intensity decay is further depicted (Kruk et al. 2010; Zhu et al. 2021a). More recently, the intensity decay period and decay distance are introduced to reflect the duration and penetration of the destructive wind force (Zhu and Collins 2021). It is also noticed that the decay period and the decay distance are not strongly related due to the variation of the storm translation speed. Furthermore, the spatial-temporal pattern of the U.S. landfalling hurricane translation speed currently remains inconclusive. Although the frequency of slow-moving hurricanes is shown to increase near the coast of the continental U.S. (Hall and Kossin 2019), the landfall events in Texas are likely to experience a shift to a faster translation speed under a warming climate (Hassanzadeh et al. 2020). This indicates the existence of both temporal and spatial variation of the storm's nearshore and inland translation speed along the U.S. coast.

While both the inland penetration and translation speed from landfalling hurricanes are key factors of their inland destructive potential, this study seeks to understand the relationship between the hurricane inland intensity decay and the nearshore translation speed. The spatial-temporal variation of the two factors is also critical in terms of future risk mitigation planning. The concept of the decay period and decay distance proposed by Zhu and Collins (2021) is employed. However, since it only considers a fixed wind threshold (from 65 kt to 35 kt) which does not represent the entire landfall process as well as the full inland damage potentials, we modify the threshold to be dynamically based on the landfall wind speed instead of the fixed 65 kt. The storm's distance of movement within the 24-h period before and after landfall is also obtained to investigate the effect of nearshore translation speed on the inland decay distance.

3.2 Materials and Methods

3.2.1 Data

We obtained the best-track dataset from the National Hurricane Center's (NHC's) North Atlantic hurricane database (HURDAT2; Landsea and Franklin 2013) as archived in the International Best Track Archive for Climate Stewardship (IBTrACS) version 4 (Knapp et al. 2010). The dataset provides tropical cyclone (TC) locations and 1 min averaged maximum sustained wind speed (MSW) intensity estimates at 6-h intervals. This study covers a 120-year period (1900–2019) for the continental U.S. landfalling hurricanes. Although satellite imagery for TCs in the North Atlantic was not available until 1966, landfalling hurricanes in the continental U.S. are considered relatively reliable since 1900 (Landsea et al. 1999; Landsea 2007; Truchelut et al. 2013). It is worth noting, however, that the accuracy of the data has steadily improved through the years due to more weather stations along the coast and improved satellite technology. Therefore, the uncertainties of the data, along with the analyses' outcomes, should be considered in this context of data reliability. This study only considers hurricanes making direct landfall in the continental United States. This excludes hurricanes making landfall in Mexico which later curved into the United States. While the 24-h pre- and post-landfall translation speed are considered, only long-lasting landfall events with at least four continuous 6-h track points on open water before landfall and another four continuous 6-h inland track points without extratropical transitions are selected. This process is determined by overlaying the TC data points with the continental U.S. map layer in ArcGIS Pro.

Since the hurricane landfall time and location could be between the 6-h synoptic interval, we also used hurricane landfall data from the Atlantic Oceanographic and Meteorological Laboratory (AOML; www.aoml.noaa.gov/hrd/hurdat/U.S.hurrs_detailed.html) to determine the landfall intensity when the center of the storm crossed the shoreline. This excludes hurricanes with MSW >64 kt wind before landfall that dissipated to tropical storm intensity when the center was over the shoreline. However, the reanalysis for the period 1970–1982 is not yet complete from either the AOML website or HURDAT2 to give landfall data actually at the coastline. To keep the consistency, allowing for this specific period to be included in the analyses, we used the last data point over the ocean and the first inland data point from HURDAT2 to linearly interpolate the intensity and time at landfall. Hurricanes that made multiple landfalls, with MSW >64 kt at each landfall, were counted as separate landfall events. The above filtering yields 90 hurricane landfall events.

3.2.2 Inland Wind Decay

The intensity decay of the inland moving hurricane can be described from different perspectives. The absolute wind changes are widely accepted as a metric quantifying the observed post-landfall intensity decay (e.g., Tuleya and Kurihara 1978; Kaplan and DeMaria 1995). Vickery (2005) later proposed an alternative version based on the change of minimum sea level pressure (MSLP). Although Klotzbach et al. (2020) suggested that the MSLP has served as a much better predictor of the observed damage than MSW in recent major U.S. landfalling hurricanes, this is typically due to a stronger relationship between MSLP and the storm surge than MSW. While the storm surge is not a major threat to inland regions, MSW may be a more useful way of describing the inland moving hurricane damage potential with the power-law relationship between the MSW and hurricane economic losses (Murnane and Elsner 2012; Zhai and Jiang 2014). More recently, Li and Chakraborty (2020) proposed the decay time scale as an indicator of how rapidly a hurricane would decay during the first day of landfall. In comparison, Zhu and Collins (2021) introduced the wind decay period and the decay distance, measuring the time and distance required for post-landfall hurricanes to decay from the hurricane wind force (with an upper threshold at MSW = 65 kt) to below tropical storm wind force (with a lower threshold at MSW = 35 kt). The decay period is defined as:

$$\text{Decay Period} = t_{lower} - t_{upper} \quad (3.1)$$

where t_{lower} is the time at the lower threshold while t_{upper} is the time at the higher threshold.

The decay distance can be obtained by summing the storm distance of movement during the decay period. This approach quantifies potential wind damage both temporally and spatially. Therefore, this study adopts Zhu and Collins' (2021) metric but sets the upper threshold to the landfall wind speed rather than fixing it at 65 kt to better represent the hurricane post-landfall damage potential. The modified decay period and decay distance describe the time and distance for a landfalling hurricane to decay from landfall MSW to below the tropical storm status. As also mentioned in Zhu and Collins (2021), landfalling hurricanes could turn back to the ocean before the intensity drops below 35 kt. Such events are mostly found along the East Atlantic and those crossing the Florida peninsula. These events (23 out of 90) are exempted from the decay period and decay distance calculation.

3.2.3 Near Shore Translation Speed

The focus of the translation speed in this study is on the 24-h period before and after the storm makes landfall. This is obtained by using the four continuous synoptic track points (6 h) from both before and after landfall to calculate the distance of movement (DM), with the reference map layer projected to the Equidistant Conic projection. Although the DM divided by 24 h yields the average translation speed, the DM alone could also serve as an alternative for the translation speed since all landfall events are measured within the same fixed 24-h period.

3.3 Effect of Translation Speed on Inland Wind Decay

We first split the landfall events into two categories indicating whether the storm has experienced acceleration (including constant; $r_{DM} \geq 1$, where $r_{DM} = DM_{post-landfall}/DM_{pre-landfall}$) or slow down ($r_{DM} < 1$) of the post-landfall translation speed relative to the pre-landfall stage (Fig. 3.1a). Landfall events from either category bring damage potential in different ways. Storms with accelerated translation speed after landfall could cause the damage to penetrate swiftly to inland regions (e.g., 2008 Hurricane Ike), while decelerated storms would stall over the coastal regions causing relatively long-lasting devastations (e.g., 2017 Hurricane Harvey). Zhu and Collins (2021) noticed that different inland moving storms could have high variations of decay distance even with the same decay period. The slope between the two describes the overall translation speed of inland moving storms during the intensity decay from the landfall MSW to below 35 kt. Here we find a significant relationship between the decay period and the decay distance with a 95% confidence interval

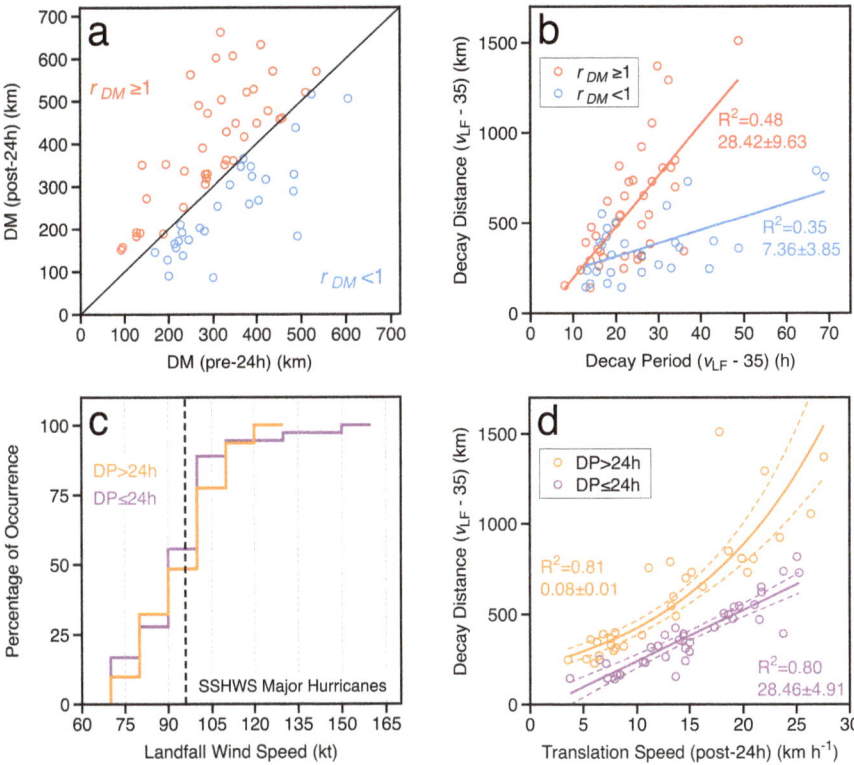

Fig. 3.1 (**a**) The hurricane distance of movement (DM) within 24 h before and after landfall. The diagonal line represents the threshold indicating the accelerated (including constant; $r_{DM} \geq 1$; red) or decelerated ($r_{DM} < 1$; blue) post-24 h translation speed relative to the pre-24 h; (**b**) Variation of the relationship between the decay period and the decay distance categorized by the r_{DM}. The solid lines from both categories represent the linear regression reported with 95% CI; (**c**) Cumulative distribution of the landfall wind speed based on whether hurricanes have decayed to tropical depression status within 24 h; (**d**) Response of the decay distance to the translation speed within 24 h of landfall. The solid lines from the two categories show the linear regression with 95% CI reported

(CI) when considering the ratio of the translation speed in the 24-h period before and after landfall (Fig. 3.1b). This implies that storms with inland deceleration tend to penetrate less but stay longer with the intensity at or above tropical storm status.

From the perspective of the decay period, we next split the landfall events into two categories based on whether the storm has maintained tropical storm status within 24 h of landfall. Figure 3.1c distinguishes the two categories with cumulative distributions on the landfall MSW. All landfall events with the Saffir-Simpson Hurricane Wind Scale (SSHWS) Category 5 intensity level (MSW $\geq$ 136 kt) have the MSW to decay below 35 kt within 24 h. This supports the empirical findings by Kaplan and DeMaria (1995), where more intense hurricanes tend to experience faster decay after landfall. However, more than half of the landfall events with less than

24 h of decay period are non-major hurricanes based on the SSHWS category. Therefore, the relationship between the landfall MSW and the inland decay is not linearly correlated when considering the wind decay period as the metric of inland intensity decay.

Since the decay distance of the storm is dependent on the change of the translation speed, understanding the response of the decay distance to the post-landfall translation speed could benefit inland hazard preparations. Figure 3.1d depicts different distributions of the decay distance over the 24-h post-landfall translation speed by the two categories indicating whether the storm has maintained tropical storm status within 24 h of landfall. For those which decayed to tropical depression status within 24 h of landfall, the decay distance can be explained by the 24-h post-landfall translation speed with a simple linear regression. This indicates that the inland penetration of TC wind from fast decaying hurricanes aligns well with the inland translation speed. In comparison, the decay distance of hurricanes with more than 24 h of decay period is better explained by the exponential fit ($R^2 = 0.81$ compared to $R^2 = 0.75$ from the linear fit). The decay distance of the long-lasting inland moving hurricanes is expected to be exponentially increased with higher 24-h post-landfall translation speed. This can be partially explained by the contributions of translation speed to the MSW (Chavas et al. 2017), as the best-track wind estimate is Earth-relative (Done et al. 2020). Although the effect of the forward motion on the MSW is considered small from intense hurricanes (Kossin 2018), it could be enlarged when TCs decay to tropical storm intensity.

It is noted that inland moving hurricanes with long decay distances contribute heavily to the exponential fit. Figure 3.2 depicts five long-lasting landfalling

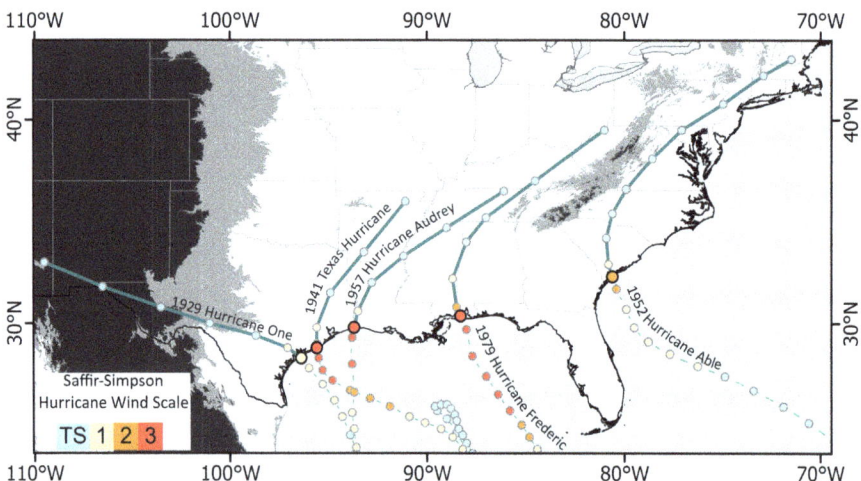

Fig. 3.2 Track of the top five landfalling hurricanes since 1900 with the longest decay distances. Track points are in 6 h intervals except for landfall and the 35 kt threshold point (last inland points). Elevation above 500 m and 1000 m is distinguished by grey and black, respectively. The topographical data are retrieved from GTOPO30 at 1 km resolution by the U.S. Geological Survey. (https://www.sciencebase.gov/catalog/item/4fb5495ee4b04cb937751d6d)

hurricanes which obtained the longest decay distance. These include Hurricane Frederic in 1979, which brought extensive damage, making it the costliest Atlantic hurricane at the time (Hebert 1980). The northward moving storms also did not encounter major terrains (e.g., 1952 Hurricane Able remained east of Appalachian Mountains (Ross 1952)). The weak topographical effects, along with higher latitudes, favor weak intensity decay and faster movement (Lanzante 2019; Yamaguchi et al. 2020). In addition, all of them have maintained their strength or intensified right before landfall, which supports the proposed positive relationship between nearshore intensification and post-landfall intensity decay (Zhu et al. 2021b).

Hurricane One in 1929 is a special case as it is marked as a very compact storm with the radius of maximum wind speed (RMS) estimated to be about 10 nm, with a 150 nm outer closed isobar (Landsea et al. 2012). The small circulation favors the storm's fast movement into New Mexico with a well-defined vortex. It is worth noting that due to the sparse observation network at the time, the inland penetration of 1929 Hurricane One was initially considered as an extratropical cyclone until the reanalysis was completed in 2010 (Landsea et al. 2012). This raised the uncertainty of not only the intensity and position but also the nature of inland moving TCs during the pre-satellite era (prior to 1970). Failure to determine whether a storm has maintained closed isobars during inland penetration or undergone extratropical transition will affect the accuracy of the decay distance calculation. Therefore, future updates of HURDAT2 data could possibly affect a small portion of the decay distance value from landfall hurricanes during the pre-satellite era, though we expect the general relationship between the translation speed and the decay distance would remain unchanged.

3.4 Long-Term Climate Variability

The impact of climate variability on TCs has been widely discussed in past decades (e.g., Webster et al. 2005; Knutson et al. 2010). However, the temporal variation of both hurricane translation speed and landfall decay has only started to be addressed in recent years. By investigating historical track positions over the period 1967–2016, Kossin (2018) argued that there is a long-term trend for TCs to decelerate, with a 20% decrease of the over land translation speed in the North Atlantic region. It was further noticed that the decadal variability of the TC translation speed over the continental U.S. is potentially linked to the Atlantic Multidecadal Oscillation (AMO) (Kossin 2019). The long-term climate mode could modify the steering flow of TCs and thus the translation speed (Chan 2019). This brings our attention to the U.S. landfall hurricane decay period (65–35 kt), which is also shown to have a potential correlation with the AMO (Zhu and Collins 2021). A considerable number of studies show the link between Atlantic hurricane activity and El Niño Southern Oscillation (ENSO; e.g., Elsner et al. 2001; Smith et al. 2007; Lin et al. 2020), where Klotzbach (2011) argues a lower U.S. continental landfalling frequency in El Niño years. Although the impact of ENSO on inland decay distance is worth investigating, the current calculation of decay period and decay distance is only suitable for long-lasting hurricanes that decay to below 35 kt after landfall before undergoing extratropical transition, either continuing to track over land or traveling back to the ocean. This reduces the sample size of landfall events and could enlarge the uncertainties when linking the decay distance to a relatively short period oscillation such as ENSO.

To compare with the long-term climate mode, we employed the AMO index provided by the NOAA Physical Sciences Laboratory (PSL; https://psl.noaa.gov/data/timeseries/AMO/) and took the average from May to November to compile the seasonal mean of the AMO index. A seven-year two-sided smoothing is applied to fill the gap years with the long-term trend removed. Similar to the findings of Zhu and Collins (2021), the decay period shows an apparent relationship with the AMO which rebounded from the low since 1980 (Fig. 3.3a), found by employing the Mann-Kendall's (non-parametric) test at the 5% significance level. This also agrees with the recent increase in decay time scale as detected by Li and Chakraborty (2020).

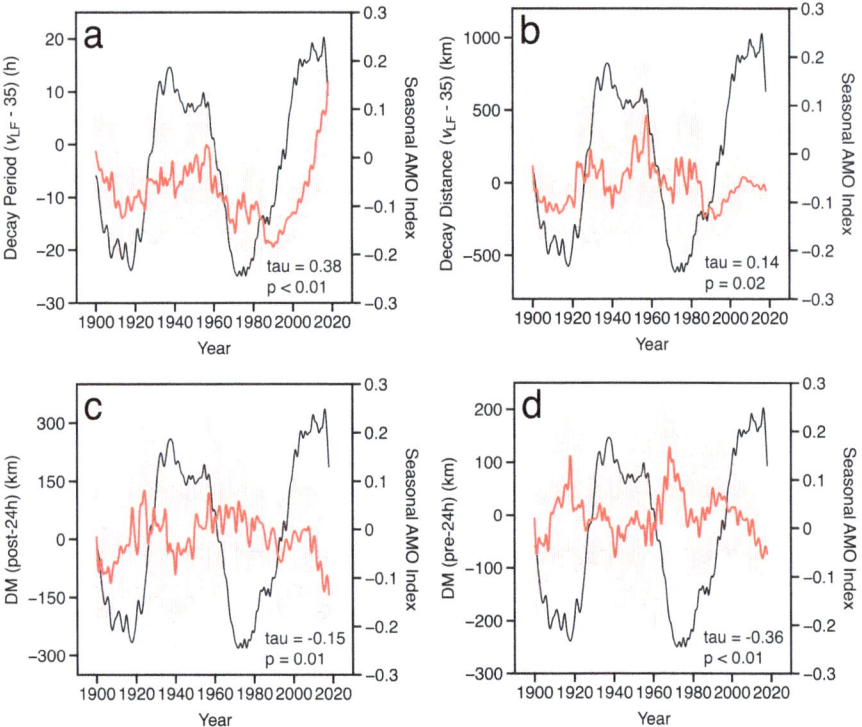

Fig. 3.3 Time series of detrended (**a**): Decay period in h; (**b**): Decay distance in km; (**c**): Distance of movement (DM) 24 h after landfall in km; (**d**): DM 24 h before landfall in km. The red line depicts the detrended 7-year two-sided moving average with standard deviations, while the black line is the detrended AMO index for reference. The correlation coefficient (tau) from the Mann-Kendall test is reported with significance (p-value)

In contrast, the relationship between the decay distance and the AMO (Fig. 3.3b) is possibly offset by the apparent negative relationship between the translation speed and the AMO (Fig. 3.3c, d), since we have shown the positive link between the 24-h post-landfall distance of movement and the decay distance (V_{LF}–35 kt). It is worth noting that the low-pass smoothing of the time series also introduces autocorrelation to the dataset and reduces the effective degrees of freedom. Therefore, the reliability of the significance of the non-parametric Kendall correlation should be considered cautiously.

3.5 Spatial Distribution

Understanding the spatial variation of hurricane inland wind decay is crucial to regional disaster mitigation. The spatial variation of hurricanes moving inland was first described by Schwerdt et al. (1979), where storms striking the Gulf Coast are claimed to have the strongest absolute wind decay and Florida the weakest. This is confirmed by a more recent study assessing the extent of the TC inland wind speed (Kruk et al. 2010). However, the absolute wind change is highly dependent on the landfall wind speed, where intense hurricanes are more likely to obtain large absolute wind change as the wind decay follows the exponential decay function (Tuleya et al. 1984; Kaplan and DeMaria 1995). By employing the landfall decay period as the metric, we present the spatial distribution of landfall events based on whether the storm has remained as a tropical storm within the first 24 h of landfall (Fig. 3.4a). No landfall event over the Florida peninsula where the system spent at least 24 h inland as a tropical cyclone has a decay period within 24 h. This supports the existing literature that landfall decay would slow down if a portion of the TC circulation remains over water (DeMaria et al. 2006). The long decay period implies hurricanes making landfall over Florida would maintain at least tropical storm strength within the first 24 h or even during the entire pathway when crossing the peninsula. Such destructive inland penetration could easily trigger statewide evacuation orders, causing large evacuation congestion and potential life and economic losses, as during Hurricane Irma in 2017 (Zhu et al. 2020; Feng and Lin 2021; Ghorbanzadeh et al. 2021).

In comparison, hurricanes landfalling over the Gulf Coast show high variations with mixed patterns. Of particular interest is the coast near Victoria in Texas (Fig. 3.4b), where all historical landfalling hurricanes in this area were obtaining greater than 24 h of decay period. Besides the unforgettable, slow-moving Hurricane Harvey in 2017, another four historical landfalling hurricanes have accelerated to the inland region. This includes Hurricane Carla in 1961, which is also one of the most intense landfalling hurricanes in the continental U.S. when taking the storm size into account (Hebert et al. 2008).

As part of the limitations of this study, landfall events that turn back to the ocean before the intensity drops below 35 kt, or while undergoing extratropical transition, are not analyzed. These typically include hurricanes crossing the Florida peninsula

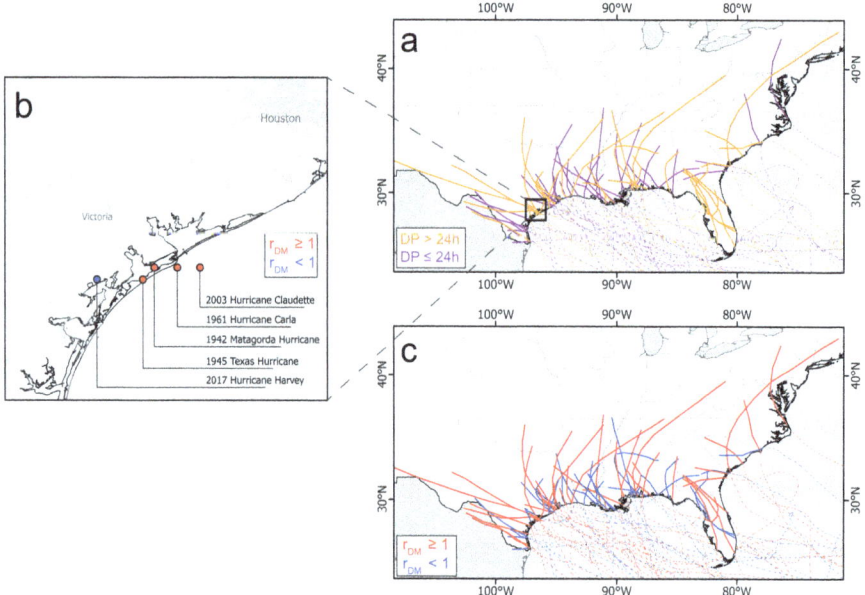

Fig. 3.4 (**a**) Spatial distribution of the landfall events distinguished by the decay period (DP); (**b**) Zoom in of the area highlighted in (**a**); (**c**) As of (**a**) but distinguished by the r_{DM}

(e.g., 2004 Hurricane Charley) or moving along the Atlantic East coast. Therefore, hurricanes making landfall while largely staying offshore off the Atlantic East coast, and those moving approximately east-west (and vice-versa) across the Florida peninsula, cannot be fully described by the landfall decay period.

The ratio of the 24-h pre- and post-landfall translation speed (r_{DM}) does not exhibit spatial patterns similar to the landfall decay period (Fig. 3.4c). Although TCs moving into higher latitude are expected to obtain larger translation speeds due to the stronger steering flow (Yamaguchi et al. 2020), the long-lived Hurricane Florence hitting North Carolina in 2018 actually maintained a slow translation speed as a result of the weakened steering current (Stewart and Berg 2019). While the TC location is projected to undergo a poleward shift in a warming climate (Kossin et al. 2014; Wang and Toumi 2021), faster translation speed overall is more likely to occur in higher latitude regions, but the r_{DM} remains uncertain.

3.6 Conclusion

In this work, the fixed decay period (65–35 kt), first introduced by Zhu and Collins (2021), is modified with the landfall decay period (V_{LF}–35 kt) to better represent the destructive force of the inland moving storms. The diverse relationship between the decay period and the decay distance can be explained by the ratio of the storm's nearshore (±24 h of landfall) translation speed. Meanwhile, for hurricanes that maintained tropical storm intensity within 24 h of landfall, the decay distance is found to be an exponential function of the translation speed within the first 24 h of landfall. This suggests that the destructive force of long-lasting inland moving storms could penetrate much farther with faster translation speed.

With the concern of the inland hurricane risk in our changing climate, we next examined the temporal variation of the key aspects regarding the hurricane inland decay. The surged landfall decay period since the 1980s reflects the slowing decay of the inland moving storms, which is in line with the literature (Li and Chakraborty 2020). The potential positive correlation found between the landfall decay period and the AMO index is similar to the result from the fixed decay period as used in Zhu and Collins (2021). Therefore, switching from the fixed decay period to the landfall decay period has minimal effect on determining the temporal variation of the inland wind decay. In comparison, the hurricane nearshore translation speed is likely to be negatively correlated with the AMO index. This helps explain the nonconcurrent pattern of the landfall decay distance with the decay period.

From the spatial perspective, except for landfalling hurricanes that underwent quick extratropical transition (e.g., 1995 Hurricane Opal) within a day of landfall, all other historical hurricanes that hit Florida during 1900–2019 had kept at least tropical storm intensity within 24 h of landfall. This supports the weak absolute wind decay found in the Florida peninsula from the existing literature (e.g., Kaplan and DeMaria 1995; Kruk et al. 2010). Landfalling hurricanes over Florida could have a portion over the ocean that helps maintain or only slowly weaken the storm (DeMaria et al. 2006). In comparison, hurricanes hitting the Gulf Coast exhibit mixed patterns with large variations among the regions. This could be a result of various topological and land surface types along the coast (Done et al. 2020). However, the coast near Victoria in Texas shows a similar pattern as in Florida (Fig. 3.3a, b), where all historical landfalling hurricanes slowly decayed with MSW maintained at or above the tropical storm status even after 24 h of landfall. Adding that 2007 Tropical Storm Erin also made landfall near this region, and even experienced a reintensification process when moving deep inland due to the soil moisture anomaly (Evans et al. 2011; Kellner et al. 2012), future studies on the land surface feedback for landfalling hurricanes in this particular region could further benefit inland risk mitigation.

References

Chan KT (2019) Are global tropical cyclones moving slower in a warming climate? Environ Res Lett 14(10):104015. https://doi.org/10.1088/1748-9326/ab4031

Chavas DR, Reed KA, Knaff JA (2017) Physical understanding of the tropical cyclone wind-pressure relationship. Nat Commun 8(1):1–11. https://doi.org/10.1038/s41467-017-01546-9

DeMaria M, Knaff JA, Kaplan J (2006) On the decay of tropical cyclone winds crossing narrow landmasses. J Appl Meteorol Climatol 45(3):491–499. https://doi.org/10.1175/JAM2351.1

Done JM, Ge M, Holland GJ, Dima-West I, Phibbs S, Saville GR, Wang Y (2020) Modelling global tropical cyclone wind footprints. Nat Hazards Earth Syst Sci 20(2):567–580. https://doi.org/10.5194/nhess-20-567-2020

Elsner JB, Bossak BH, Niu XF (2001) Secular changes to the ENSO-US hurricane relationship. Geophys Res Lett 28(21):4123–4126. https://doi.org/10.1029/2001GL013669

Evans C, Schumacher RS, Galarneau TJ Jr (2011) Sensitivity in the overland reintensification of Tropical Cyclone Erin (2007) to near-surface soil moisture characteristics. Mon Weather Rev 139(12):3848–3870. https://doi.org/10.1175/2011MWR3593.1

Feng K, Lin N (2021) Reconstructing and analyzing the traffic flow during evacuation in Hurricane Irma (2017). Transp Res Part D: Transp Environ 94:102788. https://doi.org/10.1016/j.trd.2021.102788

Ghorbanzadeh M, Burns S, Rugminiamma LVN, Erman Ozguven E, Huang W (2021) Spatiotemporal analysis of highway traffic patterns in Hurricane Irma evacuation. Transp Res Rec:03611981211001870. https://doi.org/10.1177/03611981211001870

Hall TM, Kossin JP (2019) Hurricane stalling along the North American coast and implications for rainfall. npj Climate and Atmospheric Science 2(1):1–9. https://doi.org/10.1038/s41612-019-0074-8

Hassanzadeh P, Lee CY, Nabizadeh E, Camargo SJ, Ma D, Yeung LY (2020) Effects of climate change on the movement of future landfalling Texas tropical cyclones. Nat Commun 11(1):1–9. https://doi.org/10.1038/s41467-020-17130-7

Hebert PJ (1980) Atlantic hurricane season of 1979. Mon Weather Rev 108(7):973–990. https://doi.org/10.1175/1520-0493(1980)108<0973:AHSO>2.0.CO;2

Hebert CG, Weinzapfel RA, Chambers MA (2008) The hurricane severity index–a new way of estimating a tropical cyclone's destructive potential. In: Tropical meteorology special symposium 19th conferences on probability and statistics. American Meteorological Society, New Orleans

Kaplan J, DeMaria M (1995) A simple empirical model for predicting the decay of tropical cyclone winds after landfall. J Appl Meteorol 34(11):2499–2512. https://doi.org/10.1175/1520-0450(1995)034<2499:ASEMFP>2.0.CO;2

Kellner O, Niyogi D, Lei M, Kumar A (2012) The role of anomalous soil moisture on the inland reintensification of Tropical Storm Erin (2007). Nat Hazards 63(3):1573–1600. https://doi.org/10.1007/s11069-011-9966-6

Klotzbach PJ (2011) El Niño–Southern Oscillation's impact on Atlantic basin hurricanes and US landfalls. J Clim 24(4):1252–1263. https://doi.org/10.1175/2010JCLI3799.1

Klotzbach PJ, Bell MM, Bowen SG, Gibney EJ, Knapp KR, Schreck CJ III (2020) Surface pressure a more skillful predictor of normalized hurricane damage than maximum sustained wind. Bull Am Meteorol Soc 101(6):E830–E846. https://doi.org/10.1175/BAMS-D-19-0062.1

Knapp KR, Kruk MC, Levinson DH, Diamond HJ, Neumann CJ (2010) The international best track archive for climate stewardship (IBTrACS) unifying tropical cyclone data. Bull Am Meteorol Soc 91(3):363–376. https://doi.org/10.1175/2009BAMS2755.1

Knutson TR, McBride JL, Chan J, Emanuel K, Holland G, Landsea C et al (2010) Tropical cyclones and climate change. Nat Geosci 3(3):157–163. https://doi.org/10.1038/ngeo779

Kossin JP (2018) A global slowdown of tropical-cyclone translation speed. Nature 558(7708):104–107. https://doi.org/10.1038/s41586-018-0158-3

Kossin JP (2019) Reply to: Moon IJ et al, Lanzante JR. Nature 570(7759):E16–E22. https://doi.org/10.1038/s41586-019-1224-1

Kossin JP, Emanuel KA, Vecchi GA (2014) The poleward migration of the location of tropical cyclone maximum intensity. Nature 509(7500):349–352. https://doi.org/10.1038/nature13278

Kruk MC, Gibney EJ, Levinson DH, Squires M (2010) A climatology of inland winds from tropical cyclones for the eastern United States. J Appl Meteorol Climatol 49(7):1538–1547. https://doi.org/10.1175/2010JAMC2389.1

Landsea CW (2007) Counting Atlantic tropical cyclones back to 1900. EOS Trans Am Geophys Union 88(18):197–202. https://doi.org/10.1029/2007EO180001

Landsea CW, Franklin JL (2013) Atlantic hurricane database uncertainty and presentation of a new database format. Mon Weather Rev 141(10):3576–3592. https://doi.org/10.1175/MWR-D-12-00254.1

Landsea CW, Pielke RA, Mestas-Nunez AM, Knaff JA (1999) Atlantic basin hurricanes: indices of climatic changes. Clim Chang 42(1):89–129. https://doi.org/10.1023/A:1005416332322

Landsea CW, Anderson C, Bredemeyer W, Carrasco C, Charles N, Chenoweth M, et al (2012) Documentation of Atlantic tropical cyclones changes in HURDAT. Atlantic Oceanographic and Meteorological Laboratory

Lanzante JR (2019) Uncertainties in tropical-cyclone translation speed. Nature 570(7759):E6–E15. https://doi.org/10.1038/s41586-019-1223-2

Li L, Chakraborty P (2020) Slower decay of landfalling hurricanes in a warming world. Nature 587(7833):230–234. https://doi.org/10.1038/s41586-020-2867-7

Lin II, Camargo SJ, Patricola CM, Boucharel J, Chand S, Klotzbach P et al (2020) ENSO and tropical cyclones. In: El Niño Southern Oscillation in a Changing Climate, pp 377–408. https://doi.org/10.1002/9781119548164.ch17

Lorsolo S, Uhlhorn E (2020) Managing inland hurricane wind risk: Thinking beyond the coast. Retrieved April 10, 2021, from https://www.air-worldwide.com/publications/air-currents/2020/managing-inland-hurricane-wind-risk-thinking-beyond-the-coast/

Murnane RJ, Elsner JB (2012) Maximum wind speeds and US hurricane losses. Geophys Res Lett 39(16):L16707. https://doi.org/10.1029/2012GL052740

Ross RB (1952) Hurricane able, 1952. Mon Weather Rev 80(8):138–143. https://doi.org/10.1175/1520-0493(1952)080<0138:HA>2.0.CO;2

Schwerdt RW, Ho FP, Watkins RR (1979) Meteorological criteria for standard project hurricane and probable maximum hurricane windfields, Gulf and East Coasts of the United States

Smith SR, Brolley J, O'Brien JJ, Tartaglione CA (2007) ENSO's impact on regional US hurricane activity. J Clim 20(7):1404–1414. https://doi.org/10.1175/JCLI4063.1

Stewart SR, Berg R (2019) National Hurricane Center tropical cyclone report: hurricane Florence (AL062018). National Hurricane Center 30:98

Truchelut RE, Hart RE, Luthman B (2013) Global identification of previously undetected pre-satellite-era tropical cyclone candidates in NOAA/CIRES twentieth-century reanalysis data. J Appl Meteorol Climatol 52(10):2243–2259. https://doi.org/10.1175/JAMC-D-12-0276.1

Tuleya RE, Kurihara Y (1978) A numerical simulation of the landfall of tropical cyclones. J Atmos Sci 35(2):242–257. https://doi.org/10.1175/1520-0469(1978)035<0242:ANSOTL>2.0.CO;2

Tuleya RE, Bender MA, Kurihara Y (1984) A simulation study of the landfall of tropical cyclones. Mon Weather Rev 112(1):124–136. https://doi.org/10.1175/1520-0493(1984)112<0124:ASSOTL>2.0.CO;2

Vickery PJ (2005) Simple empirical models for estimating the increase in the central pressure of tropical cyclones after landfall along the coastline of the United States. J Appl Meteorol 44(12):1807–1826. https://doi.org/10.1175/JAM2310.1

Wang S, Toumi R (2021) Recent migration of tropical cyclones toward coasts. Science 371(6528):514–517. https://doi.org/10.1126/science.abb9038

Webster PJ, Holland GJ, Curry JA, Chang HR (2005) Changes in tropical cyclone number, duration, and intensity in a warming environment. Science 309(5742):1844–1846. https://doi.org/10.1126/science.1116448

Yamaguchi M, Chan JC, Moon IJ, Yoshida K, Mizuta R (2020) Global warming changes tropical cyclone translation speed. Nat Commun 11(1):1–7. https://doi.org/10.1038/s41467-019-13902-y

Zhai AR, Jiang JH (2014) Dependence of US hurricane economic loss on maximum wind speed and storm size. Environ Res Lett 9:064019. https://doi.org/10.1088/1748-9326/9/6/064019

Zhu YJ, Collins JM (2021) Recent rebounding of the post-landfall hurricane wind decay period over the continental United States. Geophys Res Lett:e2020GL092072. https://doi.org/10.1029/2020GL092072

Zhu YJ, Hu Y, Collins JM (2020) Estimating road network accessibility during a hurricane evacuation: a case study of hurricane Irma in Florida. Transp Res Part D: Transp Environ 83:102334. https://doi.org/10.1016/j.trd.2020.102334

Zhu YJ, Collins JM, Klotzbach PJ (2021a) Spatial variations of North Atlantic landfalling tropical cyclone wind speed decay over the continental United States. J Appl Meteorol Climatol 60(6):749–762. https://doi.org/10.1175/JAMC-D-20-0199.1

Zhu YJ, Collins JM, Klotzbach PJ (2021b) Nearshore hurricane intensity change and post-landfall dissipation along the United States Gulf and East Coasts. Geophys Res Lett:e2021GL094680. https://doi.org/10.1029/2021GL094680

Chapter 4
Estimating North Atlantic Hurricane Landfall Counts and Intensities in a Non-stationary Climate

Jo Kaczmarska, Hugo C. Winter, Florian Arfeuille, and Enrica Bellone

Abstract Interest in tropical cyclone risk focuses on landfalling events, particularly on those at higher intensities, which cause the most significant damage. Modelling the frequency and intensity distributions of landfalling storms in a changing climate is extremely challenging due to the complex non-stationarities in both time and space, and the short length of the historical record. We have applied extreme value theory, in combination with non-parametric regression methods, to address the data limitations. Our method involves fitting a Poisson model for the frequency and Generalised Pareto Distribution for the intensity of landfalling U.S. storms, with the parameters of the model regressed onto a basis set of local polynomial functions (B-splines) of covariates of interest. The covariates relate to the spatial location, and to the state of the climate. We have used the model to assess the impact of different climate conditions on landfalling hurricane counts and intensities, along sections of the U.S. coastline, using annually varying climate indices.

Keywords Hurricane landfalls · P-splines · Bayes · Extremes · Climate conditioning

4.1 Introduction

Tropical cyclones are among the costliest natural catastrophes, and understanding how they depend on climate variability and trends is a highly active area of research. There are several important factors that contribute to losses within a tropical cyclone season: overall basin activity; whether storms approach land and at what intensity; the population density of affected areas; and the economic and insured values at risk. The 2005 Atlantic hurricane season is a typical case of climatic conditions favorable to tropical cyclone activity in the basin translating into high landfall counts and large

---

J. Kaczmarska (✉) · H. C. Winter · F. Arfeuille · E. Bellone
Model Development, Risk Management Solutions Ltd., London, UK
e-mail: jo.kaczmarska@rms.com; hugo.winter@rms.com; florian.arfeuille@rms.com; enrica.bellone@rms.com

economic and insured losses. However, HURDAT2 Atlantic activity data (Landsea and Franklin 2013), NOAA's list of costliest U.S. tropical cyclones (NCEI 2021), and RMS industry loss estimates provide several examples where overall basin activity does not fully explain economic and insured losses. Even the number of tropical storm and hurricane landfalls in the U.S. does not completely determine the losses in a given hurricane season. A recent example is the record season of 2020, which saw 30 named storms and 14 hurricanes in the Atlantic. Six storms made landfall in the U.S. at hurricane strength, but the storms mostly missed major urban centers. RMS real time insured loss estimates for the season total between 19 and 30 Bn USD – the order of magnitude of a single storm from 2017 (Hurricane Irma) – making it a costly but not record-breaking season. The high dependence of losses on landfall location highlights the importance of investigating impacts of climate variability and climate change at a relatively high geographical resolution.

The counts, locations, and intensities of U.S. tropical cyclone landfalls in any given year are driven by a complex and interrelated set of factors. Storm genesis counts and locations are affected by local and remote climate conditions, for example the phase of the El Niño-Southern Oscillation (ENSO), local sea surface temperatures, and the temperature profile of the atmosphere. The consequent path followed by the storm depends on the genesis location and on the steering winds, themselves affected by various climate features, such as the phase of the North Atlantic Oscillation (NAO). Local environmental conditions along the path determine the intensity evolution and thus whether the storm ultimately makes landfall and at what strength, or alternatively if it weakens and dissipates without reaching land.

Modelling the full lifecycle of tropical cyclones, whether via stochastic track models (e.g. Vickery et al. 2000; Hall and Jewson 2007) or models that combine dynamical and statistical components (e.g. Emanuel et al. 2006; Lee et al. 2018) is therefore extremely challenging. Uncertainties in every component, particularly in regions of very sparse data, are compounded over the evolution of the track, making reliable landfall estimation difficult.

Modelling the distribution of landfall counts and intensities directly, rather than taking landfalls from a track model, has the advantage of being based on more relevant data. Such a model can also be used to calibrate a basin-wide stochastic track model, which incorporates other vital information such as the storm's size, bearing, and translational speed, and maintains correlations induced by storms landfalling in multiple regions.

However, the direct modeling of landfall frequencies and intensities is not a straightforward task either, if it is to address insurance and risk mitigation requirements. The spatial resolution must be sufficiently fine to resolve the extent of urbanization around landfall locations since this will determine the level of damage and losses, as discussed. The ability to extrapolate far beyond the observation period is also important to determine insured capital requirements and engineering design criteria. The dependencies on environmental conditions of all the life stages of the cyclone result in complex temporal and spatial non-stationarities at landfall. Relationships with predictors that represent the major sources of climate variability and trends need to be incorporated and are unlikely to be linear. To our knowledge, there are no examples in the literature that fully address these requirements, although many have addressed selected parts.

To date, many of the investigations into North Atlantic landfalling tropical cyclones in the literature have segmented storms into broad categories and considered the landfall rate for each category. For example, storms have been grouped into crude regions (e.g. the whole U.S, Gulf coast, Florida coast etc.), by Saffir-Simpson intensity category (e.g. tropical storms, hurricanes and major hurricanes), and by climate condition (e.g., El Niño/la Niña or hot/cold phases).

Bove et al. (1998) studied the effect of El Niño on U.S. landfalling hurricanes, splitting years into El Niño, La Niña or neutral. Elsner and Bossak (2001) produced a climatology benchmark for hurricane counts over the whole U.S. coast and for the Gulf coast, Florida, and the East coast. Dailey et al. (2009) examined the relationship between Atlantic sea surface temperature (SST) and U.S. landfalling hurricanes by calculating the mean landfall rate for warm and cool years for the whole U.S. coast and for subregions, for tropical cyclones, hurricanes, and major hurricanes.

These approaches are useful for exploratory analysis, and benefit from their simplicity, but do not provide results at the required resolution. Relatively large groupings are required to allow enough observations for model-fitting, and considerable inhomogeneity still remains within the groups. In addition, shifts in the breakpoints of the regions and groups can lead to different results, due to their somewhat arbitrary nature. Dailey et al. (2009) addressed this last point by using a moving window of wind speeds, with results then smoothed using locally weighted scatterplot smoothing (LOWESS).

Some recent studies have acknowledged the continuous nature of climate variation and used Poisson regression for modelling the effects of the temporally varying predictors. Elsner and Jagger (2004) developed a Bayesian Poisson regression model for the mean number of coastal hurricanes, with covariates of an ENSO index, an NAO index, and their interaction. Hurricane counts for the whole U.S. coast were modelled, as well as the Southeast and Northeast coasts separately. Elsner and Jagger (2006) used a Poisson regression model for predicting U.S. coastal hurricanes for the season ahead using the NAO, Southern Oscillation Index (SOI) and Atlantic Multidecadal Oscillation (AMO) as predictors. Staehling and Truchelut (2016) used Poisson regression to relate spatially and temporally averaged upper-level divergence, relative SST, meridional wind and zonal shear vorticity to total U.S. hurricane landfall counts. A nonparametric approach was taken by Villarini et al. (2012) who used a generalized additive model for the total count of U.S. landfalling hurricanes, with predictors including tropical Atlantic SST, tropical mean SST, the NAO and SOI.

Several studies have used continuous distributions to model tropical cyclone intensities, although none that we have seen have simultaneously incorporated smooth temporal and spatial variability. Jagger and Elsner (2006) fit separate Poisson and Generalized Pareto Distribution (GPD) models for three near-coastal regions of the U.S. coast (Gulf Coast, Florida, and the East Coast), as well as a combined model for all the regions. The impact of several climate indices (a global temperature index, the AMO, the NAO and the SOI) was examined using simple binary splits of the data into above/below normal conditions. This discrete approach was acknowledged to be suboptimal, and the authors also introduced a Bayesian hierarchical model, where the Poisson rate and GPD parameters were regressed on a single climate covariate.

Only a small number of studies have modelled landfalls at a finer spatial resolution. Jagger et al. (2001) fit a Weibull distribution to yearly maximum tropical cyclone wind speeds for the coastal counties, with scale and shape parameters related to ENSO and NAO. Rumpf et al. (2010) used kernel-based smoothing and generalized nearest neighbor methods to fit an inhomogeneous Poisson process to landfall counts. However, the data were still categorized into discrete bins according to hot/cold years and strength (tropical storms, minor hurricanes, major hurricanes). Casson and Coles (1999) developed a Bayesian hierarchical model for spatial extremes where the smooth variation in the parameters of an extreme value model over the region of interest was defined by a latent spatial process. They illustrated the method with an example using simulated hurricane wind speed measurements over equal length coastal segments of the U.S. Tolwinski-Ward (2015) used a similar approach to model the annual frequency of landfalling tropical cyclones along piecewise-linear segments of the Atlantic coast of North America using a spatial generalized linear model. The Poisson rate was regressed on three climate indices (SOI, AMO and NAO), with a residual Gaussian process to capture the spatial structure.

In this chapter, we build on the existing approaches in the literature by developing a Poisson-GPD model that estimates the frequency and intensity distributions of tropical cyclones over piecewise-linear segments of the U.S. coastline with smooth variation over environmental covariates. We take a slightly different approach to that of Casson and Coles (1999) and Tolwinski-Ward (2015). Similarly to both of these studies, we use Bayesian inference, which has the advantage that return level estimates incorporate the parameter uncertainty. However, we replace the use of a Gaussian process to capture spatial structure, with P-splines, which combine a B-spline parameterization with simple roughness penalties (Eilers and Marx 1996; Lang and Brezger 2004), for both spatial and temporal covariates. This approach is preferred for its flexibility, simplicity, and its lower computational requirements.

While we have not seen this methodology used in the context of tropical cyclone wind speeds, there are examples of its application in other environmental and financial fields. Several studies have used P-splines for the modelling of extreme waves (e.g. Jonathan et al. 2014; Randell et al. 2016). Jones et al. (2016) carried out an interesting simulation study in the context of extreme wave heights that compared different nonparametric approaches relating the shape and scale parameters of a GPD to a single covariate representing wave direction. The different options considered included constant, Fourier, B-spline, and Gaussian process parameterizations, using both likelihood and Bayesian inference. They found that spline and Gaussian process parameterizations performed best. The Gaussian process parameterization was found to be computationally unwieldy unless gridding on the covariate domain was performed. In other domains, P-splines have been used to model spatial and/or temporal effects by Rodriguez-Alvarez et al. (2018) for agricultural field experiments, by Fahrmeir and Osuna (2003) for car insurance claim frequencies, and by Brezger and Lang (2006), who demonstrated applications relating to forest health and health insurance.

The rest of this paper is structured as follows. In Sect. 4.2 we discuss the data, including the climate indices considered as potential temporal predictors and the approach taken to incorporate the spatial non-stationarity. A motivating example is introduced that is followed throughout the rest of the paper. Section 4.3 describes the model methodology. Results for the motivating example are presented in Sect. 4.4, which also gives summary findings for the whole coastline. We conclude with a discussion and consideration of future improvements in Sect. 4.5.

4.2 Data

4.2.1 Track Data and the Spatial Predictor

The historical tracks are taken from the HURDAT2 reanalysis database, maintained by the National Hurricane Center (available online from https://www.nhc.noaa.gov/data/hurdat/hurdat2-1851-2020-052921.txt), from which the track positions and corresponding maximum windspeeds (Vmax) are extracted. Although historical wind speed data are subject to uncertainties when going back to the pre-satellite period, the values at landfall are considered reliable, and we use observations from 120 years, spanning 1900–2019. We have divided the North Atlantic U.S. coastline into 69 linear segments of varying length, depending on the shape of the coastline (typical length of 100 km). These segments define the resolution of our fit (i.e., we shall estimate separate frequency and intensity distributions for each of the 69 coastal segments).

Although it is possible to fit a single model to all the coastal segments at once, with some location-related predictor/s to reflect their position, we chose instead to proceed as follows. For each of the coastal segments, a set of 100 km linear extensions on each side are created, up to a maximum of 600 km (6 extensions each side), and we add the data from tracks crossing these segments to augment the relatively sparse data from the original segment. A separate model is then fitted to the data from each original segment and its extensions.

Extensions that are close to the original segment will have a much greater influence on its fitted mean rate than those that are far away. Farther extensions contribute primarily to how the rate varies in different climate conditions. This is a consequence of the model structure which is explained in Sect. 4.3. The intensity model behaves in a similar way. Figure 4.1 shows an example of a coastal segment in Texas with its 12 extensions. We define the spatial predictor, s_i, as the displacement from the center of segment i to the centre of the coastal segment of interest, with $i = 1 \ldots 13$. The same segment is shown in Fig. 4.2, with the historical tracks that cross it and any of its extensions. The windspeed values used in the fit are from the

Fig. 4.1 Example of extension segments created for a coastal segment in Texas. The predictor s_i is defined as the displacement from the center of the original segment to the center of segment i

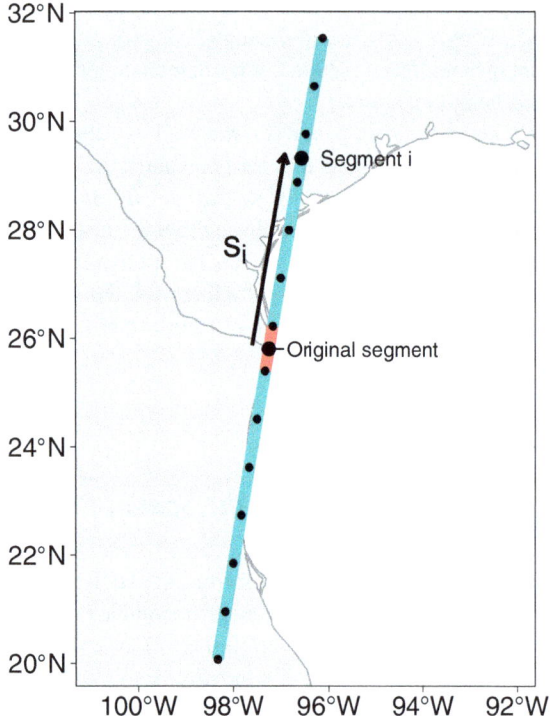

Fig. 4.2 Historical tracks crossing a costal segment in Texas (in thick blue) and its extensions (dashed blue). Colors of the tracks indicate the Saffir Simpson category of the track maximum windspeed at the 6-hourly point before crossing the segment/extension (cat5: red, cat 4: dark orange, cat3: orange, cat2: yellow, cat1: cream, tropical storm: cyan)

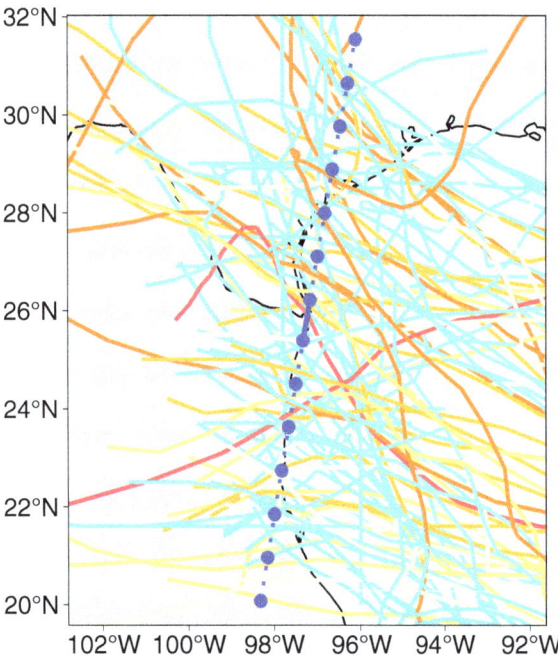

last over-water 6 hourly point before crossing a segment, and the number of extensions is reduced in situations where we considered that the climatic conditions would not be similar over the region covered. As an example, the extensions of a few segments in the Northeastern section of the Gulf coast are truncated to the east, in order to exclude any in the Atlantic. The segment smoothing methodology is further described in Sect. 4.3.

Although fitting multiple models in this way is a more onerous undertaking than using a single model for the whole coastline, we believe this disadvantage is outweighed by the advantages. In particular, it means that storms included in a given fit are expected to respond similarly to a given set of climate conditions, without the need to include interaction terms between the spatial and temporal predictors, as would be required for the single model approach. The chosen method also ensures that information from storms having similar characteristics to the ones landfalling at a specific segment can always be included, even where the orientation of the segment makes its experience markedly different to its coastal neighbors. A potential concern with the approach is that the reliability of the data tends to reduce with distance from land. However, none of the segments are far from the coast and, as discussed, the influence of the data in the fits also reduces with distance from the original coastal segment.

4.2.2 Climate Data and Temporal Predictors

Landfalls along the North Atlantic coastlines exhibit variation that can be associated with changes in climate conditions. Although there is considerable debate in the literature as to the historical drivers of some of the changes, this is not of concern for the purposes of this study. Our motivation is rather to develop a model which relates tropical cyclone landfall frequencies and intensities to the main environmental factors that impact them, and which can be used to model potential future scenarios. We have chosen three climate indices to use as temporal predictors in our model. The selection of these indices was based on a balance between a number of factors: the strength of their relationship to physical processes which impact tropical cyclone genesis, intensification and paths; the reliability with which they can be estimated, and the extent to which stakeholders are familiar with them.

Although local sea surface temperature is widely used in empirical relationships for tropical cyclone activity, it is not ideal for climate impact studies, since the potential for intensification depends not only on SST, but also on the thermodynamic imbalance within the troposphere and the outflow temperature (Emanuel et al. 2013). Using the SST without accounting for these other effects is expected to lead to an overstatement of the levels of future activity as the climate continues to warm (Vecchi and Soden 2007). The combined influence from SST, vertical temperature and humidity profiles, and outflow temperature can be estimated by calculating the Potential Intensity (PI), which represents the thermodynamic limit of the intensity a storm can reach given its environment (Bister and Emanuel 1998). Rather than using an index based on windspeed, we calculate the Minimum Sustainable Pressure (MSP), in pressure units. The code used to estimate the MSP is provided by Kerry Emanuel and can be downloaded from Emanuel (2021).

For the years 1900–2015, the input data used for the MSP calculation is taken from the Twentieth Century Reanalysis project (20CR, Compo et al. 2001) version 3 (Slivinski et al. 2019, 2021). 20CR is a product based on the assimilation of surface pressure using observed SSTs as boundary conditions and has a horizontal resolution of 1 degree. One of the main considerations in the choice of reanalysis was the relative temporal homogeneity of 20CR, due to the absence of drastic changes in assimilated observations over the study period. The years 2016–2019, which are not available in 20CR, are computed using the NCEP reanalysis dataset, adjusted to reflect the bias between the two reanalysis sets. To mitigate the impact of storms themselves on the SSTs (cold wake), we first calculate the tenth quantile of the MSP daily distribution for each month (i.e. we use the day with the third lowest MSP/highest intensity potential, which is unlikely to have been significantly impacted by any storm activity), and then use the June–November average of these values for each year (1900–2019). The values are finally averaged spatially over the hurricane main development region and standardized to create an annual index (MDR: 10 N–20 N, 70 W–15 W).

The MSP index does not account for processes that can inhibit storm intensification such as vertical wind shear, for which another proxy is required. ENSO is a well-known climate pattern associated with SST anomalies in the Pacific. It influences storms in the Atlantic through modulation of the vertical wind shear and tropospheric humidity/temperature, La Niña/El Niño conditions having a positive/negative influence on storm activity respectively. We use the Oceanic Niño Index (ONI) to represent ENSO, based on the averaged SST over the Niño 3.4 region (5 S–5 N, 120–170 W), for which 3 months' running mean anomalies are computed based on 30-year periods updated every 5 years. SST data is taken from HadISST (Rayner et al. 2003) for years 1900–2019.

The final index included is the NAO, which is potentially important for landfall activity due to its impact on storm paths. It is characterized by modifications in the strength of the Bermuda high pressure system (enhanced/reduced during the NAO positive/negative phase). We use an NAO index based on a principal component analysis of sea level pressure anomalies in the Atlantic, as defined in Hurrell et al. (2003). There is no established consensus in the literature on the months used to create an annual NAO index. It is generally thought that the NAO values in the early hurricane season months have a larger influence on tropical cyclone landfalls, thus here we use anomalies over the April–August season.

Figure 4.3 shows that all three of these indices have an effect on the example Texas coastal segment and its extensions, at least in the storm frequencies. For this segment, over the historical period considered, ENSO has the most significant impact, with over five times the rate of landfalling storms in La Niña conditions, compared with El Niño. The MDR MSP index has the next highest impact, with over

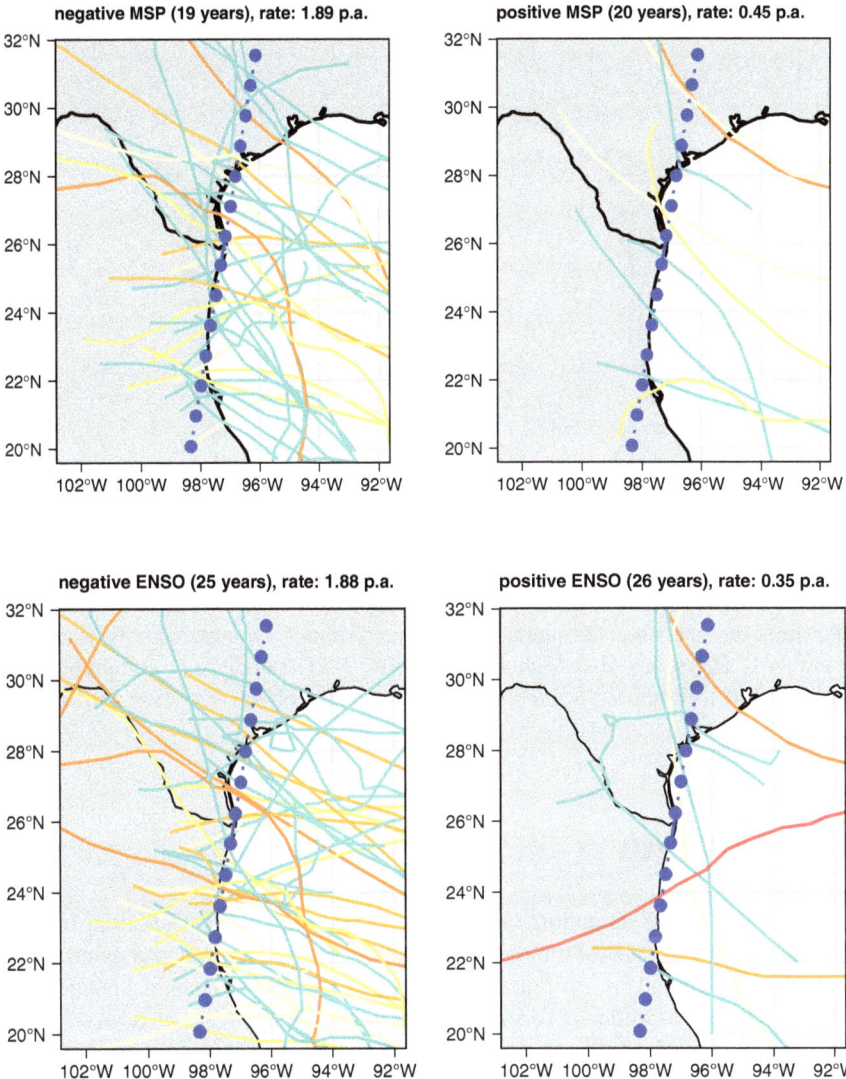

Fig. 4.3 Historical tracks crossing a costal segment in Texas (in thick blue) and its extensions (dashed blue) in different climate conditions. The rate shown is the total across all the segments. Colors of the tracks are as in Fig. 4.2 and indicate the Saffir Simpson category at the 6-hourly point before crossing the segment/extension. MSP and NAO negative and positive conditions are defined as greater than one standard deviation from the mean. ENSO negative/positive phases are defined as greater than 0.5 and less than −0.5 respectively

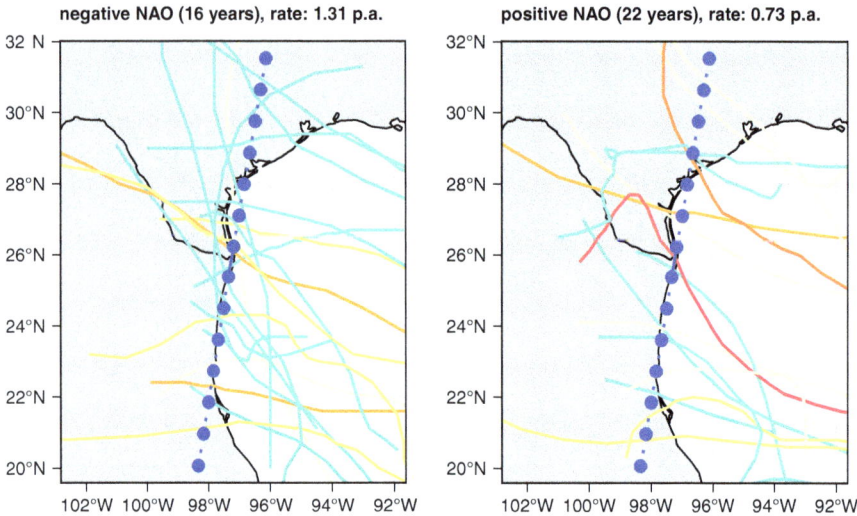

Fig. 4.3 (continued)

four times the rate when the standardized index is above 1 compared with when it is below −1. Rates are also higher in negative NAO conditions, compared with positive, but this impact is much smaller. Intensity effects are more difficult to judge visually.

4.3 Methodology

An important aim of modelling the frequency and intensity of landfalling tropical cyclones is to understand how the probability of exceeding particular wind speed values (often plotted as exceedance frequency (EF) curves) could change with different climate conditions. To this end, we now outline the statistical model that we use to estimate this type of quantity. In this section we explain how our method is applied to the single coastal segment and its extensions described in Sect. 4.2. For illustrative purposes, the predictors are taken as the displacement from the mid-point of the coastal segment, and a single climate index. Ultimately, we fit models with different combinations of the four potential predictors described in Sect. 4.2. Since all terms in the model are assumed to be additive, any equations shown here can readily be extended to include more predictors. Section 4.3.1 outlines the Poisson-GPD model and Sect. 4.3.2 introduces the P-splines approach used to relate predictors to the model parameters. This model is fitted in the Bayesian framework, as described in Sect. 4.3.3. Finally, our model selection approach is explained in Sect. 4.3.4.

4.3.1 Poisson-GPD Model

The intensity of landfalling tropical cyclones is assumed to follow a GPD, which is a two-parameter distribution, suitable for modelling the tails of other distributions (see Coles 2001 for an introduction to extreme value modelling). It requires a threshold value, above which the distribution can be assumed to hold. In this study, it is necessary to provide inferences for all extreme levels of tropical storm intensity and stronger (Vmax >17 m/s). Therefore, we fit the GPD model with a threshold of 17 m/s for each segment. However, for some segments, standard threshold diagnostics (such as the mean residual life plot and parameter stability plot, see Coles 2001) would suggest a higher threshold. In this case we fit an additional GPD model at a second higher threshold and smooth the resulting exceedance frequency curves in between the two thresholds.

Following Chavez-Demoulin and Davison (2005) and other authors, we re-parameterize the GPD to have orthogonal parameters, ξ (shape) and $\nu = \beta(1 + \xi)$ (adjusted scale). The adjusted scale parameter ν is assumed to depend on both the displacement and the climate index, but the shape parameter does not depend on predictors.

Conditional on a tropical cyclone crossing segment i in year j with intensity above a threshold value, u, the intensity y_{ij}, follows a GPD distribution with density:

$$f_Y(y_{ij}|s_i, x_j, u) = \begin{cases} \dfrac{(1+\xi)}{\nu(s_i, x_j)} \left[1 + \dfrac{\xi(1+\xi)}{\nu(s_i, x_j)}(y_{ij} - u)\right]^{-\left(1+\frac{1}{\xi}\right)}, & \xi \neq 0 \\ \dfrac{1}{\nu(s_i, x_j)} \exp\left(\dfrac{-(y_{ij} - u)}{\nu(s_i, x_j)}\right), & \xi = 0 \end{cases}$$

for $y_{ij} > u$ when $\xi \geq 0$, and $0 < y_{ij} < u - \dfrac{\nu(s_i, x_j)}{\xi(1+\xi)}$ when $\xi < 0$. The parameter $\nu(s_i, x_j)$ is a non-parametric function of s_i, the displacement of the mid-point of segment i from the midpoint of the coastal segment being fitted (as shown in Fig. 4.1), and x_j, the value of the climate index in year j.

The rate of occurrence of storms crossing segment i in year j with intensity above the threshold u is assumed to have a Poisson distribution with rate $\lambda(s_i, x_j)$ per 100 years and 100 km:

$$f_N(n_{ij}|s_i, x_j, l_i) \sim \text{Poisson}\left[\dfrac{l_i NY}{100^2} \lambda(s_i, x_j)\right],$$

where NY is the number of years in the data, l_i is the length of the i th segment in km.

The nature of the non-parametric functions $\nu(s_i, x_j)$ and $\lambda(s_i, x_j)$ is explained in Sect. 4.3.2. Briefly, we assume that there is no interaction between s_i and x_j, so the effect of the climate predictor does not vary with distance from the original costal segment.

As already mentioned, a key goal of the analysis is to estimate exceedance frequencies (EF). The EF (per 100 km and 100 years) for windspeed value y, conditional on climate index x, and spatial displacement s is defined as:

$$EF(y|s,x) = \lambda(s,x)[1 - F_Y(y|s,x,u)],$$

where $\lambda(s,x)$ is the rate parameter of the Poisson distribution and F_Y is the distribution function of the GPD. The EF is calculated at a set of values y using the parameter estimates from the fitted Poisson-GPD model to create an EF curve.

4.3.2 P-splines

The relationships between the climate and displacement predictors and Poisson rate/GPD scale parameter are not necessarily linear. Regressing a response on a set of basis functions of a predictor is one way of allowing for non-linearity in a relationship. One could, for example, regress a response on the polynomial basis $(1, x, x^2, \ldots, x^p)$ of a predictor, x, to fit a non-linear relationship between the response and the predictor. A drawback of such a basis, however, is its global nature, which makes it difficult to achieve a good fit over the whole domain of x simultaneously. B-splines are local polynomial basis functions that are computed recursively as differences of truncated power functions (de Boor 1978). They make an attractive and much more numerically stable basis.

Eilers and Marx (1996) introduced a curve fitting method that combines regression on a large set of equally spaced B-splines (of the order of 30 to 50) with a penalty to control the smoothness of the function, based on the differences between adjacent coefficients. They called the method P-splines to reflect the B-spline basis and the simple difference penalties.

The penalty term is multiplied by a smoothing parameter, which acts in a similar way to the bandwidth in kernel-smoothing. As the multiplier is reduced, the function is allowed to become less and less smooth. As it increases to infinity the limiting function is constant if using first order differences for penalties, and linear if using second order differences. Figure 4.4 shows an example function, smoothed with second-order penalties, with different values of the smoothing parameter.

The order of the penalties also affects the extrapolation, which is assumed constant for first-order and linear for second-order penalties. Second order penalties are generally preferred and have been used for the GPD and Poisson model fits. However, given the sparsity of observations in the tails of the climate predictor distributions, linear extrapolation was considered undesirable, so we chose to moderate the extrapolation by incorporating a fixed additional first-order penalty on the outer few coefficients only (Eilers and Marx 2010).

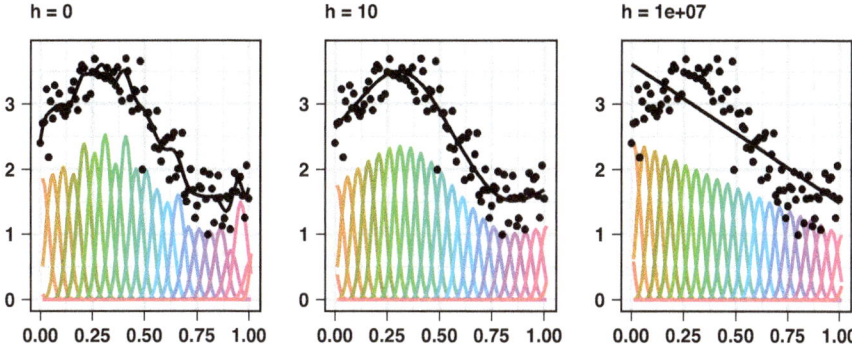

Fig. 4.4 Examples of P-spline regression with 20 B-splines and second order penalties with different values of the smoothing parameter, h

4.3.3 Bayesian Model and Implementation

A Bayesian version of P-splines was introduced by Lang and Brezger (2004). It replaces the difference penalties with their stochastic analogues, which are Gaussian random walk priors for the regression coefficients.

From a Bayesian perspective, parameters are treated as random variables, for which we estimate posterior distributions, given the likelihood and our prior beliefs about the parameters.

Poisson Component
For the Poisson component, the parameters for which we estimate posterior distributions are the coefficients $\beta_{\lambda 1}, \beta_{\lambda 2}$, and the smoothing parameters, $h_{\lambda 1}, h_{\lambda 2}$, and their joint posterior distribution is given by:

$$f(\beta_{\lambda 1}, \beta_{\lambda 2}, h_{\lambda 1}, h_{\lambda 2} \mid n, s, x, l) \propto f_N(n \mid s, x, l, \beta_{\lambda 1}, \beta_{\lambda 2}) f(\beta_{\lambda 1} \mid h_{\lambda 1}) f(\beta_{\lambda 2} \mid h_{\lambda 2})$$
$$f(h_{\lambda 1} \mid a_{\lambda 1}, b_{\lambda 1}) f(h_{\lambda 2} \mid a_{\lambda 2}, b_{\lambda 2}).$$

The first component on the right-hand side, $f_N(n \mid s, x, l, \beta_{\lambda 1}, \beta_{\lambda 2})$ is the Poisson likelihood, conditional on the covariates s and x and rate λ, and given segment lengths l, with λ given by:

$$\lambda = \exp(B_{Ns}\beta_{\lambda 1} + B_{Nx}\beta_{\lambda 2}),$$

where B_{Ns} is the design matrix of B-splines based on the displacement from the center of the segment to the center of the coastal segment being fitted, and B_{Nx} is the design matrix of B-splines for the climate index predictor. The dimension of both basis matrices is $(NY \times NS) \times 50$, where NY is the number of years as before (120 in this study), NS is the number of linear segments (typically 13 as shown in Fig. 4.1), and we have used 50 B-splines.

The second and third terms are the prior densities for the coefficient vectors $\beta_{\lambda 1}$ and $\beta_{\lambda 2}$ respectively, given the smoothing parameters, $h_{\lambda 1}$ and $h_{\lambda 2}$. With first order penalties, these would be given by:

$$\beta_{\lambda 1,0} \sim N(0,1),$$

$$\beta_{\lambda 1,k} \sim N\left(\beta_{\lambda 1,k-1}, \frac{1}{h_{\lambda 1}}\right), k = 1 \ldots 50,$$

and similarly for $\beta_{\lambda 2}$. This construction sets the mean for a coefficient to the previous coefficient's value with the precision parameter, $h_{\lambda 1}$, controlling how different it is permitted to be ($h_{\lambda 1}$ high means low variance, so coefficients will be very similar to each other i.e. a high level of smoothing).

With second order penalties, we have instead,

$$\beta_{\lambda 1,0} \sim N(0,1),$$

$$\beta_{\lambda 1,1} \sim N\left(\beta_{\lambda 1,0}, \frac{1}{h_{\lambda 1}}\right), \quad \beta_{\lambda 1,k} \sim N\left(2\beta_{\lambda 1,k-1} - \beta_{\lambda 1,k-2}, \frac{1}{h_{\lambda 1}}\right), \quad k = 2 \ldots 50.$$

Finally, the last two terms $f(h_{\lambda 1} | a_{\lambda 1}, b_{\lambda 1})$ and $f(h_{\lambda 2} | a_{\lambda 2}, b_{\lambda 2})$ are hyper-priors for the smoothing parameter. These were both specified as uninformative Gamma priors, with shape parameters $a_{\lambda 1} = a_{\lambda 2} = 1$ and rate parameters $b_{\lambda 1} = b_{\lambda 2} = 0.0005$.

GPD Component

For the GPD component the parameters for which we estimate posterior distributions are the shape parameter ξ, the coefficients $\beta_{\nu 1}, \beta_{\nu 2}$, and their respective smoothing parameters, $h_{\nu 1}, h_{\nu 2}$, and their joint posterior distribution is given by:

$$f(\xi, \beta_{\nu 1}, \beta_{\nu 2}, h_{\nu 1}, h_{\nu 2} | y, s, x, u) \propto f_Y(y|s, x, \xi, \beta_{\nu 1}, \beta_{\nu 2}, u) f(\beta_{\nu 1}|h_{\nu 1}) f(\beta_{\nu 2}|h_{\nu 2})$$
$$f(h_{\nu 1}|a_{\nu 1}, b_{\nu 1}) f(h_{\nu 2} | a_{\nu 2}, b_{\nu 2}) f(\xi|a_\xi, b_\xi).$$

The setup for the coefficients is exactly the same as for the Poisson component. However, in this case we used Gaussian priors for the $\log_{10}$ of the smoothing parameters, with mean and standard deviation of 2, instead of the more usual Gamma priors. These were found to perform better in terms of the convergence of the Markov chain Monte Carlo (MCMC) algorithm. Finally, a fairly informative Gaussian prior was used for the shape parameter with mean -0.35 and standard deviation of 0.1. This was based on our belief that the windspeed distribution should be bounded.

In this study, we use a recently developed approach called the No-U-Turn Sampler (NUTS) within the framework of Hamiltonian Monte Carlo (HMC) (Hoffman and Gelman 2014) to sample from the posterior distribution. The chosen

MCMC scheme is implemented in Stan interface to the R statistical software (Stan Development Team 2021a, b; Vehtari 2017).

4.3.4 Model Selection

The previous sections describe the model fitting for a single coastal segment and its extensions. We repeat this process for all the coastal segments covering the U.S. coastline (see Sect. 4.2.1) and retain the posterior parameter distributions and corresponding EF curves only for the original segments, discarding the results for the extensions. These outputs can also be combined to give estimates at lower resolutions of interest (e.g. at a region composed of multiple separate segments).

Several Poisson-GPD models with different predictor structures (i.e. combinations of the potential predictors) are fitted and evaluated separately at each coastal segment. The choice of model selection technique aims to balance the minimization of segment-level out-of-sample prediction errors with the need to maintain consistency between neighboring segments along the coastline.

Standard approaches to model selection (see McElreath (2020) for a review) include the Akaike Information Criterion (AIC), the Bayesian Information Criterion (BIC), the Deviance Information Criterion (DIC), and cross-validation approaches (such as leave one out cross validation, LOO).

Applying these standard approaches separately at each of the 69 original segments has the drawback that small differences in model performance are likely to lead to different predictor structures at coastal segments in close proximity, even when the physical dissimilarities are limited. As an example, consider two neighboring segments: in the first the model with only the ENSO predictor is marginally preferred over one with only the NAO, and in the second this slight preference is reversed. The ENSO-only model would be selected for the first, and the NAO-only for the second, leading to very different results at two neighboring and climatically similar segments during certain phases of the ENSO and NAO.

An alternative approach is to choose a set of fixed regions (e.g. Texas, Central Gulf, Florida, etc.) and to aggregate the model selection criteria from the separate models within each region to decide upon a predictor structure for the whole region. This approach leads to consistent inferences within each region. However, the definition of regions is open to subjectivity and could lead to discontinuities at the edges.

The approach used here avoids the need to define specific regions subjectively and ensures a smooth transition between the coastal segments. It achieves this by fitting all possible predictor structures at each coastal segment and combining them using Bayesian model averaging (BMA). We considered two approaches for calculating model weights: (i) the Watanabe-Akaike Information Criterion (WAIC) (Watanabe 2010); (ii) stacking of predictive distributions (Yao et al. 2018).

WAIC and stacking both aim to assess the out-of-sample prediction error but take slightly different approaches. WAIC is calculated on the within-sample results with an adjustment for potential overfitting whereas stacking utilizes LOO cross validation instead. Both approaches can be used in the Bayesian context as they look across the whole posterior distribution, as opposed to focusing on point estimates (as do AIC, BIC and DIC).

WAIC is less computationally intensive than stacking, since stacking requires each model to be fitted n times (where n is the number of data points in the fit). However, the use of Pareto smoothed importance sampling leave-one-out cross validation (PSIS-LOO) helps to reduce the computational complexity (Vehtari et al. 2017). Ultimately, we have chosen to use stacking, primarily because it performs better when comparing several models with similar predictors (Yao et al. 2018).

The final model is defined by exceedance frequency curves for each segment. These are obtained by first averaging over the samples from the MCMC output for each of the considered predictor structures, and then combining these into a single model using the model averaging weights. In this way, both parameter and model uncertainty are incorporated.

4.4 Results

The results section is organized in the following way. We first illustrate the model averaging approach, including the list of predictor structures for the frequency and intensity models, and an assessment of the model weights across the U.S. coast. Then, we take a single segment within Texas (the same segment as in Sect. 4.2) to provide model fit diagnostics and illustrate how the proposed model works. Finally, we piece together results from all the segments around the Atlantic U.S. coast and describe some dominant themes.

4.4.1 Model Selection and Model Averaging

Section 4.3 outlined the approach for combining the fits from multiple Poisson-GPD models with different predictor structures, and this section describes the results. Initial exploratory analysis was undertaken to limit the number of potential models and Table 4.1 shows the final set of predictors used for the frequency and intensity models.

Differences between frequency and intensity predictor structures relate to the displacement (Disp) and NAO covariates. The NAO is included in the set of potential frequency predictors, as it is observed to have an impact on tropical cyclone occurrence in some regions within the North Atlantic basin. On the other hand, the NAO does not appear to have a noticeable impact on the storm intensity, so it has not been included in any of the intensity models. The displacement predictor is included in all the frequency models, while we allow for intensity models with only climate predictors, only displacement, or both. This choice for the intensity framework stems

Table 4.1 List of different predictors used within the frequency and intensity models

Frequency predictors	Intensity predictors
Disp + MSP	MSP
Disp + ENSO	ENSO
Disp + NAO	MSP + ENSO
Disp + MSP + ENSO	Disp
Disp + MSP + NAO	Disp + MSP
Disp + ENSO + NAO	Disp + ENSO
	Disp + MSP + ENSO

The final set of combined frequency-intensity models is created by combining all possible predictor structures from each column to give 42 different combinations

from the difficulty in reliably estimating a large number of parameters with a small number of observations. By including models without the displacement predictor, it becomes easier to explore the impact of climate indices.

Figure 4.5 shows model averaging weights for the Poisson frequency model at each coastal segment using the stacking method. The segments are numbered from 1 to 69, starting at the border with Mexico and finishing in Maine. The plot shows that the fitted model has the desired property that the weights change fairly smoothly along the coastline, particularly in the areas of higher risk. Any discontinuities typically reflect different segment orientations. The weights give an indication of the relative importance of the different predictors in different parts of the coastline. For example, the model averaging identifies the NAO as an important predictor within the Central Gulf region (segments 9–20) as well as the Southeast (39–48) and Northeast (49–69) Atlantic coast, while in Texas (1–8) and the Florida peninsula (21–38) it assigns higher weights to models including ENSO and MSP. A similar plot can be constructed for the intensity model (not shown); the corresponding variation of the weights along the coastline is fairly smooth, although less so than in Fig. 4.5.

Exceedance frequency curves for each of the considered predictor structures are combined into a single EF curve for each segment, using the model averaging weights from the frequency and intensity models. This model average will be referred to as the fitted model in the following sections and all plots and results are taken from this model.

4.4.2 Poisson-GPD Results with Predictors

Results from a single segment are first presented as an example of a Poisson-GPD model fit. Figure 4.6 shows the rates of observed tropical cyclone landfalls and the fit from the Poisson model for the chosen segment in Texas and the extensions either side. The data and fit for the coastal segment are shown at the zero displacement value. To construct this plot, the climate indices are set to their observed values. The P-spline methodology captures the predictor relationship well, with the smoothing

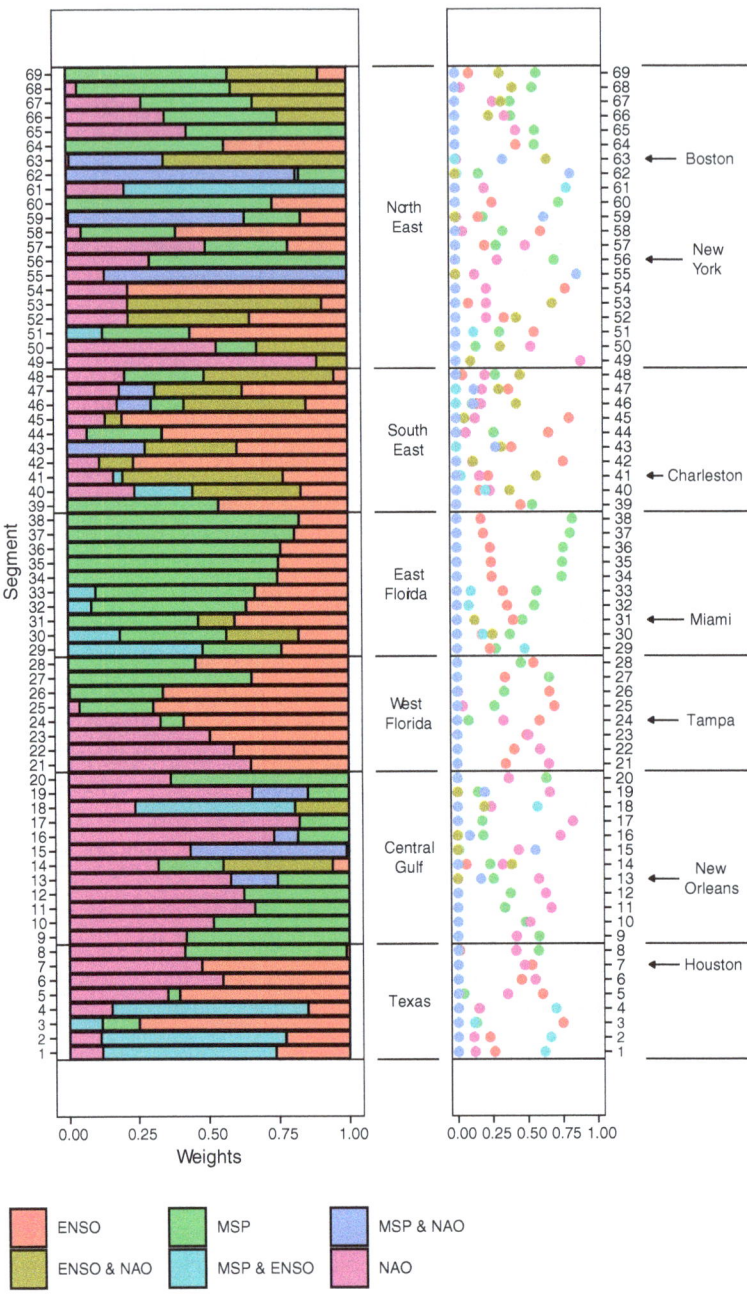

Fig. 4.5 BMA weights for the Poisson frequency model using the stacking method incorporating six different models with varying covariate structures. The same data has been presented in two different ways (stacked barplot, and simple point plot) to bring out different features. Segments follow the U.S. mainland coast from 1 at the Texas/Mexico border to 69 on the Northeast U.S./Canada border; region names and locations of some major cities have also been included

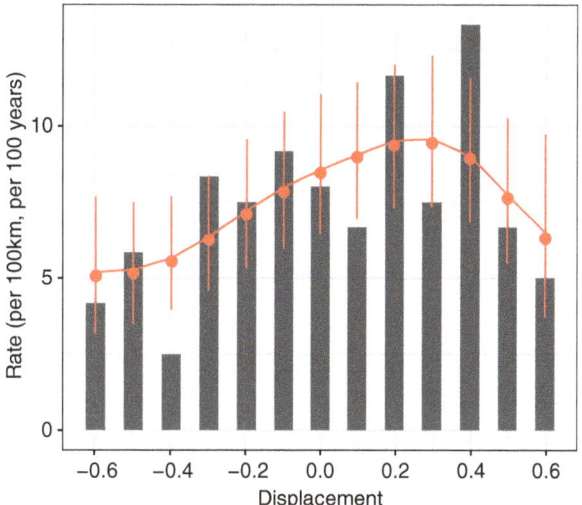

Fig. 4.6 Tropical cyclone landfalling rate (per 100 km and per 100 years) across segment 2 in Texas (displacement = 0) and extension gates for observations (bar plot) and for the Poisson model fitted with P-splines (red line). The posterior mean is given by the red dot and a 90% credible interval is shown by the red intervals around the mean

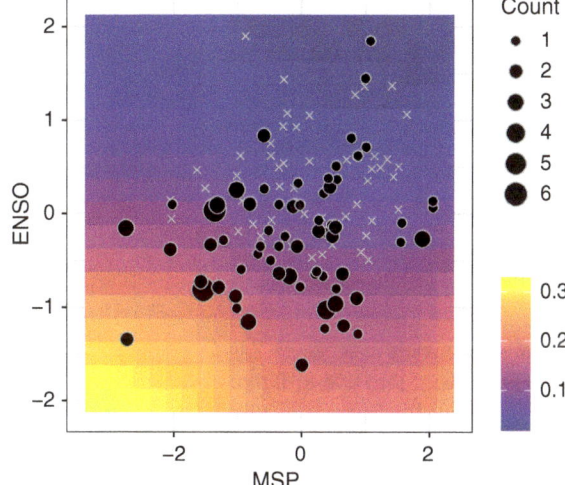

Fig. 4.7 Summaries of the model fit to observed landfalls for segment 2 in Texas. Observed counts (dots) given for each year over the segment and its extensions (years with no events are represented by grey crosses) with the fitted Poisson rate parameter for the segment itself (expressed per annum) given by background shading. The NAO is assumed to be neutral in the fitted values

resulting in a slight increase in the rate at the chosen segment, compared to its observed value.

Figure 4.7 shows how the tropical cyclone landfall counts and fitted rates vary with the ENSO and MSP indices at the same segment in Texas. Each point in the plot represents the count over one of the years in the period 1900 to 2019. The background shading represents the fitted parameter, λ, expressed as a rate per annum. We observe that the rate tends to be larger during negative MSP conditions and during La Niña (the negative phase of the ENSO index), which broadly agrees with the observed data.

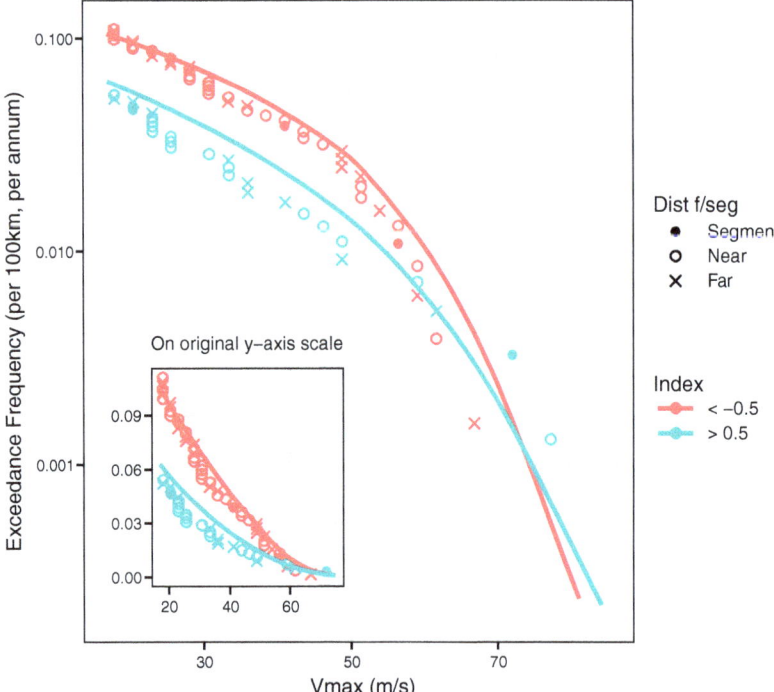

Fig. 4.8 Observed (points) and fitted (lines) exceedance frequency curves for segment 2 in Texas for different MSP conditions: MSP < −0.5 (red); MSP > 0.5 (blue). Observations include storms crossing segment 2 itself (filled points) and its extensions (near: < 300 km away, circles; far: 300–600 km away, crosses). The y-axis of the main plot is on a log scale to highlight fit in larger extremes, the inset plot is on the original scale to highlight the model fit at smaller extremes

Figure 4.8 combines the Poisson and GPD components into Vmax exceedance frequency curves at segment 2 in Texas and compares them with observations. The curves are split into two different MSP conditions, grouping together years in which the MSP index is less than −0.5, and those where it is greater than +0.5. The model EF curves are created from fits at the observed values of the predictors. The plot shows the sparsity of data at the segment itself (three observations for each of the MSP conditions) and highlights the importance of incorporating additional data from the extensions into the fit. The contribution of extensions decays with distance; for visual ease the observations are binned by distance from the original segment and plotted accordingly. The model reflects the impact of MSP conditions reasonably well, given inherent uncertainties at long return periods. Positive MSP leads to fewer and generally weaker tropical cyclone landfalls, both in the model and observations. At rarer exceedance frequencies, our estimates of Vmax tend to be larger for positive MSP than for negative MSP, a feature which is also evident in the underlying data at this segment.

4.4.3 Overall Results and Dominant Themes

Results such as those shown in Figs. 4.7 and 4.8 have been produced for all the segments of the U.S. coast, for different climate conditions. In this section we discuss the dominant themes that emerge from this analysis.

Table 4.2 summarizes results for the U.S. coast as a whole and gives the landfall rate and proportion of tropical cyclones that landfall as major hurricanes (Category 3 and above on the Saffir-Simpson scale) in selected climate conditions. The results that relate landfall activity to the MSP index are of most interest in terms of understanding the impact of climate change. The values of the MSP as the climate continues to warm are expected to reduce (Ting et al. 2019), and our analysis indicates that this would lead to enhanced U.S landfall frequency for tropical cyclones, hurricanes, and major hurricanes. For reference, there is no clear consensus in the literature on whether tropical storm and hurricane numbers increase or decrease with climate change, while an intensity increase is generally accepted (Knutson et al. 2020; Emanuel 2020). However, the results in this section are not directly comparable to these studies, as most climate change research had focused on basin counts rather than landfall.

The Poisson-GPD model also reflects climate variability impacts on landfall activity across the whole U.S. coast: Table 4.2 shows ENSO affects frequencies and intensities, while the NAO has a somewhat smaller effect.

Although overall U.S. results are of interest, it is also important to investigate regional differences in impacts. Figure 4.9 shows maps of the difference in modeled landfalling tropical cyclone rates per 100 years between negative, positive, and neutral values of ENSO, and negative and neutral values of the MSP and NAO indices. In each plot, the values of the other climate covariates have been set to zero. Figures 4.10 and 4.11 combine landfall frequency and intensity for example regions,

Table 4.2 Fitted tropical cyclone landfall rates and intensity ratios in different climate conditions for the whole U.S. coast

Predictor	Value	Fitted landfall rate p.a.	Intensity ratio
ENSO	−2	5.74	0.18
	0	4.43	0.13
	+2	3.50	0.12
MSP	−2	5.66	0.14
	0	4.43	0.13
	+2	4.42	0.10
NAO	−2	4.61	0.15
	0	4.43	0.13
	+2	4.02	0.12

The intensity ratio is equal to the fitted rate of major hurricanes (category 3+) divided by the fitted rate of tropical cyclones. In each row, the corresponding predictor is set to the Value column, with all other covariates set to zero (e.g., the top row provides results from our fitted model with the ENSO index set at −2 and the MSP and NAO covariate set at 0)

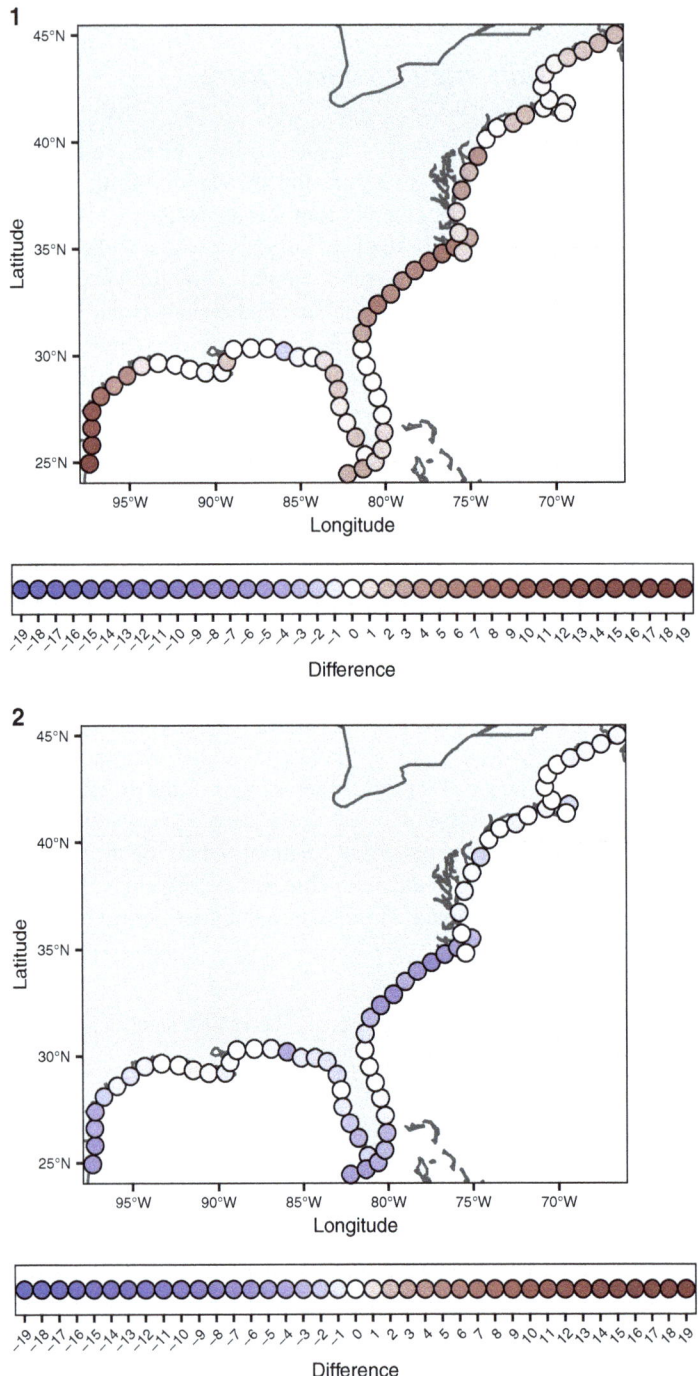

Fig. 4.9 Map of the difference in the landfall rate per 100 years from the fitted model between: ENSO = −2 and ENSO = 0 (1st plot); ENSO = +2 and ENSO = 0 (2nd plot); MSP = −2 and MSP = 0 (3rd plot); NAO = −2 and NAO = 0 (4th plot). For each plot, all other covariates within the fitted model have been set to neutral (i.e. set to zero)

4 Estimating North Atlantic Hurricane Landfall Counts and Intensities in...

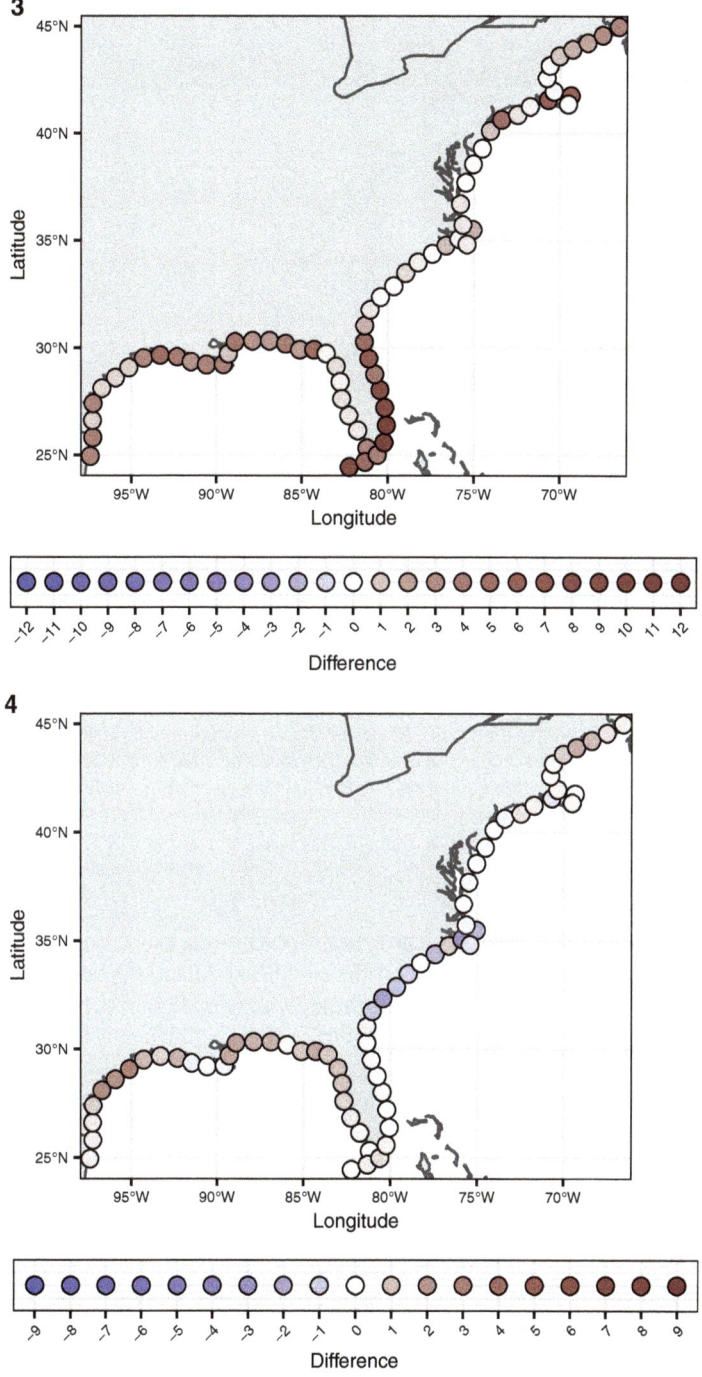

Fig. 4.9 (continued)

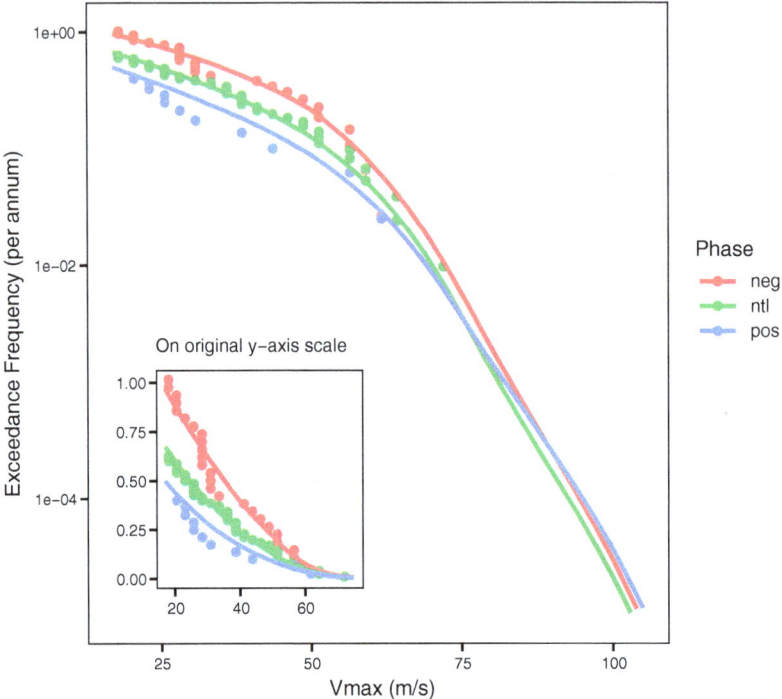

Fig. 4.10 EF curves split by the ENSO phase for the Texas region with historical observations (points) and fitted model (lines). The y-axis of the main plot is on a log scale to highlight fit in larger extremes, the inset plot is on the original scale to highlight the model fit at smaller extremes. ENSO positive/negative phases are defined as index values greater than 0.5 and less than −0.5 respectively

showing Vmax exceedance frequency curves split by phases of the climate covariates.

The first two maps in Fig. 4.9 show that ENSO has a major impact on tropical cyclone landfalling rates in Texas and the Southeast Atlantic coast, with La Niña conditions leading to more landfalls in these two areas. Figure 4.10 compares the fitted Poisson-GPD model against observed landfalls for different ENSO phases in Texas, and shows that our model captures observed distributions well in this region. Jagger and Elsner's (2006) analysis suggested that at longer return periods, return levels were higher in El Niño conditions than in La Niña in the Gulf region. This effect can be seen in Fig. 4.10 in Texas to some extent, and more clearly in the Central Gulf region (not shown). We would caution, however, that estimation at high return levels in conditions of data sparsity is uncertain. The south facing Gulf region (from Eastern Texas to the Florida panhandle) only shows limited dependence on ENSO in terms of tropical cyclone frequency. This can be explained by a shift in genesis locations during El Niño conditions, with relatively more storms originating in the northern part of the Gulf of Mexico and moving northwards, compensating for the reduced activity in the MDR.

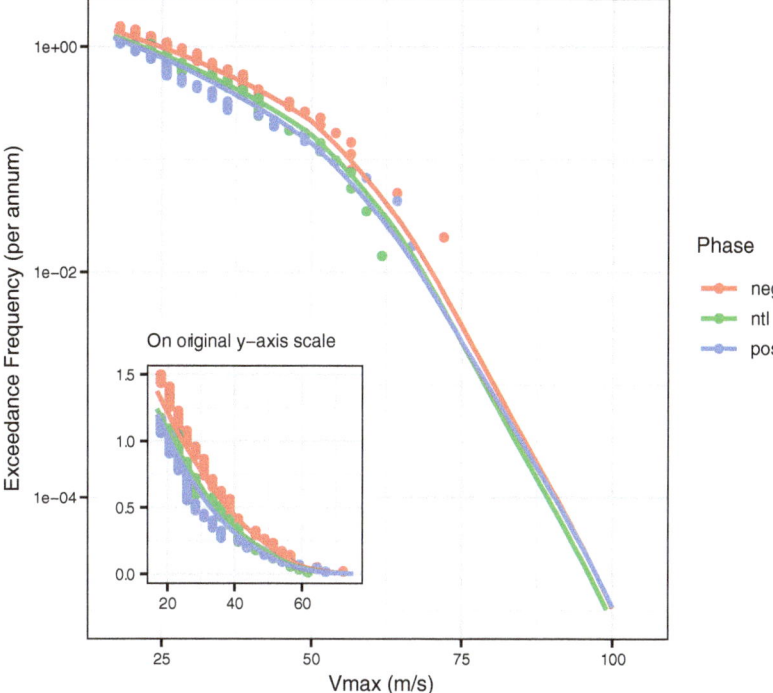

Fig. 4.11 EF curves split by MSP conditions in the Central Gulf region for historical observations (points) and fitted model (lines). The y-axis of the main plot is on a log scale to highlight fit in larger extremes, the inset plot is on the original scale to highlight the model fit at smaller extremes. MSP negative and positive conditions are defined as greater than one standard deviation from the mean

Landfall frequencies along the Florida East coast show some dependence on ENSO conditions, mainly in the South (Fig. 4.9). In general, the ENSO dependence, with overall more frequent landfalls in the U.S., confirms previous findings (e.g. Bove et al. 1998).

The MSP climate predictor is found to have the greatest effect in Eastern Florida and the Central Gulf, with negative conditions leading to more landfalling tropical cyclones, as shown in Fig. 4.9 (3rd plot). Figure 4.11 shows the agreement between observed and modeled Vmax exceedance frequencies in the Central Gulf, split by different MSP phases. Higher intensities are associated with the negative phase, especially at short and medium return periods. At higher return periods, this effect is less clear in the observations, and the fitted curves for the different phases are closer together. In the Southeast region of the U.S. (between Northeast Florida and Virginia), there is a tendency for stronger vertical wind-shear off the coast when SST conditions are favorable in the MDR (Kossin et al. 2017). This wind shear-SST dipole translates into a relatively limited (and non-linear) landfall frequency dependency on the MSP index (Fig. 4.9 3rd plot). In general, most of the U.S. coastline shows an increased landfall frequency in negative MSP conditions.

The impact of the NAO on landfall activity in the U.S. is generally smaller than the effect of ENSO and MSP. There is no clear consensus in the literature as to the physical explanations for the regional NAO impacts. Figure 4.9 (4th plot) shows that the modeled landfall variability due to the NAO is mostly limited to the Gulf region and the Southeast Atlantic coast. Positive NAO conditions are linked to decreased landfall frequencies in the Gulf, and increased landfall frequencies at the Southeast Atlantic coast. The behavior is reversed in negative NAO conditions and may be partly due to the tendency for storms to track straighter into the Gulf of Mexico during NAO negative conditions, and recurve more towards the Atlantic coast during positive conditions (e.g., Elsner 2003). Other potential explanations include the prevalence of different Atlantic SST patterns in positive and negative NAO phases, with conditions in the negative phase being generally favorable to TC landfalls along the U.S., with the exception of the Southeast region.

4.5 Discussion and Conclusion

In this paper, we have estimated the frequency and intensity distributions of tropical cyclones over piecewise-linear segments of the U.S. coastline under different climate conditions. The relatively sparse data for each segment has been enhanced by the addition of storms that cross climatically similar linear extensions to either side. The methodology is based on semi-parametric Poisson-GPD models, fitted in a Bayesian framework, with P-splines used for both the spatial smoothing and to allow non-linear relationships with several climate indices. Finally, models with different sets of predictors have been combined using Bayesian Model Averaging. This modelling strategy has allowed us to model tropical cyclone landfalls with continuous, relatively smooth variation along the coastline, making effective use of all the available data.

The results broadly confirm findings from earlier studies. Both ENSO and the MSP indices are found to have major impacts on U.S. tropical cyclone landfalling rates, with a lesser, but nevertheless important, effect from the NAO. Rates are found to increase significantly in La Niña conditions in Texas and the Southeast Atlantic coast, but little dependence on ENSO is predicted in the south facing Gulf region. The latter result can be explained from a physical perspective by the shift in genesis location in El Niño conditions, with relatively more storms originating in the northern part of the Gulf and tracking northward compared to La Niña conditions, compensating for the reduced activity in the MDR. The MSP predictor was found to have the greatest impact in the Central Gulf, and Eastern Florida with negative index values (hotter conditions) leading to more tropical cyclones. The tendency to see a stronger wind-shear off the Atlantic coast when SSTs are high in the MDR is thought to explain the limited impact in the Southeast region. The NAO impact is seen as a decrease in landfall frequencies in the Gulf and an increase in the Southeast Atlantic during its positive phase. This is consistent with a tendency for storms to recurve more in the Atlantic during positive conditions.

Intensity effects are less clear. The methodology allows the spatial variation in the intensity distribution to be modelled smoothly along the coastline. In terms of the climate dependence, a moderate increase in the ratio of major storms to all tropical cyclones is estimated between strong la Niña and strong El Niño conditions, and similarly between strongly negative and strongly positive MSP conditions.

The complexity of climate signals makes it necessary to exercise caution when interpreting the results of modelling studies. Different climatic effects combine to influence landfall rates and intensities in different regions in a basin. Any statistical method will struggle to reproduce all of these signals, especially given the relative sparsity of landfall data and the correlation between some of the predictors. These issues make it particularly important to relate model results to potential physical explanations and to highlight agreement or disagreement with published literature. As in all statistical methods applied to climate data, there is an implicit assumption that the relationships will hold in a changing climate. Known concerns with SST as a predictor (Sobel et al. 2019) have been addressed by replacing it with MSP, but nevertheless care should be taken if projecting far into the future. For example, the strength of the dipole between SST (and hence MSP) and wind-shear, discussed in the Results section, may decrease as the climate warms.

We conclude with some suggestions for further research. In this study, the shape parameter of the GPD is assumed to be independent of covariates. Initial exploratory analysis suggested that relationships between the shape parameter and climate predictors could not be estimated robustly, due to data sparsity. As a result, regional variation is incorporated by fitting the shape parameter separately at each coastal segment, but no temporal variability is included in the model. Investigation of the EF curves along the U.S. coastline, in different climate conditions, suggests that further research into the shape parameter and its potential dependence on climate predictors may be warranted.

The threshold above which the GPD is assumed to hold is another area of future research. The current method assumes a threshold of 17 m/s at all coastal segments. Where needed, it uses a second higher threshold and smooths the two resulting EF curves in between. A potential improvement would be to incorporate a non-extreme value distribution below a single threshold (e.g. Randell et al. 2016). More robust methods of threshold selection could also be incorporated (e.g. Northrop and Coleman 2014; Northrop and Attalides 2017).

Another potential area worth investigating further is the model selection approach. Data at each coastal segment is sparse, so including all potential predictors in every model would lead to over-fitting. Here we chose to address this issue by fitting multiple models with subsets of the potential predictors and using Bayesian Model Averaging to combine them. An alternative might be to use slab and spike priors. This approach would include all potential predictors in a single model for each segment and force the coefficients towards zero for the weaker signals to avoid overfitting (Scheipl et al. 2012).

References

Bister M, Emanuel KA (1998) Dissipative heating and hurricane intensity. Meteorog Atmos Phys 65:233–240

Bove MC, Elsner JB, Landsea CW, Niu X, O'Brien JJ (1998) Effect of El Niño on U.S. landfalling hurricanes, revisited. Bull Am Meteorol Soc 79(11):2477–2482

Brezger A, Lang S (2006) Generalized structured additive regression based on Bayesian P-splines. Comput Stat Data Anal 50(4):967–991

Casson E, Coles S (1999) Spatial regression models for extremes. Extremes 1:449–468

Chavez-Demoulin V, Davison AC (2005) Generalized additive modelling of sample extremes. J R Stat Soc Ser C Appl Stat 54(1):207–221

Coles SG (2001) An introduction to statistical modeling of extreme values. Springer, London

Compo GP, Whitaker JS, Sardeshmukh PD, Matsui N, Allan RJ, Yin X, Gleason BE, Vose RS, Rutledge G, Bessemoulin P, Brönnimann S, Brunet M, Crouthamel RI, Grant AN, Groisman PY, Jones PD, Kruk MC, Kruger AC, Marshall GJ, Maugeri M, Mok HY, Nordli Ø, Ross TF, Trigo RM, Wang XL, Woodruff SD, Worley SJ (2001) The twentieth century reanalysis project. Q J R Meteorol Soc 137:1–28

Dailey PS, Zuba G, Ljung G, Dima IM, Guin J (2009) On the relationship between North Atlantic sea surface temperatures and U.S. hurricane landfall risk. J Appl Meteorol Climatol 48(1): 111–129

De Boor C (1978) A practical guide to splines. Springer, New York

Eilers PH, Marx BD (1996) Flexible smoothing with B-splines and penalties. Stat Sci 11(2):89–121

Eilers PH, Marx BD (2010) Splines, knots, and penalties. WIREs Comput Stat 2(6):637–653

Elsner JB (2003) Tracking hurricanes. Bull Am Meteorol Soc 84(3):353–356

Elsner JB, Bossak BH (2001) Bayesian analysis of U.S. hurricane climate. J Clim 14(23): 4341–4350

Elsner JB, Jagger TH (2004) A hierarchical Bayesian approach to seasonal hurricane modelling. J Clim 17(14):2813–2827

Elsner JB, Jagger TH (2006) Prediction models for annual U.S. hurricane counts. J Clim 19(12): 2935–2952

Emanuel KA (2020) Response of global tropical cyclone activity to increasing CO2: results from downscaling CMIP6 models. J Clim 34:57–70. https://doi.org/10.1175/JCLI-D-20-0367.1

Emanuel K (2021) Products | Kerry Emmanuel. https://emanuel.mit.edu/products. Accessed 9 July 2021

Emanuel K, Ravela S, Vivant E, Risi C (2006) A statistical deterministic approach to hurricane risk assessment. Bull Am Meteorol Soc 87(3):299–314

Emanuel K, Solomon S, Folini D, Davis S, Cagnazzo C (2013) Influence of tropical tropopause layer cooling on Atlantic hurricane activity. J Clim 26:2288–2301

Fahrmeir L, Osuna L (2003) Structured count data regression. Collaborative Research Center 386, Discussion Paper 334

Hall TM, Jewson S (2007) Statistical modelling of North Atlantic tropical cyclone tracks. Tellus A Dyn Meteorol Oceanogr 59(4):486–498

Hoffman MD, Gelman A (2014) The no-U-turn sampler: adaptively setting path lengths in Hamiltonian Monte Carlo. J Mach Learn Res 15:1593–1623

Hurrell JW, Kushnir Y, Ottersen G, Visbeck M (2003) An overview of the North Atlantic oscillation. Geophys Monogr Am Geophys Union 134:1–36

Jagger TH, Elsner JB (2006) Climatology models for extreme hurricane winds near the United States. J Clim 19(13):3220–3236

Jagger TH, Elsner JB, Niu X (2001) A dynamic probability model of hurricane winds in coastal counties of the United States. J Appl Meteorol 40(5):853–863

Jonathan P, Randell D, Wu Y, Ewans K (2014) Return level estimation from non-stationary spatial data exhibiting multidimensional covariate effects. Ocean Eng 88:520–532

Jones M, Randell D, Ewans K, Jonathan P (2016) Statistics of extreme ocean environments: non-stationary inference for directionality and other covariate effects. Ocean Eng 119:30–46

Knutson TR, Camargo SJ, Chan JCL, Emanuel KA, Chang-Hoi H, Kossin JP, Mohapatra M, Satoh M, Sugi M, Walsh KJE, Wu L (2020) Tropical cyclones and climate change assessment: part II: projected response to anthropogenic warming. Bull Am Meteorol Soc 101(3):E303–E322. https://doi.org/10.1175/BAMS-D-18-0194.1

Kossin JP, Hall T, Knutson T, Kunkel KE, Trapp RJ, Waliser DE, Wehner MF (2017) Extreme storms. In: Wuebbles DJ, Fahey DW, Hibbard KA, Dokken DJ, Stewart BC, Maycock TK (eds) Climate science special report: fourth National Climate Assessment, volume I. U.S. Global Change Research Program, Washington, DC, pp 257–276

Landsea CW, Franklin JL (2013) Atlantic hurricane database uncertainty and presentation of a new database format. Mon Weather Rev 141:3576–3592

Lang S, Brezger A (2004) Bayesian P-splines. J Comput Graph Stat 13(1):183–212

Lee C, Tippett MK, Sobel AH, Camargo SJ (2018) An environmentally forced tropical cyclone hazard model. J Adv Model Earth Syst 10:223–241. https://doi.org/10.1002/2017MS001186

McElreath R (2020) Statistical rethinking: a Bayesian course with examples in R and STAN, 2nd edn. Chapman and Hall/CRC, Boca Raton

NOAA National Centers for Environmental Information (NCEI) (2021) U.S. billion-dollar weather and climate disasters. https://www.ncdc.noaa.gov/billions/. Accessed 9 July 2021

Northrop P, Attalides N (2017) Cross-validatory extreme value threshold selection and uncertainty with application to ocean storm severity. J R Stat Soc Ser C 66(1):93–120

Northrop PJ, Coleman CL (2014) Improved threshold diagnostic plots for extreme value analyses. Extremes 17(2):289–303

Randell D, Turnbull K, Ewans K, Jonathan P (2016) Bayesian inference for nonstationary marginal extremes. Environmetrics 27(7):1180–4009

Rayner NA, Parker DE, Horton EB, Folland CK, Alexander LV, Rowell DP, Kent EC, Kaplan A (2003) Global analyses of sea surface temperature, sea ice, and night marine air temperature since the late nineteenth century. J Geophys Res 108(D14):4407

Rodriguez-Alvarez MX, Boer MP, van Eeuwijk FA, Eilers PH (2018) Correcting for spatial heterogeneity in plant breeding experiments with P-splines. Spat Stat 23:52–71

Rumpf J, Weindl H, Faust E, Schmidt V (2010) Structural variation in genesis and landfall locations of North Atlantic tropical cyclones related to SST. Tellus A Dyn Meteorol Oceanogr 62:243–255

Scheipl F, Fahrmeir L, Kneib T (2012) Spike-and-slab priors for function selection in structured additive regression models. J Am Stat Assoc 107(500):1518–1532

Slivinski LC, Compo GP, Whitaker JS, Sardeshmukh PD, Giese BS, McColl C, Allan R, Yin X, Vose R, Titchner H, Kennedy J, Spencer LJ, Ashcroft L, Brönnimann S, Brunet M, Camuffo D, Cornes R, Cram TA, Crouthamel R, Domínguez-Castro F, Freeman JE, Gergis J, Hawkins E, Jones PD, Jourdain S, Kaplan A, Kubota H, Le Blancq F, Lee T-C, Lorrey A, Luterbacher J, Maugeri M, Mock CJ, Kent Moore GW, Przybylak R, Pudmenzky C, Reason C, Slonosky VC, Smith CA, Tinz B, Trewin B, Valente MA, Wang XL, Wilkinson C, Wood K, Wyszyński P (2019) Towards a more reliable historical reanalysis: improvements for version 3 of the twentieth century reanalysis system. Q J R Meteorol Soc 145:2876–2908

Slivinski LC, Compo GP, Sardeshmukh PD, Whitaker JS, McColl C, Allan RJ, Brohan P, Yin X, Smith CA, Spencer LJ, Vose RS, Rohrer M, Conroy RP, Schuster DC, Kennedy JJ, Ashcroft L, Brönnimann S, Brunet M, Camuffo D, Cornes R, Cram TA, Domínguez-Castro F, Freeman JE, Gergis J, Hawkins E, Jones PD, Kubota H, Lee TC, Lorrey AM, Luterbacher J, Mock CJ, Przybylak RK, Pudmenzky C, Slonosky VC, Tinz B, Trewin B, Wang XL, Wilkinson C, Wood K, Wyszyński P (2021) An evaluation of the performance of the twentieth century reanalysis version 3. J Clim 34(4):1417–1438

Sobel AH, Camargo SJ, Previdi M (2019) Aerosol versus greenhouse gas effects on tropical cyclone potential intensity and the hydrological cycle. J Clim 32:5511–5527

Staehling EM, Truchelut RE (2016) Diagnosing United States hurricane landfall risk: an alternative to count-based methodologies. Geophys Res Lett 43(16):8798–8805

Stan Development Team (2021a) Stan modeling language users guide and reference manual, 2.27. https://mc-stan.org. Accessed 9 July 2021

Stan Development Team (2021b) RStan: the R interface to Stan, R package version 2.21.2. http://mc-stan.org/. Accessed 9 July 2021

Ting M, Kossin JP, Camargo SJ, Li C (2019) Past and future hurricane intensity change along the U.S. East Coast. Sci Rep 9(1):7795. https://doi.org/10.1038/s41598-019-44252-w

Tolwinski-Ward SE (2015) Uncertainty quantification for a climatology of the frequency and spatial distribution of North Atlantic tropical cyclone landfalls. J Adv Model Earth Syst 7: 305–319

Vecchi G, Soden B (2007) Effect of remote sea surface temperature change on tropical cyclone potential intensity. Nature 450:1066–1071

Vehtari A (2017) Extreme value analysis and user defined probability functions in Stan, Stan Case Studies. https://mc-stan.org/users/documentation/case-studies/gpareto_functions.html. Accessed 9 July 2021

Vehtari A, Gelman A, Gabry J (2017) Practical Bayesian model evaluation using leave-one-outcross-validation and WAIC. Stat Comput 27:1413–1432

Vickery PJ, Skerlj PF, Twisdale LA (2000) Simulation of hurricane risk in the U.S. using empirical track model. J Struct Eng 126(10):1222–1237

Villarini G, Vecchi GA, Smith JA (2012) U.S. landfalling and North Atlantic hurricanes: statistical modeling of their frequencies and ratios. Mon Weather Rev 140(1):44–65

Watanabe S (2010) Asymptotic equivalence of Bayes cross validation and widely applicable information criterion in singular learning theory. J Mach Learn Res 11:3571–3594

Yao Y, Vehtari A, Simpson D, Gelman A (2018) Using stacking to average Bayesian predictive distributions (with discussion). Bayesian Anal 13(3):917–1007

Chapter 5
Analysis of the Future Change in Frequency of Tropical Cyclone-Related Impacts Due to Compound Extreme Events

Patrick A. Harr, Antoni Jordi, and Luke Madaus

Abstract Tropical cyclone-related hazards are often comprised of compound, connected events that individually amplify the total impacts. Often, hazard risk assessments focus on one factor rather than the compound nature of multiple forcing mechanisms. It is possible that extreme event analysis in a univariate context may underestimate the probabilities and impacts of extreme events. In this study, a framework addresses multivariate analysis of risk due to compound hazards related to tropical cyclone characteristics. Combinations of observations and simulations are used to identify possible frequencies of annual chance extreme events forced via connected individual events. The framework emphasizes the statistical dependence of multiple physical variables that contribute to extreme compound events when individual events are not extreme.

To make the analysis clear, specific locations are analyzed using both univariate and a joint analysis. The joint analysis is conducted using copula functions to remove the restriction that marginal distributions need come from the same family of probability functions. The primary results suggest that univariate and joint return periods for key tropical cyclone-related hazards could shorten in the future and univariate frequency analysis may underestimate the magnitude of the extreme events because the univariate analyses do not account for the dependence structure between the paired environmental factors.

Keywords Tropical cyclones · Extreme value analysis · Compound risk · Climate change

P. A. Harr (✉) · A. Jordi · L. Madaus
Jupiter Intelligence, San Mateo, CA, USA
e-mail: pat.harr@jupiterintel.com; Antoni.jordi@jupiterintel.com; luke.madaus@jupiterintel.com

© The Author(s) 2022
J. M. Collins, J. M. Done (eds.), *Hurricane Risk in a Changing Climate*, Hurricane Risk 2, https://doi.org/10.1007/978-3-031-08568-0_5

5.1 Introduction

The most recent understanding of possible changing tropical cyclone (TC) characteristics under climate change (including natural variability as well as anthropogenic factors) has been comprehensively summarized by Knutson et al. (2019, 2020). Based on review of global climate model (GCM) output, Knutson et al. (2020) provided measures of confidence to projected changes in TC frequency, track, intensity, and storm surge globally and over ocean basins that experience TC activity. The studies on which the Knutson et al. (2020) conclusions are based represent explicit numerical simulations plus dynamically and stochastically downscaled results to identify historical and possible future TC characteristics. While most of the Knutson et al. (2020) studies are based on models from the Coupled Model Intercomparison Project – Version 5 (CMIP5), with marginally appropriate spatial resolution (though augmented using downscaling methods) to properly resolve TC characteristics. The summary by Roberts et al. (2020) provides measures of projected future changes in TC characteristics using the more recent CMIP6 and higher-resolution models that are much more capable of representing TC frequency, motion, and structure. Consistent projections of possible future changes in TC characteristics from GCM, dynamically and stochastically downscaled, and high-resolution model simulations are leading to increased understanding in the "what" and "why" of changing TC characteristics.

Although the magnitude of projected sea-level rise is less than TC-induced storm surge, Knutson et al. (2020) note that increasing coastal flood impacts due to the combination of storm surge and sea-level rise is a very high confidence result. Knutson et al. also suggest that moderate-to-high-confidence results are increasing TC intensity and TC-associated rainfall. The latter changes are also evident in the high-resolution model simulations summarized in Roberts et al. (2020). Changes in TC frequency and track are of less confidence as differences exist among more recent model simulations that vary over geographic region (Bhatia et al. 2018) and, in some cases, vary with downscaling methodology (Emanuel 2013, 2020). As projected future changes in TC-related impacts to society depend on the nature of changes in TC characteristics of frequency, track, intensity, rainfall, and storm surge, it is critical to establish how projected changes in these TC characteristics impact the overall TC hazard. For this reason, several full TC hazard model systems have been developed (e.g., Lee et al. 2018; Jing and Lin 2020) to assess future TC-related hazards. In these systems, all the primary TC characteristics are examined in the historical record to assess the current TC-hazard impacts. Using these impacts as a baseline, these comprehensive TC-hazard model systems can eventually be applied to the model-projected future environmental states.

In this study, a more basic approach is taken in which changing TC-related hazards are based on the most confident conclusions of Knutson et al. (2020) on increases in TC intensity and precipitation. Here, the distributions of historical TC frequency and tracks, defined through a stochastically downscaled approach, are applied in future environments, and TC intensity and rainfall changes are estimated in relation to the future environmental states.

Historical impacts of TC-related hazards are often placed in a framework of a hazard frequency analysis (e.g., Jagger and Elsner 2006; Emanuel and Jagger 2010; Irish et al. 2011) that estimates return periods for key thresholds of wind, precipitation, surge, etc. However, TC-related hazards are often complex and multivariate in nature – a mix of wind, precipitation, storm surge, and tornadic activity. The area receiving the impacts from these hazards is determined by the storm structure and motion. The impacts also depend on various characteristics of the landfall location such as population density and distribution, building codes, topography (land and sea), tidal state, and inland watersheds. This study does not account for all of these TC- and non-TC-related factors. Rather, projected changes in TC hazards are solely based on a joint frequency analysis to assess changing characteristics of hazards in which joint wind and precipitation thresholds are exceeded.

The concurrent or consecutive occurrence of multiple extremes is referred to as a compound event (Wahl et al. 2015). Examples include a heat wave and drought or ocean and fluvial flooding. Compound events can be categorized as several types. One such type is successive or simultaneous extreme events such as extreme precipitation or storm surge. A second is an indirect amplification of an impact from one or more individual extremes (e.g., drought and heat wave). A third type is one in which individual events are not extreme by themselves, but rather the combination of events results in extreme events or impacts. This may occur when a moderate storm surge occurs at an above average high tide.

Many studies (Kopp et al. 2017; Field 2012) have identified the potential for the occurrence and severity of compound events to increase in the near future. This requires advances in recognition, understanding, and modeling of interacting physical processes to better quantify the risk that compound events pose to infrastructure and society. A primary challenge to assessing risk associated with compound events is that dependencies among environmental factors, which comprise a particular hazard (i.e., flood, drought, fire, etc.), can cause the estimation of event probabilities to be much more difficult than if the factors are independent (Salvadori et al. 2016). For a given landfall event, poor representation of the dependencies among physical factors that define compound events can often lead to underestimation of the risk of such events provided that the impacts of dependent factors are larger than the impacts of the independent combination of individual factors.

In this study, the joint hazards to be addressed are TC intensity and precipitation. A basic methodology, which is a simplified version of the full TC hazard models of Lee et al. (2018) and Jing and Lin (2020), is used to estimate the bivariate joint distribution of intensity and precipitation. Although simpler than the full hazard models defined above, the system employed here better captures the heterogeneity in environmental conditions associated with TC activity over the North Atlantic to derive estimates of future TC intensity and precipitation based on climate model-projected conditions. Once derived, the bivariate distribution is used to assess the return probabilities of the joint occurrence of the hazards. The joint probability is compared to the univariate probability of the separate occurrence of each hazard and their variations in time.

5.2 Method

To investigate the compound hazards of TC intensity and precipitation, a sequence of statistical and physical models is applied to define current and future joint distributions of TC intensity and precipitation. For TC intensity, which is defined by the maximum 10-m sustained wind, a regression-based model is trained with historical TC data and then used to estimate intensity over historical and future synthetic TC events. Based on the estimated TC intensity, a parametric wind profile (Chavas and Lin 2015; Chavas et al. 2015) is used to obtain a two-dimensional gradient-level-sustained wind field, which is then converted to a 10-m-sustained wind using a planetary boundary layer model (Kepert 2001; Kepert and Wang 2001). For precipitation, a physical model of TC-induced precipitation (Lu et al. 2018; Xi et al. 2020) is applied based on the three-dimensional low-level TC winds and environmental winds and their interaction with the environmental moisture. The estimated historical and future statistical character of TC winds and precipitation is then subjected to a joint probability analysis to determine the compound return levels for specified return periods. Computation of TC intensity and precipitation plus univariate and joint return periods are conducted at approximately 5 km spatial resolution over global regions where such hazards occur. While this analysis is conducted globally, only the North Atlantic basin, which includes the North Atlantic Ocean, the Gulf of Mexico, and the Caribbean Sea, is considered in this study. Details of each step are provided below.

(a) *Synthetic tracks*

To compensate for the low occurrence of TCs, synthetic TCs are generated using the North Atlantic Stochastic Hurricane Model (NASHM, Hall and Jewson 2007; Hall and Yonekura 2013). NASHM consists of three components that determine the formation, motion, and decay of each generated TC and three components that determine the TC structure in terms of its maximum wind speed, central pressure, and radius of maximum winds. For the historical period, climate covariates are included to capture the El Niño Southern Oscillation (ENSO) and vertical wind shear (VWS) influences. Use of climate covariates in each NASHM component is based on the physical relationships between the climate factors and North Atlantic TC characteristics (e.g., formation, motion, intensity). NASHM is used to generate multiple realizations of the pre-defined historical time period such that thousands of simulations are used to define the statistical character of TC activity over the North Atlantic. Re-sampling of the tracks generated with the historical environmental conditions is used to place tracks into future environments. The formation and motion characteristics of each track are unchanged, but the TC intensity model is used to compute a new intensity (i.e., Vmax) using the future environment conditions. Therefore, the statistical character of TC frequency, formation location, and track are held constant in future years, but, as defined below, intensity changes using the regression-based relationships among environment and TC predictors, and the TC intensity change.

(b) *TC Intensity model*

Regression-based methods for TC intensity prediction have been developed and used in operational settings (DeMaria and Kaplan 1994; DeMaria et al. 2005; Knaff et al. 2003) and in support of studies of risk due to hazards related to TCs (Vickery et al. 2000; Vickery and Lavelle 2014; Lee et al. 2015, 2016; Lin et al. 2017). The TC intensity model in the current study follows from the developments of Lee et al. (2015, 2016) and Lin et al. (2017) but differs in set up by including a set of singular-value decomposition (SVD, Cherry 1996) modes as a classifier to be applied in a finite mixture model (FMM)-based framework. The predictor pool is comprised of environment characteristics (Table 5.1), which are drawn from the European Centre for Medium-Range Weather Forecasts (ECMWF) reanalysis (ERA-5) for the historical period of 1950–2019, and future years are defined by a full, non-selective ensemble constructed from the full suite of CMIP6 models using the Shared Socio-economic Pathways-5 8.5 (SSP585) scenario. While field variables are available directly from the CMIP6 output, the maximum potential intensity (MPI) is calculated following Bister and Emanuel (1998). The historical TC characteristics used as predictors (Table 5.2) are from the International Best Track Archive for Climate Stewardship (IBTrACS, Knapp et al. 2010) during the period 1950–2012 as the dependent data set, and 2013–2019 as an independent set.

Following Lee et al. (2015, 2016), the TC intensity model is of the form:

$$\Delta V\max = V\max{}_{t+6h} - V\max{}_t = L\{S_t, S_{t-6h}, E_t, E_{t-6h}\} + \varepsilon_{t+6h} \quad (5.1)$$

Table 5.1 Environmental predictors

Parameter	Notation	Calculation
Midlevel relative humidity	MIDRH	Averaged from 700 to 400 hPa within 200 to 800 km annulus around TC center
Vertical wind shear	VWS	200–850 hPa wind speed difference averaged within 200–800 km annulus around the TC center
Maximum potential intensity of sustained wind	MPI (Vmax)	Averaged over 500 km-radius area around the TC center
Delta MPI	Δ(MPI)	The difference between the MPI and the actual intensity (Vmax).
Relative TC intensity	Vmax/MPI	TC intensity divided by MPI
Relative change in TC intensity	Δ(Vmax/MPI)	6-h change in relative TC intensity

Table 5.2 TC predictors

Parameter	Notation
TC intensity	Vmax
Translation speed	Vtr
Latitude/longitude	Lat./Lon.
Precious 6-h change in TC intensity	Δ(Vmax)

Where S_t represents TC-related predictors at time t, E_t represents environmental predictors at time t, and L represents a general linear operator. The term, ε, is a stochastic forcing term applied at each evaluation and is defined by sampling from the distribution of errors from the dependent data set. Initially, L represented an ordinary least squares (OLS) operator, but was modified to be of a linear mixture model type using the SVD modes defined below. In practice, the SVD modes are used to separate TC intensities into subgroups (Alqahtani and Kalantan 2020), and based on the statistical character of the grouped predictand and predictors, significant predictors are chosen and the model is developed for each group. Group construction and its dependence on the SVD modes are defined below.

Once the TC intensity model coefficients are defined based on IBTrACS data, the model is re-applied using the synthetic track set defined within the historical period and re-sampled to apply over future years. As in the training of the model, ERA-5 data are used to define the environmental predictors while TC predictors are derived from the synthetic track sets. In future years, the environment predictors come from CMIP6 SSP585 ensemble and the TC predictors again come from the synthetic track set re-sampled into the future years.

Use of SVD modes stems from the realization that a regression relationship based on assuming that the predictor-predictand relationships are homogenous over all observations might not capture changing environmental conditions that could occur on a variety of time and space scales. Based on developed regression-based models of TC intensity (e.g., DeMaria and Kaplan 1994; Lee et al. 2015, 2016; Lin et al. 2017), MPI, VWS, and MIDRH are consistently found to be significant environmental predictors. Therefore, these three parameters are placed into a combined SVD analysis that analyzes the ERA-5 depiction of the parameters from 1950 to 2020. As defined above, monthly environmental fields from the ERA-5 are used and further combined into three-month averages for May–June–July (MJJ) and August–September–October (ASO). Therefore, the SVD loadings represent interannual variability in each three-month interval that covers the early (i.e., MJJ) and peak (i.e., ASO) portions of the TC season over the North Atlantic. For all three fields, the SVD analysis is applied to standardized anomalies that have been detrended and processed with a cosine (latitude) weighting.

In a positive orientation, the leading combined SVD mode for ASO (Fig. 5.1a–c) represents higher than normal MPI (Fig. 5.1a) over the main development region (MDR, defined here as 10°N–20° N and 20°W–80°W; Goldenberg and Shapiro 1996; Bell and Chelliah 2006) that extends from the eastern Atlantic into the Caribbean Sea. Over the western Atlantic along the North American coastline and eastward, the MPI would be below normal. Also, in the positive orientation, the SVD mode 1 represents enhanced MIDRH (Fig. 5.1b) over the subtropical Atlantic and western portion of the MDR (i.e., Caribbean Sea), and lower than average VWS (Fig. 5.1c) over the eastern MDR (i.e., eastern tropical Atlantic). Therefore, the ASO SVD positive mode 1 indicates high MPI, high MIDRH, and low VWS throughout the MDR. Alternatively, the negative orientation of SVD mode 1 would represent low MPI, low MIDRH, and high VWS over the region.

5 Analysis of the Future Change in Frequency of Tropical Cyclone-Related... 93

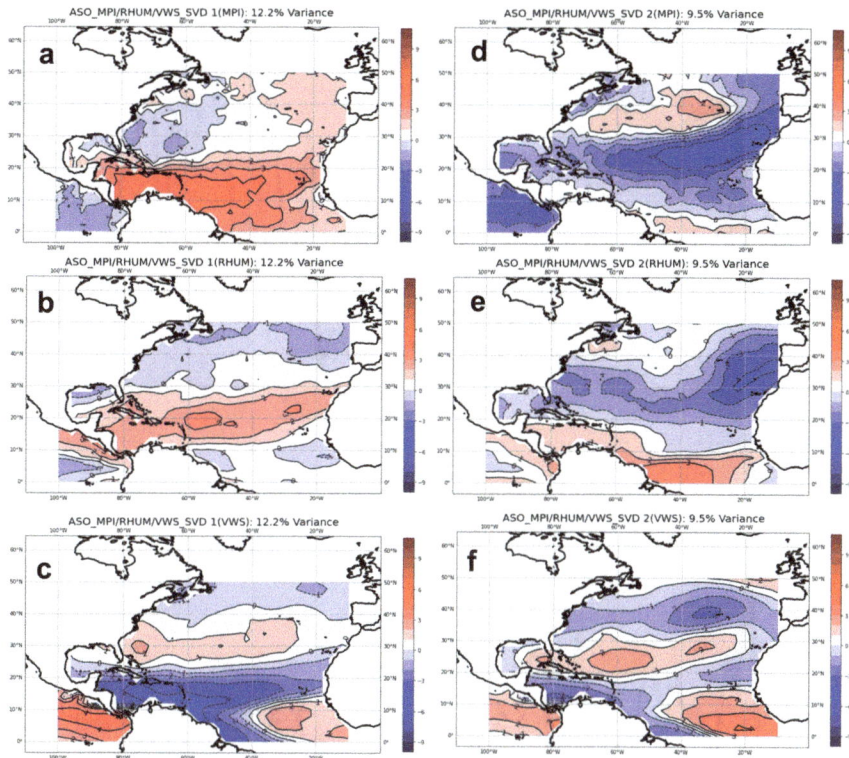

Fig. 5.1 The first (**a–c**) and second (**d–f**) SVD patterns for the August–September–October average during the period 1950–2018 using ERA-5 data with (**a, d**) MPI(Vmax), (**b, e**) MIDRH, and (**c, f**) VWS. The displayed patterns represent the positive sense of each pattern

The coefficients for SVD mode 1 (Fig. 5.2a) capture the interannual and some multidecadal variability (Bell and Chelliah 2006) that is similar to impacts from variations in atmospheric aerosol concentrations and the ocean circulations during low TC count years in the 1970s and 1980s. As defined, the spatial SVD mode 1 pattern defines conditions somewhat related to the well-defined ENSO conditions that affect TC characteristics over the North Atlantic (Gray 1984; Goldenberg and Shapiro 1996). SVD mode 1 has a negative relationship with the multivariate El Niño Index (MVEI) (Fig. 5.3a) and a positive relationship with seasonal accumulated cyclone energy (ACE) (Fig. 5.3b). Therefore, positive SVD mode 1 defines patterns in MPI, MIDRH, and VWS connected to La Niña conditions and increases in ACE. A negative SVD mode 1 would represent patterns indicative of El Niño and decreases in ACE. The combined SVD mode 1 patterns explain approximately 12% of the total variance among the three environmental fields. However, the correspondence among SVD mode 1, ENSO, and ACE suggests the mode represents an important factor in the interannual variability among North Atlantic large-scale environmental conditions and TC characteristics, which must be viewed with

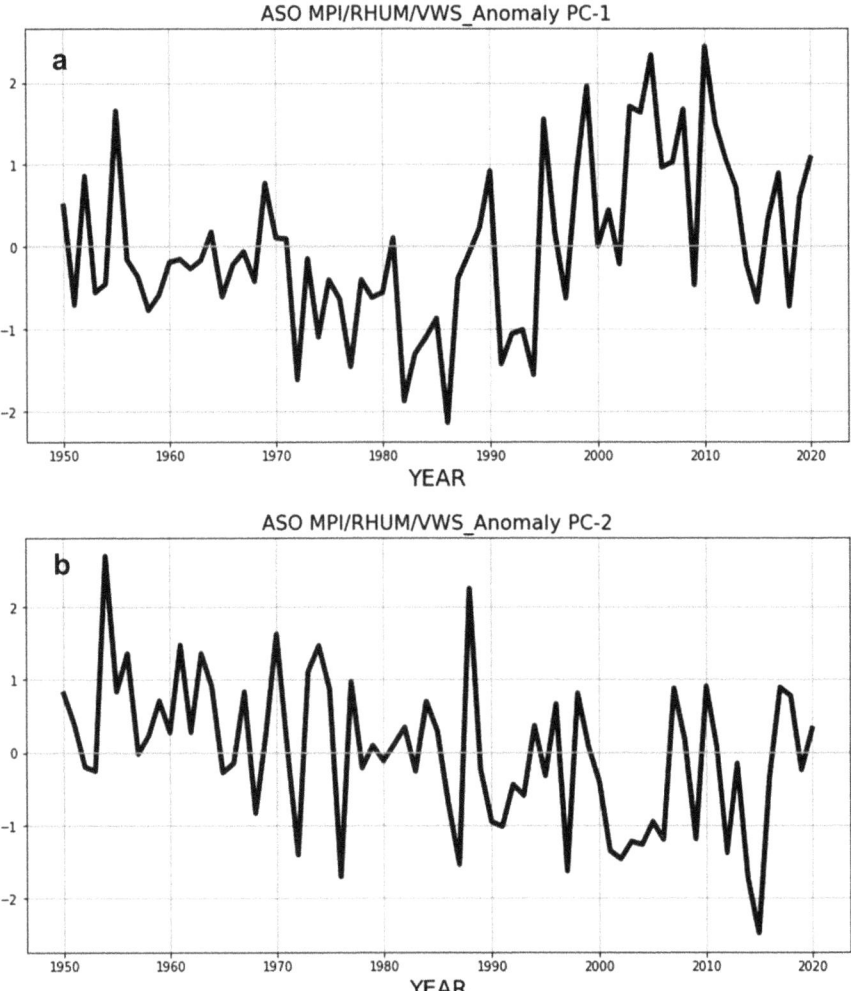

Fig. 5.2 The SVD scores for the first two SVD patterns in Fig. 5.1

knowledge of the likely underestimation of ACE in the pre-satellite era. The second ASO SVD mode (Fig. 5.1d–f) also represent large-scale variations in the three environmental factors. Generally, primary centers in each pattern are shifted poleward of the mode 1 centers with maximum amplitude over the eastern North Atlantic. While the SVD mode 2 scores (Fig. 5.2b) define interannual variability that does not vary with ENSO or other established large-scale circulation modes, the coherent structures of each pattern (Fig. 5.1d–f) require examination beyond that required for use in this study, in which the SVD analysis is meant to serve as a dimension-reduction and classification method.

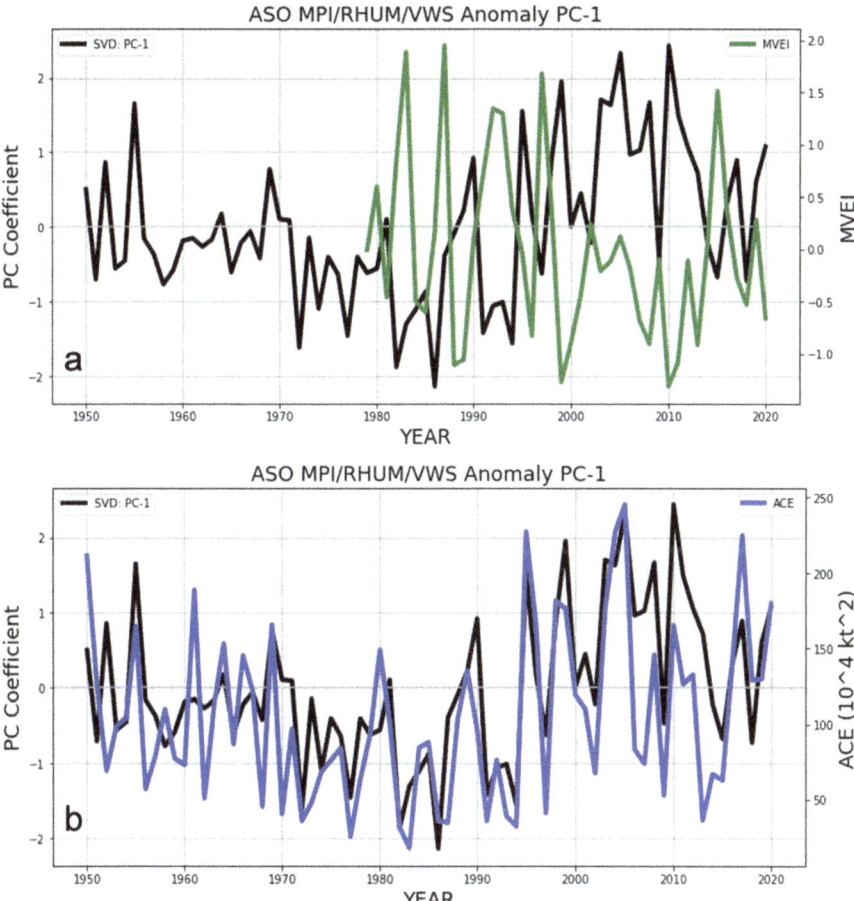

Fig. 5.3 The SVD scores as in Fig. 5.2 with (**a**) the Multivariate Enso Index (MVEI), and (**b**) the Accumulated Cyclone Energy (ACE) included with the SVD scores

Based on the characteristics of the SVD mode 1 patterns, the time variation of the mode 1 coefficients, and their association with ENSO and ACE, SVD mode 1 is used to define the heterogeneity in the predictor and predictand subpopulations for the TC intensity regression model. Essentially, separate regression models are based on the sign of the SVD mode 1 coefficient during the 1950–2012 training period and applied during the 2013–2019 independent data period.

As defined above, the TC intensity model is applied to the NASHM tracks that define thousands of realizations of TC seasons re-sampled from the historical period into future years. Therefore, the SVD analysis was applied to the same three parameters for future years to classify years for the mixture-based regression model. This would only be a viable option if the SVD structures of the future years resembled those of the historical years. The first SVD mode over 2015–2050

(Fig. 5.4) using CMIP6 SSP585 output is similar in pattern to SVD mode 1 from 1950 to 2019 and ERA-5 data. In a positive sense, the mode represents above average MPI over the MDR, plus above average MIDRH and reduced VWS over the same region as the enhanced MPI. Time SVD coefficients (Fig. 5.5) define the interannual variability with a suggestion of some multidecadal variability as present in SVD mode 1 from 1950 to 2019. Therefore, the similarity among the spatial and loading patterns of SVD mode 1 in the historical data and future projections justify use of the SVD mode as a classifier in both periods when applying the TC intensity regression model as will be illustrated in Sect. 5.3.

(c) *TC Wind Model*

For each of the thousands of NASHM tracks re-sampled for the historical period and placed into the future environment, an estimated TC intensity is calculated from the TC intensity model defined in Sect. 5.2a. The new intensity (Vmax) at each interval along the track is then used to construct a two-dimensional gradient-level sustained wind field using a parametric model (Chavas et al. 2015; Chavas and Lin 2015). The gradient level wind is then used to generate a surface-level (i.e., 10 m) sustained wind via a linear version of the planetary boundary layer model developed by Kepert (2001). The use of a parametric wind model followed by a boundary layer model to fully define the surface winds and their asymmetries is similar in construct to that of Done et al. (2020), except for their use of the full nonlinear boundary layer model of Kepert and Wang (2001). The final surface-level winds include asymmetries due to storm motion and variations in land/sea boundaries, terrain, and surface roughness, which are used to compute a spatially-varying drag coefficient for a neutrally-stratified atmosphere in a manner similar to that employed by Feldman et al. 2019. The combination of the efficiency of a parametric model and a boundary layer model that is capable of representing interaction of the flow with varying terrain and surface roughness characteristics in line with the requisite dynamics provides realistic horizontal and vertical winds over ocean, coastal, and land regions.

(d) *TC Precipitation Model*

The surface-level winds output from the TC wind model are used to in a TC precipitation model as defined in Lu et al. (2018) and employed in Feldman et al. (2019) and Xi et al. (2020). The model is based on a partitioning of the vertical motion at the top of the boundary layer into contributions from terrain, friction, stretching, baroclinic effects, and radiation. A vertical vapor flux is computed using the net vertical velocity and the saturation specific humidity. The vertical vapor flux is then modified by a specified precipitation efficiency. Lu et al. (2018) define the TC precipitation rate as:

$$\text{Prate} = \epsilon_p \frac{\rho_{\text{air}}}{\rho_{\text{liquid}}} q_s \{w_f + w_h + w_t + w_s + w_r\} \qquad (5.2)$$

where ϵ_p is the precipitation efficiency, ρ is the density of air and water, q_s is the saturation specific humidity, and the vertical velocity components are due to friction

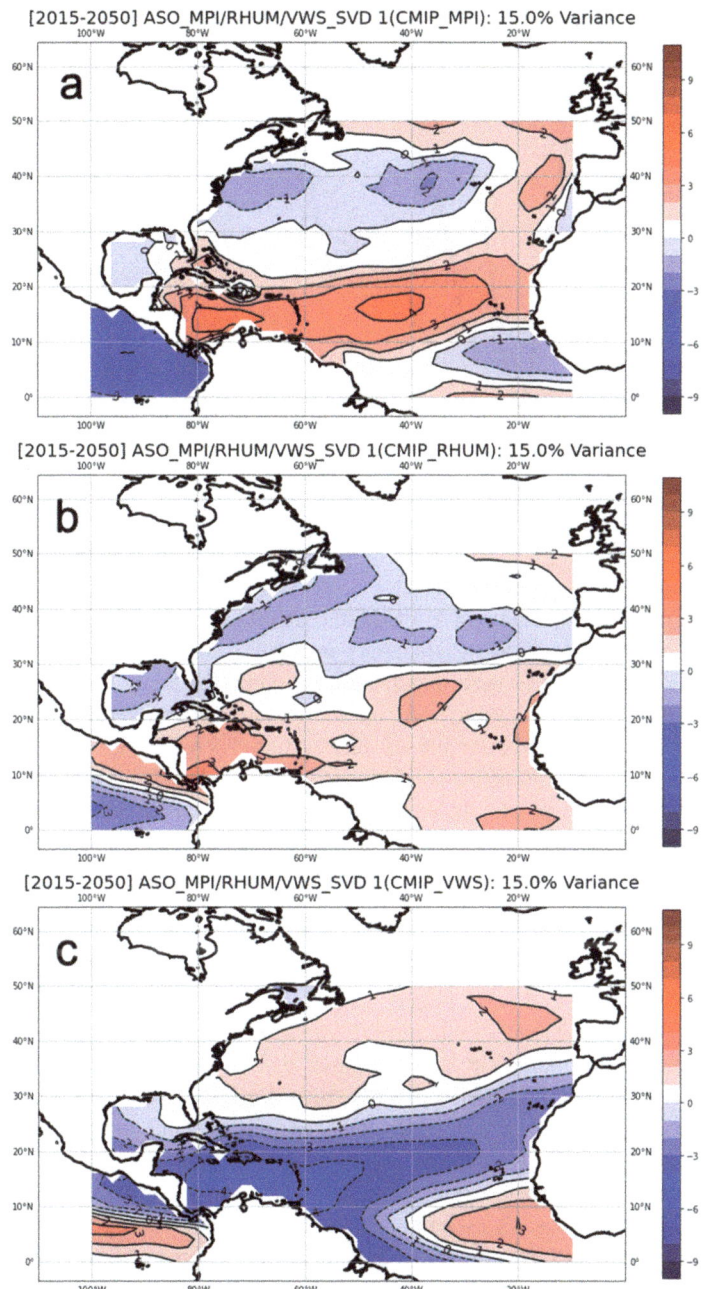

Fig. 5.4 The first SVD pattern for August–September–October over the years 2015–2050 using an ensemble of CMIP6 model output. The displayed patterns represent the positive sense of each mode

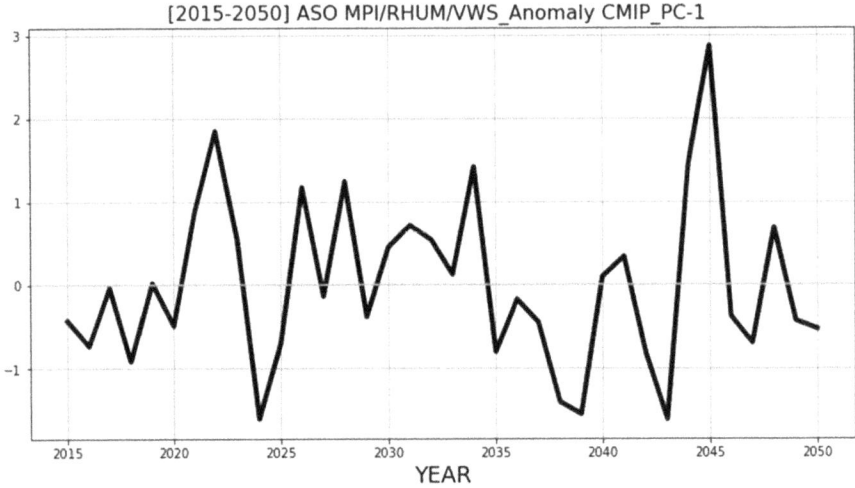

Fig. 5.5 The SVD scores for the 2015–2050 SVD mode 1 pattern in Fig. 5.4

(w_f), terrain (w_h), stretching (w_t), baroclinic effects (w_s) and radiation (w_r). The baroclinic contribution represents changes in the TC vorticity field that are driven by interaction of the TC vortex and the environmental vertical wind shear. All environmental fields required to compute the various vertical velocity components plus the saturation specific humidity are defined from ERA-5 for the historical period and from the CMIP6 SSP585 ensemble for the period 2015–2050. As defined above, monthly data are interpolated to the time and location of each TC position. Application of the winds from the boundary layer model provides high-resolution coverage of vertical motion in the TC environment.

(e) *Joint Probabilistic Wind and Precipitation Analysis*

As discussed above, impacts from a TC are typically multivariate in the sense that combined (i.e., wind and precipitation, wind and storm surge) may contribute most to total impact rather than single variable impacts such as wind, precipitation, or surge. While TCs contain multivariate hazards, the probabilistic estimation of critical return levels for specified return periods (in years) is often done in a univariate setting. In this study, a multivariate analysis of the wind and precipitation is undertaken to obtain the probability of joint events and the joint return periods.

Many studies have been conducted to define the univariate relationship between environmental factors such as wind, precipitation or storm surge, and flood frequency (e.g., Chu and Wang 1998; Elsner et al. 2008; Emanuel and Jagger 2010; Malmstadt et al. 2010; Irish et al. 2011). In a univariate framework, a probability distribution is fit to the environmental data and extreme value statistical methods (Coles 2001) are employed to identify probabilities associated with levels of forcing (i.e., precipitation amount or storm surge level). Often, the univariate analysis is expanded to define joint distributions of environmental hazards (Sackl and

Bergmann 1987; Yue et al. 2001). While an empirical counting method (Mudd et al. 2017) may be employed, the joint distribution analyses are often designed such that the same family of marginal distributions is assumed for all environmental forcing. For example, a bivariate Gumbel (Yue et al. 2001), bivariate normal (Sackl and Bergmann 1987), or bivariate gamma (Yue 2001) have all been used in various studies.

However, recent investigations, which mainly focused on the compound nature of flood frequency analysis, have shown that the best fit marginal distribution of individual environmental forcing often belongs to different families of probability distributions (Salvadori et al. 2016; Favre et al. 2004; Grimaldi and Serinaldi 2006; Zhang and Singh 2006). These studies rely on the concept of the copula (Nelson 1999) to model the dependence structure among multiple environmental forcing factors without regard to their respective marginal distributions. It has been shown (Zhang and Singh 2006) that copula-based analyses result in a more realistic assessment of the multivariate environmental forcing mechanisms of a compound event because the resulting copula, which defines the dependence structure of the multivariate distribution, fits the empirical distribution of data better than a distribution established around a joint parametric distribution and marginal distributions assumed to be of the same family for all forcing factors. Therefore, the copula construct provides for models of the dependent structure of compound events regardless of the distribution to which the univariate marginal distributions belong. Importantly, joint return periods can be estimated from copulas, which is a primary benefit in TC hazard analysis.

In the bivariate case, the copula maps the marginal distributions to a joint cumulative distribution (CDF) through the expression:

$$H(x, y) = C\{W(x), P(y)\} \tag{5.3}$$

where H is the joint CDF for random variables of TC intensity, x, and TC precipitation, y, $W(x)$ and $P(y)$ are the marginal CDFs for TC intensity and TC precipitation, respectively, and C defines the copula function that maps the marginal distributions to the joint CDF.

An initial step in the analysis of compound hazards is to test for tail dependence between the marginal distributions. For this purpose, the Spearman's rho and Kendall's tau rank-based non-parametric measures of dependence (Genest and Favre 2007) are used and significance probability values are estimated to examine the null hypothesis that no dependence between the two variables exists. Because of the importance of the upper (positive) tails of the intensity and precipitation marginal distributions, and given a tail dependence, Archimedean copulas are examined as they only fit the upper tails of the joint distribution (Zhang and Singh 2007). While three Archimedean copulas are calculated, the Gumbel copula was in all cases examined here (not shown) most applicable based on its greater dependence on the positive tails of the marginal distributions.

From Eq. (5.3), the joint probability density function (PDF) is defined by:

$$h_{x,y} = c[W(x), P(y)]w(x)p(y) \tag{5.4}$$

where w and p are the PDFs of TC intensity and precipitation, and c is the PDF of the copula function defined as:

$$c(u,v) = \frac{\partial^2 C(u,v)}{\partial u \partial v}. \tag{5.5}$$

Here, u,v denotes the marginal CDFs, W and P. A copula parameter, θ, defines the degree of association between u and v. The form of the Gumbel copula is then:

$$C(u,v) = \exp\left\{-\left[(-\ln u)^\theta + (-\ln v)^\theta\right]^{\frac{1}{\theta}}\right\} \tag{5.6}$$

The parameter, θ, is estimated using a method based on the inversion of the non-parametric dependence measure, Kendell's tau (Requema et al. 2013), which allows use of the empirical marginal distributions rather than parametric distributions. A goodness-of-fit test is applied to examine the Gumbel copula versus other standard Archimedean copulas by comparing the quantiles of the parametric copula with a nonparametric distribution that is based on the empirical non-exceedance probabilities of the joint occurrences of intensity and rainfall. The resulting copula Q-Q plots can be used to graphically assess the ability of each copula to capture the tail dependencies in the marginal distributions. The Q-Q plots were produced (not shown) for each analysis described below, and in all cases the Gumbel copula fit was deemed best.

An important aspect of the copula application is computation of the joint return periods. The primary return period, which is also defined as the "OR" return period is when a threshold intensity or precipitation may be exceeded (Fig. 5.6). Therefore, the primary return period examines the probability of wind exceeding a threshold or precipitation exceeding a threshold (i.e., $P(\text{Wind} > x$ OR Precipitation $> y$). A secondary return period is labeled the "AND" return period in which the threshold intensity and precipitation are exceeded (i.e., $P(\text{Wind} > x$ AND Precipitation $> y$). The primary return period is computed using the copula function as:

$$R_{X,Y}^{\vee} = \frac{\mu_T}{P(X > x \vee Y > y)} = \frac{\mu_T}{1 - C(W(x), P(y))} \tag{5.7}$$

where $\vee$ signifies "OR" and μ_T is the mean inter-arrival time between successive events (i.e., $\mu_T = 1$ for annual events). The "AND" return period is defined as:

$$R_{X,Y}^{\wedge} = \frac{\mu_T}{P(X > x \wedge Y > y)} = \frac{\mu_T}{1 - W(x) - P(y) + C(W(x), P(y))} \tag{5.8}$$

where $\wedge$ signifies "AND". The inter-arrival time for each location is defined based on the IBTrACS data during the period 1950–2019. In this application, the primary

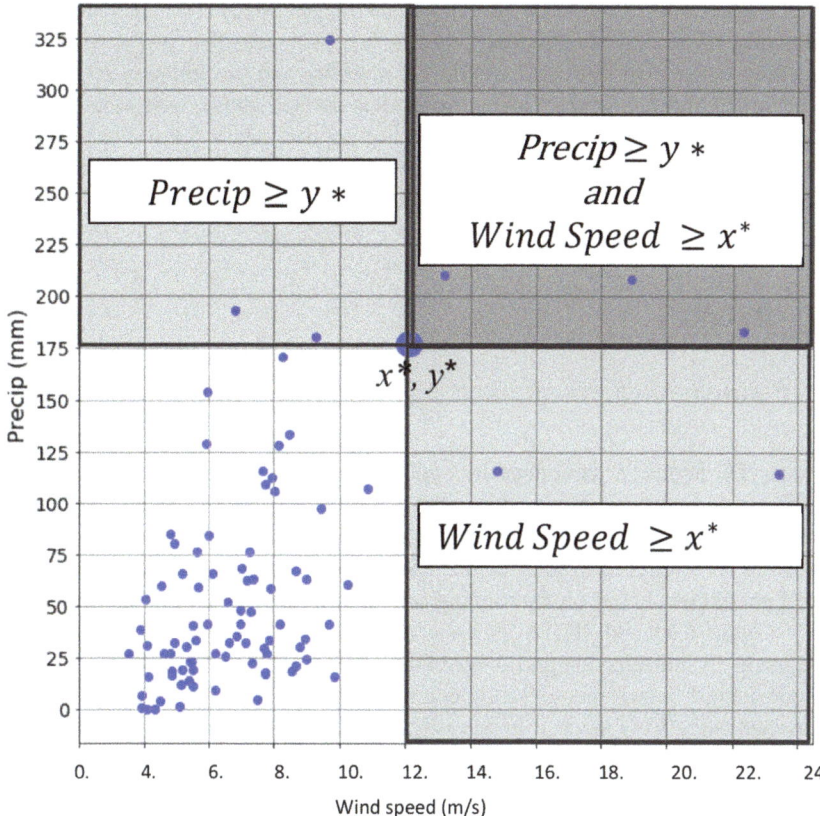

Fig. 5.6 Schematic of the scatter in the joint distribution of wind speed (m s^{-1}) and precipitation (mm) with the regions labeled as to their interpretation in specifying joint return periods related to a specified wind threshold (x^*) and precipitation threshold (y^*). In the OR return period, probabilities (or return periods) are P(Wind Speed> x^* OR Precipitation >y^*) and in the AND return period probabilities (or return periods) are P(Wind Speed > x^* AND Precipitation >y^*)

return period is mainly examined as severe TC impacts can occur due to wind or precipitation and the occurrence of extremes in both is not necessary.

To summarize, the following steps are taken to obtain joint return periods:

1. Compute Kendall's rank correlation, τ, from the observed data;
2. Determine the copula parameter, θ, from τ;
3. Compute the generating function, ϕ, for each copula; and
4. Obtain the copula function from ϕ and θ.

The analyses presented below also examines the marginal distributions and return periods of wind and precipitation to compare with the joint return periods. Analyses of the univariate distributions are based on application of the generalized extreme value (GEV) distribution (Coles 2001), which has been shown to be applicable to TC winds (Jagger and Elsner 2006; Emanuel and Jagger 2010). The GEV defines the probability of having at least one event in a given year that has a wind speed exceeding a specified threshold. While TC intensity and precipitation are obtained for historical and future years using the models defined above, univariate and joint hazard analyses are applied to a spatial grid of approximately 5 km. For this study, selected points of interest will be highlighted here to narrow the focus and interpretation of future changes in TC intensity, precipitation, and joint impacts.

5.3 Results

(a) *TC intensity*

Because TC intensity, as defined by application of the regression-based model to historical and future conditions, drives the computation of the TC wind fields, which then drives the TC precipitation distribution, the characteristics of the estimated TC intensities are examined in some detail. The significance and sign of the TC intensity model parameters based on the training set consisting of the ERA-5 environmental fields (Table 5.1) and IBTrACS track data (Table 5.2) indicate that there are differences in the selection and order of key predictors based on whether the data are considered homogeneous (ALL column in Table 5.3) or if a classification is performed using SVD mode 1. In all cases, the most recent change in intensity is one of the two most significant predictors. For all data and for the SVD 1+ classification, the recent change normalized by MPI is also a key predictor. However, this relative change in Vmax is not significant in the SVD 1− classification as it is replaced by the difference between Vmax and MPI, which is labeled Δ(MPI). In the ALL case and in SVD 1+, Δ(MPI) appears as the third-most significant predictor. While MPI alone is not a significant predictor, the use of MPI in normalizing Vmax or in the difference between Vmax and MPI does appear in all classes (e.g., Merrill 1987; DeMaria and

Table 5.3 TC intensity model variables in order of selection when applied to the historical period using ERA-5 and IBTrACS data followed by the sign of their coefficient

Ordered variables	All		SVD 1+		SVD 1−	
1	Δ(Vmax)	+	Δ(Vmax/MPI)	+	Δ(Vmax)	+
	Δ(Vmax/MPI)	+	Δ(Vmax)	+	Δ(MPI)	+
	Δ(MPI)	+	Δ(MPI)	+	Vmax/MPI	−
	Vmax/MPI	−	Vmax/MPI	−	Vmax	−
	Vmax	−			Latitude	−
	Latitude	−			Vtr	−
	Vtr	−				
R^2	0.43		0.54		0.48	

The R^2 value is at the bottom of each column. The column labeled "All" is for all cases with no classification based on the sign of SVD mode 1

Kaplan 1994). All three of these predictors relate to TC intensity change in a positive sense, which indicates that intensifying TCs will be estimated to continue intensifying (persistence factor) apart from unfavorable environmental factors. Also, the positive difference between current MPI and Vmax signifies continued intensification. Without classification, TC-related parameters are significant in that Vmax, latitude, and translation speed all have negative association with intensity change. In the SVD 1+ classification, no TC-based predictors are chosen while all three TC-based predictors are chosen in the SVD 1− classification. Finally, neither VWS nor MIDRH appear as significant predictors, though they are contained in the classification using SVD mode 1. These large-scale environmental fields likely do not appear as individual predictors due to the use of monthly data interpolated to the time and location of each TC position, which may not fully represent the relationship between VWS, MIDRH, and TC intensity. Nevertheless, the R^2 values around 0.5 indicate a reasonable fit to the observations in all three versions of the regression-based model though the R^2 value is highest in the SVD 1+ classification.

To better understand the TC intensity model results and the effect of classification using SVD mode 1, the model estimations of historical and future tracks that pass through the NE and GoM regions (Fig. 5.7) are examined. The model training is based on classifying ERA-5 environmental fields and IBTrACS data based on SVD 1+ and SVD 1−. As defined in Table 5.3, a type of null classification is included by using all data (the ALL classification) to compare with the classifications. Here, the model is then applied to the NASHM synthetic tracks from the same historical period, and to future years via re-sampling of NASHM tracks into future years. Application of the TC intensity model to re-sampled synthetic tracks scaled into future years from 2020 to 2050 provides a measure of the change in intensity due to future environments as defined by the CMIP6 SSP585 ensemble and the relationships defined by the TC intensity model.

Recall that SVD mode 1+ represents higher than normal MPI and MIDRH, and lower than normal VWS over the MDR, while SVD mode 1− represents opposite conditions. The track density for the ALL classification tracks NASHM tracks simulated for the historical period of 1950–2019 (Fig. 5.8a) that enter the NE region (Fig. 5.7) at some point during their lifetime exhibit a typical recurving track profile as they move poleward along the east coast of North America after passing through the MDR or eastern GoM. The intensity of such TCs (Fig. 5.8b) is highest when they move toward the northwest in the very western and northern portion of the MDR. As an example, the difference between the TC intensities in the year 2050 and the same tracks taken from the historical period (Fig. 5.8c) indicates that there is a modest increase of 2–5 m s^{-1} in intensity throughout their path, except in the region where the historical tracks typically reach their lifetime maximum intensity. In that region, the TC intensity model indicates a decrease in 2050, which is likely caused by a bias in the intensity model and a lag related to the importance of intensity change persistence as a key predictor. Since this bias does not seem to affect tracks in the key regions where the joint hazard analysis is conducted, it is not considered a serious issue, though efforts are being taken to reduce this bias. The general range of TC intensity increases of 2–5 m s^{-1} is in line with medium – high confidence in

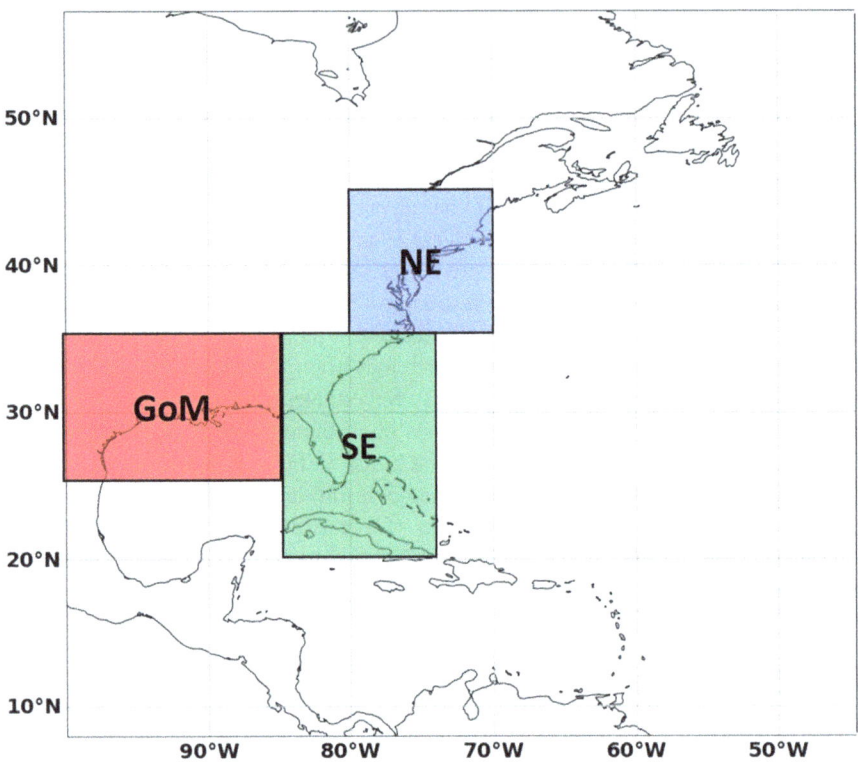

Fig. 5.7 Regions over which the joint probability analysis of compound TC intensity and precipitation hazards are computed

future changes of TC intensity suggested by the CMIP5 archive (Knutson et al. 2020). The range of TC intensity increase is also consistent with the magnitudes found in Roberts et al. (2020) using the CMIP6 high-resolution model intercomparison project suite.

To compare estimated TC intensities based on the SVD classifications, the year 2045 is chosen as an SVD 1+ year and 2050 is chosen as an SVD 1− year (Fig. 5.5). In the SVD 1+ classification (Fig. 5.8d–f), the track density (Fig. 5.8d) along the east coast of North America is less and the density throughout the MDR is greater than the ALL classification. This is likely related to the increase in MPI, MIDRH, and reduced VWS represented by the SVD 1+ patterns. While the density of tracks entering the NE region during the SVD 1+ classification is less, the lifetime maximum intensities (Fig. 5.8e) are higher and higher intensities cover a larger area as more intense storms approach the SE region from the MDR. Again, these traits are consistent with the SVD mode 1+ patterns. When the SVD 1+ classification regression model is applied to the synthetic tracks placed into the environment of 2045 (Fig. 5.8f), the negative bias still exists where observed intensities are highest but there is an increase in intensities located along what appears like a typical

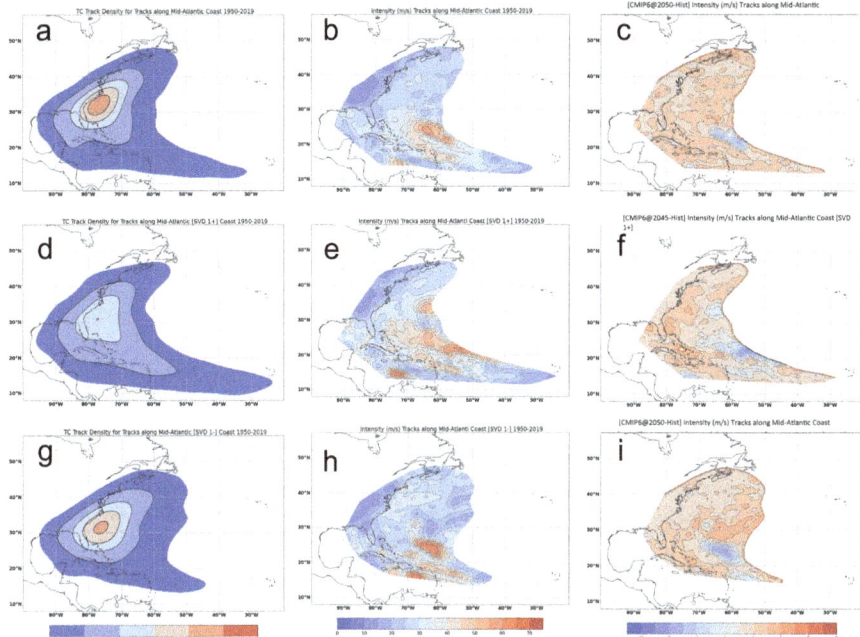

Fig. 5.8 TC track characteristics for all TCs in the IBTrACS set from 1950 to 2019 that enter the NE region of Fig. 5.7 for the (**a–c**) ALL data set with no classification by SVD mode 1, (**d–f**) SVD mode 1+ classification, and (**g–i**) SVD mode 1− classification. The left column contains the track densities (tracks per half degree lat./lon. per year). The middle column contains the IBTrACS TC intensities along the tracks, and the right column contains the difference between estimated TC intensity in 2050 using the TC intensity model and the IBTrACS intensities. All intensity values are m s^{-1}

recurring track followed by a TC that exits the MDR and moves northwestward toward the North American coastline. Additionally, the increase in intensities during the SVD 1+ classification is greater than the ALL class and is consistent with the SVD 1+ patterns of MPI, MIDRH, and VWS.

In the SVD 1− classification (Fig. 5.8g–i), there is a higher track density along the east coast of North America, but lower density over the MDR. This suggests that TCs entering the NE region during the SVD 1− classification have shorter tracks in that they either have less time in the MDR or form farther north. Both of these conditions would be consistent with the SVD mode 1− patterns of a poleward shifted increase in MPI, MIDRH, and VWS such that these factors are less favorable over the MDR but more favorable poleward of the MDR (Goldenberg and Shapiro 1996). The maximum track intensities during SVD 1− (Fig. 5.8h) do not extend northwestward toward the eastern coastline of North America as found with the intensities in the SVD mode 1+ classification (Fig. 5.8e). However, the increase in intensities in 2050 (Fig. 5.8i) exhibits a pattern similar to that of SVD mode 1+, with

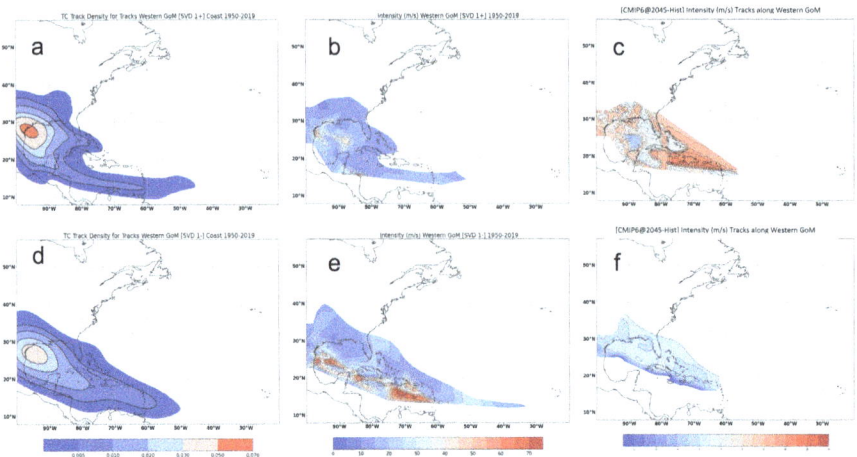

Fig. 5.9 As in Fig. 5.8, except only for the (**a–c**) SVD mode 1+ classification and (**d–f**) SVD mode 1− classification and for all TCs that enter the very western portion of the GoM region of Fig. 5.7

a swath of increasing intensities leading to the Florida peninsula from the southeast. However, the magnitudes are less than in mode SVD 1+.

A similar look at the TC intensity results is conducted for TCs that enter the very western portion of the GoM region of Fig. 5.7, and only the SVD mode classification results are shown (Fig. 5.8). During SVD 1+ (Fig. 5.9a), the track density over the western GoM is higher than it is for SVD mode 1− (Fig. 5.9d). However, along the very southern portion of the track density, the estimated intensities during the historical period are higher in SVD 1− (Fig. 5.9e) than in SVD 1+ (Fig. 5.9b). This may be due to the relatively fewer number of cases in the SVD 1− in this region. For the SVD 1+ classification, the estimated future TC intensities (Fig. 5.9c) increase throughout the Caribbean Sea prior to when a typical track would enter the GoM. In the SVD 1− classification, the estimated future intensities are lower than historical values throughout the entire region. All of these characteristics, with exception for high intensities of the historical SVD 1− data, are consistent with the SVD mode 1 positive and negative patterns of MPI, MIDRH, and VWS that relate to an increase (positive mode) or decrease (negative mode) in favorable conditions in the MDR over which many TCs travel prior to entering the GoM.

Overall, a mixed regression based on classifications using SVD mode 1 does not appreciably alter the fit of each model as R^2 values don't differ among the ALL and two SVD mode classifications (Table 5.3). However, the partitioning improves the physical representation of TC tracks and intensities as they could be possibly altered in future environments when a particular year may be in SVD mode 1+ or SVD mode 1− state. Additionally, the predictor selection reflects the changing aspect of TC characteristics in that during SVD mode 1− classifications, the recurving-type of track density is more dominant and TC-type predictors related to latitude and translation speed are significant in that class whereas they do not appear in SVD mode 1+ when recurving tracks may be less frequent.

While there is certainly room for improvement in the approach to a combined physical-statistical model of TC intensity, the goal here is to provide a careful attempt at estimating future TC intensities when the environment is changing but track frequencies or locations are held constant. Additionally, the character of the future TC intensity estimates over the near- and on-shore (landfall) regions (Fig. 5.7) over which the joint hazard analysis is applied is quite reasonable.

(b) *Joint Frequency Analysis of Wind and Precipitation*

Using estimated historical and future TC intensities applied to the large number of synthetic tracks for each period, the wind and precipitation models are applied to produce gridded fields of each parameter that provide the statistical basis for the univariate and joint TC wind and precipitation frequency analysis. For this application, the gridded fields were defined at 5 km horizontal resolution, but that can be set to a higher or lower resolution as needed. Beginning with 2020 and proceeding through 2050 in 5-year increments, the individual wind and precipitation data for each year at each of the grid points are analyzed using the GEV distribution to define the marginal distributions and return levels of wind and total storm precipitation for return periods of 2–1000 years. Examples of such analysis for 2020 (Fig. 5.10a–b) and 2050 (Fig. 5.10c–d) indicate the spatial distribution and magnitude of precipitation return levels for return periods of 100 (Fig. 5.10a, c) and 500 years (Fig. 5.10b, d) for the years 2020 (Fig. 5.10a, b) and 2050 (Fig. 5.10c, d). In the following, 2050 will continue to be used to typify the future change in TC wind and intensity. While 2050 is an SVD 1− year, it is chosen because SVD 1− years are indicative of El Niño and low ACE (Fig. 5.3), so it is expected that the combination of the SVD 1− model and track characteristics with the most future year will yield a rather conservative estimate of the change in TC characteristics in both the univariate and joint sense.

To provide more detail on the time-varying return levels of wind and precipitation in the univariate setting, and to compare those with the joint return levels, analysis for the NE area will focus over the New York City (NYC) and Long Island (LI) region (Fig. 5.10) and another area north of Cape Hatteras and near the mouth of Chesapeake Bay (Fig. 5.10). A third area (not shown) is over the western GoM near the Houston metro area. These areas are chosen to provide a sense of the variation in the univariate and joint return levels for a subtropical GoM region, a mid-Atlantic region, and a midlatitude region, which each encompass varying types of TC characteristics. For each region, the univariate return levels using the GEV distribution are provided with uncertainties based on a bootstrap of 1000 iterations to define a 0.95 confidence level about the return level curve. In each display of return level, a subsample of 500 points chosen randomly from the approximate average of 2000 tracks per region are included to give a sense of the reasonable use of the GEV distribution. Also, on each display of univariate return level, the 100- and 500-year return periods and levels are marked to allow for comparison among years and locations.

For the region around NYC and LI (Fig. 5.11), the slope of the wind return-level curve (Fig. 5.10a–b) becomes shallower for longer return periods, which indicates

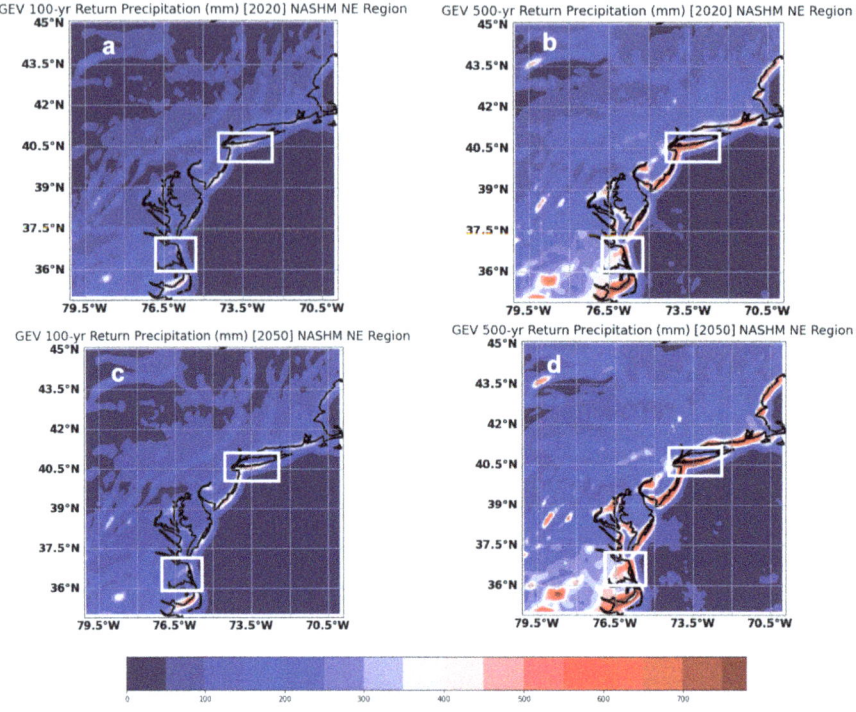

Fig. 5.10 Spatial distribution of TC total precipitation return levels (mm) for return periods of 100 years (left column) and 500 years (right column) for (**a**, **b**) 2020 and (**c**, **d**) 2050. The white boxes define two regions over which the univariate and joint frequency analyses are conducted

the limiting character of TC intensity over this region as TCs move into the midlatitude environment with decreasing intensity and the possible start of extratropical transition. The GEV fit to the data is adequate as only one value lies outside the uncertainty bounds in the tail of the distribution for 2050. As defined in the guidelines for 100- and 500-year return periods, there is a slight increase in wind return levels from 2020 to 2050. Similar characteristics are observed in the return levels of total TC precipitation but the increase in total storm precipitation return levels from 2020 to 2050 (Fig. 5.11c, d) is larger than the increase in wind return levels.

The non-parametric test for tail dependency between the marginal distributions of wind and precipitation yields a p value of 0.002, which allows one to reject the null hypothesis of no dependency. The dependency between the wind and precipitation is qualitatively clear in the scatter of the 500 randomly chosen data points plotted on the joint return periods in Fig. 5.12. The joint OR (Fig. 5.12a, c) return levels plus AND return levels (Fig. 5.12b, d) are displayed over the range of the estimated wind and precipitation data to clearly see differences between the two estimates. Differences between some chosen wind and precipitation return levels from the univariate and joint analyses are compared in Table 5.4 using the two blue circles placed on the

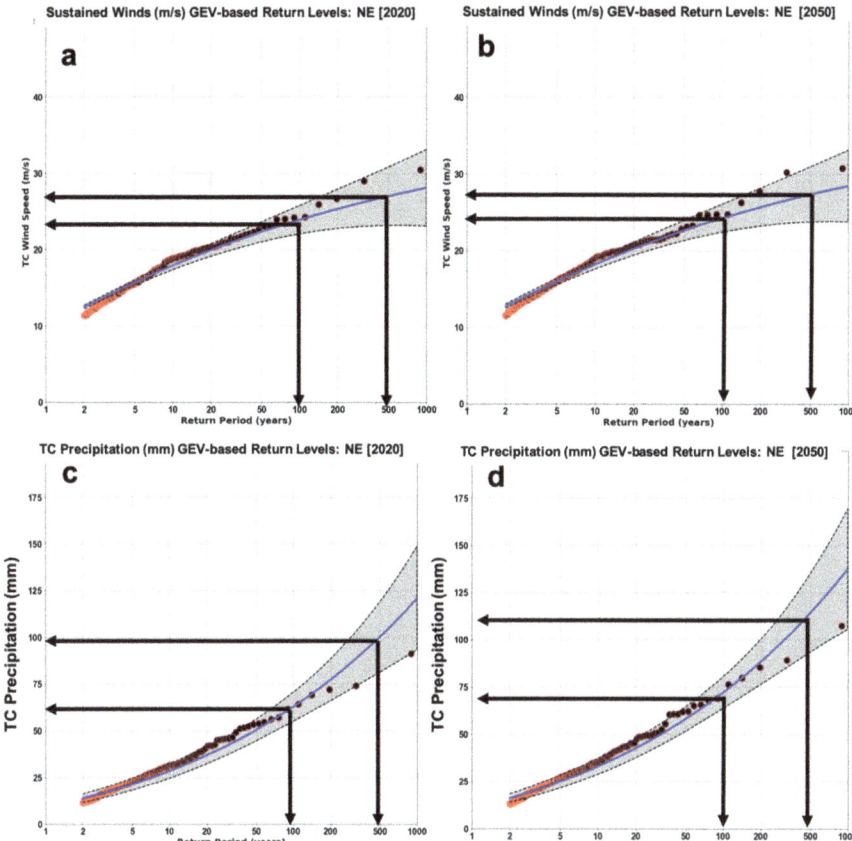

Fig. 5.11 Univariate return levels for (**a, b**) TC winds (m s^{-1}) and (**c, d**) TC total precipitation (mm) for the NYC-LI region and (**a, c**) 2020 and (**b, d**) 2050. Black arrows mark the 100- and 500-year return period values. The shaded area defines the 0.95 uncertainty level and circles define 500 samples from the NASHM synthetic tracks that passed through the region

joint return levels in Fig. 5.12 as reference. For given levels of univariate wind and precipitation values, the return periods decrease in 2050 from what they were in 2020. This is also true for the OR plus AND joint return levels. There are two additional important, more subtle characteristics defined by the univariate and joint analyses. One is that the shorter return period for a given return level increases with increasing return period, and this is more evident in the joint return periods (Fig. 5.12). Also, the joint OR return periods for comparable values of wind and precipitation are shorter than the univariate return periods. Requema et al. (2013) and Kwon and Lall (2016) describe the inequality relationship that defines the decrease in the joint return OR period compared to univariate return periods, which are in turn lower than the AND return periods. This is an important characteristic as it implies that ignoring joint characteristics could lead to incorrect estimation of return periods, which can impact certain mitigation policies.

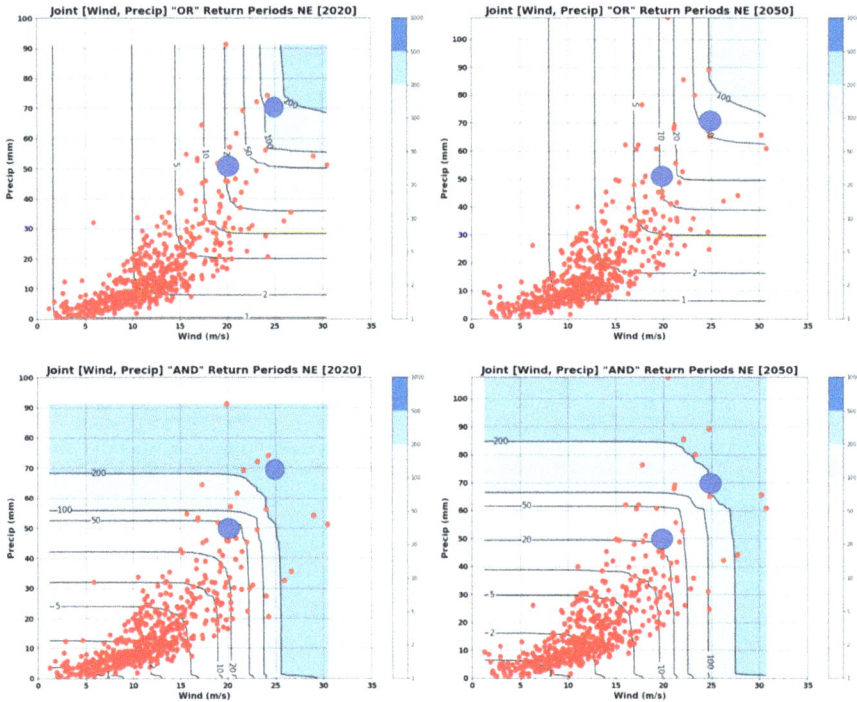

Fig. 5.12 Joint TC total precipitation and winds for the NYC-LI region in 2020 (top) and 2050 (bottom). Contours define return periods (years) and the red circles are 500 samples taken from the NASHM synthetic tracks. The blue circles mark wind and precipitation values used in Table 5.4 to compare joint and univariate return levels for 2020 and 2050

Table 5.4 Comparison of univariate and joint return levels and return periods for selected values of wind and TC total precipitation for the NYC-LI region and for years 2020 and 2050

New York City and Long Island region		
Wind (m s^{-1})	2020 return period (years)	2050 return period (years)
25	120	100
20	20	20
Precipitation (mm)		
70	155	100
50	55	35
Joint wind, precipitation		
[Wind >25 OR Precip >70]	150	70
[Wind >20 OR Precip >50]	20	10
[Wind >25 AND Precip >70]	270	200
[Wind >20 AND Precip >50]	45	25

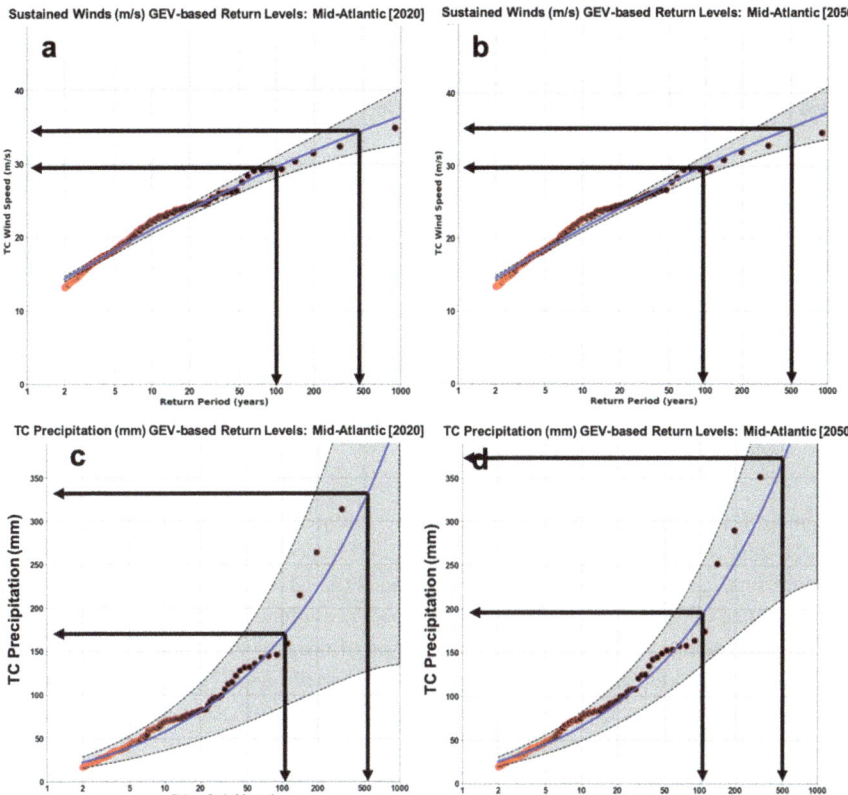

Fig. 5.13 Univariate return levels for (**a, b**) TC winds (m s^{-1}) and (**c, d**) TC total precipitation (mm) for the mid-Atlantic region in (**a, c**) 2020 and (**b, d**) 2050. Black arrows mark the 100- and 500-year return period values. The shaded area defines the 0.95 uncertainty level and circles define 500 samples from the NASHM synthetic tracks that passed through the region

For the region of the mid-Atlantic coastline where TC activity increases and is more intense in general, the wind return-level curves for 2020 (Fig. 5.13a) and 2050 (Fig. 5.13b) have greater slopes than the curves for the NYC and LI region owing to the increase in frequency of more extreme TC wind and precipitation events. Decreases in return periods between 2020 and 2050 occur (Table 5.5), but they are smaller than found for the NYC-LI region. This is probably due to the choice of an SVD 1− year and the unfavorable large-scale environmental patterns associated with that mode (Figs. 5.1a–c, 5.8g–i). The GEV fit to the precipitation data (Fig. 5.13c, d) is not optimal as there is a long tail due to several extreme precipitation events, and the uncertainty in return levels beyond a 100-year return period is quite large for both years. However, return level values of total precipitation are comparable to those in Feldman et al. (2019) for radar locations over the mid-Atlantic region.

Table 5.5 Comparison of univariate and joint return levels and return periods for selected values of wind and TC total precipitation for the Mid-Atlantic region and for years 2020 and 2050

Mid-Atlantic region		
Wind (m s^{-1})	2020 return period (years)	2050 return period (years)
30	140	100
25	20	20
Precipitation (mm)		
350	570	500
250	240	200
Joint wind, precipitation		
[Wind >30 OR Precip >350]	170	100
[Wind >25 OR Precip >250]	40	20
[Wind >30 AND Precip >350]	300	240
[Wind >25 AND Precip >250]	270	180

The test for tail dependency between TC intensity and precipitation over the mid-Atlantic region results in rejection of the null hypothesis of no dependency, but the significance is not as high as found with the NYC-LI wind and precipitation data. The influence of the extreme precipitation events is evident in the scatter of wind and precipitation points (Fig. 5.14) and in the shape of the two-dimensional joint return period curves and strong gradient along the wind-speed axis. Regardless, joint OR return periods for corresponding values of wind and precipitation are much shorter than the univariate return periods in Table 5.5.

For the western GoM region, the return level curves are indicative of a region over which TCs have a higher intensity and produce extreme precipitation, often due to sufficiently high sea-surface temperatures and deep, warm water eddies. The use of the GEV distribution for the GoM region wind (Fig. 5.15a, b) is more appropriate than for precipitation (Fig. 5.15c, d) data. Some extreme precipitation events lead to a long right tail, which exhibits high uncertainty in return levels. Also, as found in the regions along the Atlantic coast, the differences in wind return levels between 2020 and 2050 are small (Table 5.6). This is again likely due to the use of an SVD 1− year in which the characteristics of TC tracks over the western GoM are quite different than in an SVD 1+ year. However, the tendencies for shorter return periods in 2050 for a given return level are still evident, as are the shorter return periods for joint wind and precipitation values than the univariate return levels.

Tail dependence between TC winds and precipitation over the western GoM region is also significant and visually evident in the scatter of data displayed on the joint return period curves (Fig. 5.16). As found for the Atlantic regions, for given return levels the joint OR plus AND return periods decrease from 2020 to 2050. Also, the OR joint return period for given return levels is less than corresponding univariate return periods.

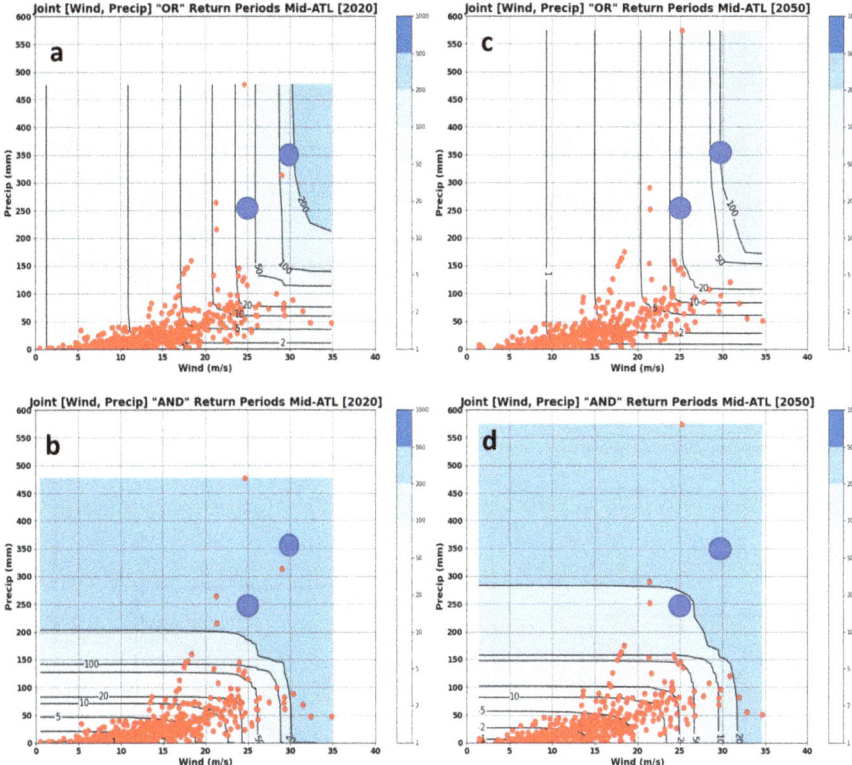

Fig. 5.14 Joint TC total precipitation (mm) and winds (m s^{-1}) for the Mid-Atlantic region in 2020 (top) and 2050 (bottom). Contours define return periods (years) and the red circles are 500 samples taken from the NASHM synthetic tracks. The blue circles mark wind and precipitation values used in Table 5.5 to compare joint and univariate return levels for 2020 and 2050

Overall, three tendencies are found in the variation of return periods for each region. For given return levels of TC intensity and precipitation, univariate and joint OR plus AND return periods are shorter in 2050 than in 2020. The differences between the 2 years increase with increasing return period. The OR return periods of joint events are consistently shorter than univariate events over the three regions with some variation. As found in the analysis of the TC intensity model, changes in intensity are modest between 2020 and 2050 and that is reflected by the differences in univariate and joint return periods. This is clear over the mid-Atlantic and western GoM regions where differences between wind return levels in 2050 and 2020 are small and differences in precipitation are larger, which affects the joint return periods such that there is a rather sharp return-period gradient along the wind speed axis. This is not as evident in the joint return periods over the NYC-LI region (Fig. 5.12).

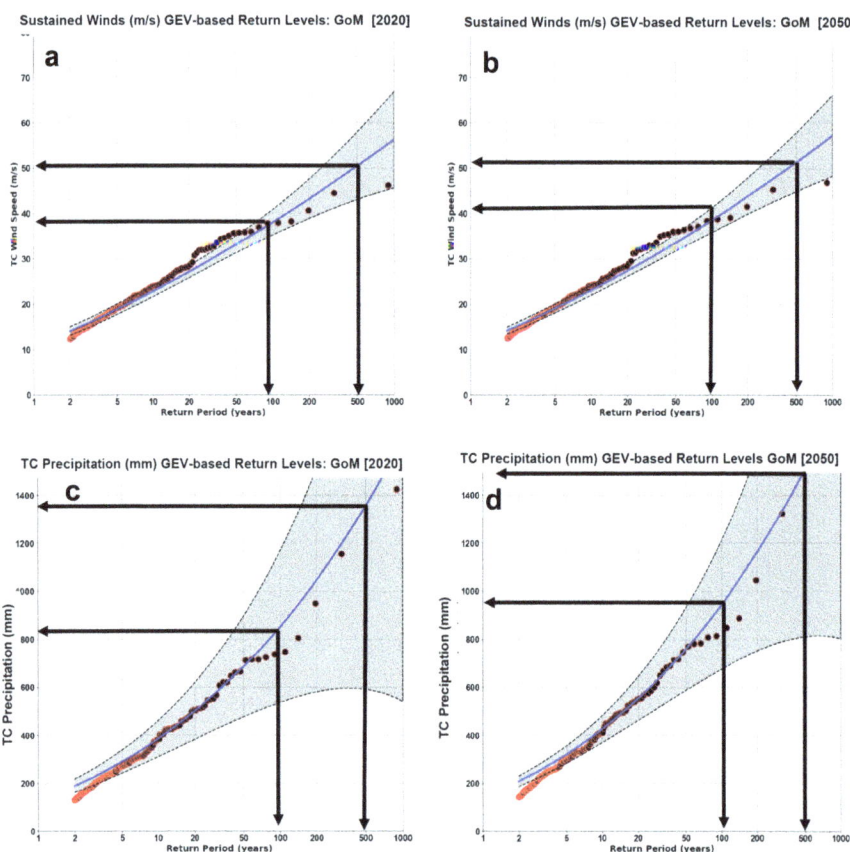

Fig. 5.15 Univariate return levels for (**a**, **b**) TC winds (m s^{-1}) and (**c**, **d**) TC total precipitation (mm) for the western GoM region in (**a**, **c**) 2020 and (**b**, **d**) 2050. Black arrows mark the 100- and 500-year return period values. The shaded area defines the 0.95 uncertainty level and circles define 500 samples from the NASHM synthetic tracks that passed through the region

5.4 Discussion and Conclusions

To understand possible changes in the character of impacts from TC wind and precipitation into the future, a combined system of models is constructed to estimate TC intensity, TC winds, and TC maximum precipitation. The analysis strategy assumes that the frequency and tracks of future TCs will not differ greatly from historical characteristics, but TC intensity might change. This is roughly based on consensus views of the state of climate change impacts to TC characteristics as summarized in Knutson et al. (2020) and references therein. However, uncertainties in climate model projections and in statistical estimation of key distributions remain important and require further analysis and understanding. Using historical TC tracks and reanalysis depiction of the historical environment, a regression-based model is

5 Analysis of the Future Change in Frequency of Tropical Cyclone-Related...

Table 5.6 Comparison of univariate and joint return levels and return periods for selected values of wind and TC total precipitation for the western GoM region and for years 2020 and 2050

Western GoM region		
Wind (m s^{-1})	2020 return period (years)	2050 return period (years)
35	50	48
25	15	11
Precipitation (mm)		
1000	200	150
700	60	45
Joint wind, precipitation		
[Wind >35 OR Precip >1000]	100	40
[Wind >25 OR Precip >700]	15	9
[Wind >35 AND Precip >1000]	265	210
[Wind >25 AND Precip >700]	75	40

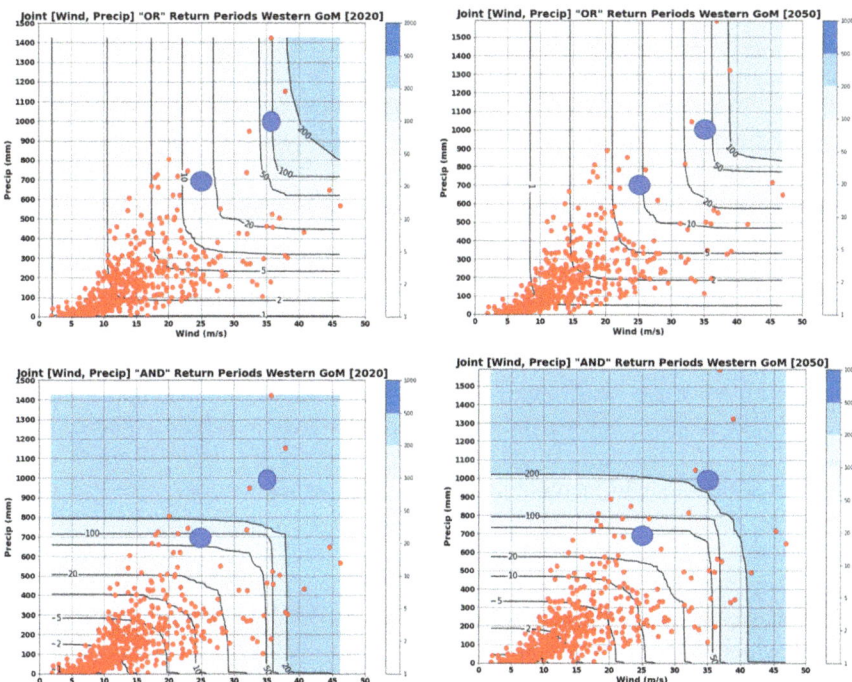

Fig. 5.16 Joint TC total precipitation (mm) and winds (m s^{-1}) for the western GoM region in 2020 (top) and 2050 (bottom). Contours define return periods (years) and the red circles are 500 samples taken from the NASHM synthetic tracks. The blue circles mark wind and precipitation values used in Table 5.6 to compare joint and univariate return levels for 2020 and 2050

trained to estimate TC intensity using past TC characteristics and environmental parameters. That model is then applied to simulated synthetic tracks under the same historical environment and then to future environments from 2020 to 2050 using an ensemble of CMIP6 models and the SSP585 scenario. Also, since the environment under which TCs form, move, and intensify changes on an interannual (and shorter time scales) basis, a regression model format was derived to classify predictors based on representation of the changes in environment rather than treat the environment as being homogenous in time. The estimated historical and future TC intensities are then used in a parametric wind model and boundary-layer model to produce a two-dimensional low-level wind field. The low-level winds reflect land-sea differences, terrain, and surface roughness and are then used in a model of TC precipitation to produce TC total precipitation swaths and also maximum precipitation rates. The wind and precipitation data are produced over regional grids at a specified resolution for thousands of synthetic track realizations. These data then form the basis of a univariate and joint frequency analysis of wind and precipitation return levels for varying return periods and for the years 2020–2050.

Because global climate models are unable to robustly represent TC wind and precipitation structures, and because biases may be introduced due to methods employed to recognize TCs in global climate model output, the construction of a simplified model sequence and synthetic track model set is a basic step to understanding potential changes in the statistical character of TC-related hazards over time. Also, since TC impacts are often multivariate in nature, a key aspect of this study was to construct a joint frequency analysis of the impacts of wind and precipitation. Additionally, a goal was to compare the return levels and periods of potential impactful conditions when analyzed in a univariate method versus jointly.

While the frequency analyses are applied over all grid points, for clarity, three regions of varying TC character and environments were chosen to analyze. While changes in univariate-based return levels of wind and precipitation between 2020 and 2050 are modest, they are similar to those identified in the CMIP6-based high-resolution model intercomparison project multi-model ensemble (Roberts et al. 2020) in which a slight increase in average 10-m winds was detected. It is noted that these slight increases up to 2050 are below the current margin of error in observations of TC intensity and model projections indicate an increase in intensity trends beyond 2050, which is beyond the time scope of this study. Additionally, while the year 2050 is farthest into the future of years analyzed here, it is an SVD 1– year in which environmental conditions are less than optimal for TC frequency and possibly TC intensity, though it is important to note that years with a below-average number of TC events do not correspond to years with TCs of less intensity.

Application of the model sequence, univariate frequency analysis, and joint frequency analysis yields three primary conclusions:

(i) For comparable return levels of wind and TC total precipitation, return periods in 2050 are shorter than they are in 2020;

(ii) The difference between return periods in 2050 from those in 2020 increase as the return period increases. Therefore, low probability events in 2020 occur with greater probability in 2050; and
(iii) For comparable levels of wind and precipitation, joint OR return periods are shorter than univariate return periods, which has implications for underestimating extreme events and that can potentially impact mitigation policies.

While the analyses presented here focus on three specific regions and one TC ocean basin, application globally can be examined to define regional differences in the character of joint and univariate return periods and their possible change into future years. Additionally, further efforts to estimate uncertainties due to climate model projections and estimation of distribution parameters will improve the understanding of differences in return periods over time. Plus, it is generally acknowledged (Knutson et al. 2020; Murakami et al. 2020) that changing climate conditions prior to the mid twenty-first century are not too sensitive to emission scenarios, and thus differences identified in return periods between 2020 and 2050 may amplify beyond 2050, which was the latest year addressed in this study. Finally, the joint frequency analysis is not limited to a bivariate scenario and inclusion of other high-impact factors such as storm surge can be used to extend the analysis to the three primary TC hazards of wind, precipitation, and surge.

Acknowledgments The authors acknowledge the very constructive comments provided by the anonymous reviewers and editors, which contributed to an improved manuscript.

References

Alqahtani NA, Kalantan ZI (2020) Gaussian mixture models based on principal components and applications. Math Probl Eng 2020:Article ID 1202307. https://doi.org/10.1155/2020/1202307
Bell GD, Chelliah M (2006) Leading tropical modes associated with interannual and multi-decadal fluctuations in North Atlantic hurricane activity. J Clim 19:590–612
Bhatia K, Vecchi G, Murakami H, Underwood S, Kossin J (2018) Projected response of tropical cyclone intensity and intensification in a global climate model. J Clim 31:8281–8303. https://doi.org/10.1175/JCLI-D-17-0898.1
Bister M, Emanuel KA (1998) Dissipate heating and hurricane intensity. Meterol Atmos Phys 50: 233–240
Chavas DR, Lin N (2015) A model for the complete radial structure of the tropical cyclone wind field. Part II: wind field variability. J Atmos Sci 27:3093–3113. https://doi.org/10.1175/JAS-D-15-0185.1
Chavas DR, Lin N, Emanuel K (2015) A model for the complete radial structure of the tropical cyclone wind field. Part I: comparison with observed structure. J Atmos Sci 72:3647–3662. https://doi.org/10.1175/JAS-D-15-0014.1
Cherry S (1996) Singular value decomposition analysis and canonical correlation analysis. J Clim 9: 2003–2009. https://doi.org/10.1175/1520-0442(1996)
Chu P, Wang J (1998) Modeling return periods of tropical cyclone intensities in the vicinity of Hawaii. J Appl Meteorool Climatol 37:951–960
Coles S (2001) An introduction to statistical modeling of extreme values. Springer, London. isbn:978-1-85233-459-8

DeMaria M, Kaplan J (1994) A statistical hurricane intensity prediction scheme (SHIPS) for the Atlantic basin. Weather Forecast 9:209–220. https://doi.org/10.1175/1520-0434(1994) 009<0209:ASHIPS>2.0.CO:2

DeMaria M, Mainelli M, Shay LK, Knaff JA, Kaplan J (2005) Further improvements to the statistical hurricane intensity prediction scheme (SHIPS). Weather Forecast 20:531–543. https://doi.org/10.1175/WAF862.1

Done JM, Ge M, Holland GJ, Dima-West I, Phibbs S, Saville GR, Wang Y (2020) Modelling global tropical cyclone wind footprints. Nat Hazards Earth Syst Sci 20:567–580. https://doi.org/10.5194/nhess-20-567-2020

Elsner JB, Jagger TH, Liu KB (2008) Comparison of hurricane return levels using historical and geological records. J Appl Meteorol Climatol 47:368–374

Emanuel KA (2013) Downscaling CMIP5 climate models shows increased tropical cyclone activity over the 21st century. Proc Natl Acad Sci 110(30):12219–12224. https://doi.org/10.1073/pnas.1301293110

Emanuel K (2020) Response of global tropical cyclone activity to increasing CO2: results from downscaling CMIP6 models. J Clim 34(1):57–70. https://doi.org/10.1175/JCLI-D-20-0367.1

Emanuel K, Jagger T (2010) On estimating hurricane return periods. J Appl Meteorol Climatol 49:837–844

Favre A-C, El Adlouni S, Perreault L, Thiémonge N, Bobeé B (2004) Multivariate hydrological frequency analysis using copulas. Water Resour Res 40:W01101. https://doi.org/10.1029/2003WR002456

Feldman M, Emanuel K, Zhu L, Lohmann U (2019) Estimation of Atlantic tropical cyclone rainfall frequency in the United States. J Appl Meteorol Climatol 58:1853–1866. https://doi.org/10.1175/JAMC-D-19-0011.1

Field CB (2012) Managing the risks of extreme events and disasters to advance climate change adaption: special report of the intergovernmental panel on climate change. Cambridge University Press, Cambridge

Genest C, Favre A-C (2007) Everything you always wanted to know about copula modeling but were afraid to ask. J Hydrol Eng ASCE 12:347–368

Goldenberg SB, Shapiro LJ (1996) Physical mechanisms for the association of El Niño and West African rainfall with Atlantic major hurricanes. J Clim 9:1169–1187

Gray WM (1984) Atlantic seasonal hurricane frequency: part I: El Niño and 30-mb quasi-bienniel oscillation influences. Mon Weather Rev 112:1649–1668

Grimaldi S, Serinaldi F (2006) Asymmetric copula in multivariate flood frequency analysis. Adv Water Resour 29:1155–1167

Hall TM, Jewson S (2007) Statistical modeling of North Atlantic tropical cyclone tracks. Tellus 59A:486–498

Hall T, Yonekura E (2013) North American tropical cyclone landfall and SST: a statistical model study. J Clim 26:8422–8439

Irish JL, Resio DT, Divoky D (2011) Statistical properties of hurricane surge along a coast. J. Geophys Res Oceans 116:1–15

Jagger T, Elsner JB (2006) Climatology models for extreme hurricane winds near the United States. J Clim 19:3220–3236

Jing R, Lin N (2020) An environment-dependent probabilistic tropical cyclone model. J Adv Model Earth Sys 12:e2019MS001975. https://doi.org/10.1029/2019MS001975

Kepert JD (2001) The dynamics of boundary layer jets within a tropical cyclone core. Part I: linear theory. J Atmos Sci 58:2469–2484

Kepert JD, Wang Y (2001) The dynamics of boundary layer jets within a tropical cyclone core. Part II: nonlinear enhancements. J Atmos Sci 58:2485–2501

Knaff JA, DeMaria M, Sampson CR, Gross JM (2003) Statistical, 5-day tropical cyclone intensity forecasts derived from climatology and persistence. Weather Forecast 18:80–92. https://doi.org/10.1175/1520-0434(2003)018<0080:SDTCIF>2.0.CO:2

Knapp KR, Kruk MC, Levinson DH, Diamond HJ, Neumann CJ (2010) The international best track archive for climate stewardship (IBTrACS). Bull Am Meteorol Soc 91:363–376. https://doi.org/10.1175/2009BAMS2755.1

Knutson T et al (2019) Tropical cyclones and climate change assessment: part I: detection and attribution. Bull Am Meteorol Soc 100:1987–2007. https://doi.org/10.1175/BAMS-D-18-0189.1

Knutson T et al (2020) Tropical cyclones and climate change assessment: part II: projected response to anthropogenic warming. Bull Am Meteorol Soc 101:E303–E322. https://doi.org/10.1175/BAMS-D-18-0194.1

Kopp R, Easterling D, Hall T, Hayhoe K, Horton R, Kunkel K, LeGrande, A (2017) Potential surprises – compound extremes and tipping elements. Climate science special report: a sustained assessment activity of the U.S. Global Change Research Program, pp 608–635

Kwon H-H, Lall U (2016) A copula-based nonstationary frequency analysis for the 2012–2015 drought in California. Water Resour Res 52:5772–5675. https://doi.org/10.1002/2016WR018959

Lee C-Y, Tippett MK, Camargo SJ, Sobel AH (2015) Probabilistic multiple linear regression modeling for tropical cyclone intensity. Mon Weather Rev 143:9330954. https://doi.org/10.1175/MWR-D-14-00171.1

Lee C-Y, Tippett MK, Sobel AH, Camargo SJ (2016) Autoregressive modeling for tropical cyclone intensity climatology. J Clim 29:7815–7830. https://doi.org/10.1175/JCLI-D-15-0909.1

Lee C-Y, Tippett M, Sobel AH, Camargo SJ (2018) An environmentally forced tropical cyclone hazard model. J Adv Model Earth Syst 10:223–241. https://doi.org/10.1002/2017MS001186

Lin N, Jing R, Wang Y, Yonnekura E, Fan J, Xue L (2017) A statistical investigation of the dependence of tropical cyclone intensity change on the surrounding environment. Mon Weather Rev 145:2813–2831. https://doi.org/10.1175/MWR-D-16-0368.1

Lu P, Lin N, Emanuel K, Chavas D, Smith J (2018) Assessing hurricane rainfall mechanisms using a physics-based model: hurricanes Isabel (2003) and Irene (2011). J Atmos Sci 75:2337–2358. https://doi.org/10.1175/JAS-D-17-0264

Malmstadt JC, Elsner JB, Jagger TH (2010) Risk of strong hurricane winds to Florida cities. J Appl Meteorol Climatol 49:2121–2132

Merrill RT (1987) An experiment in statistical prediction of tropical cyclone change. NOAA Tech. Memo. NWS NHC-34

Mudd L, Rosowsky D, Letchford C, Lombardo F (2017) Joint probabilistic wind–rainfall model for tropical cyclone Hazard characterization. J Struct Eng 143(3). https://doi.org/10.1061/(ASCE)ST.1943-541X.0001685

Murakami H, Delworth TL, Cooke WF, Zhao M, Xiang B, Hsu P-C (2020) Detected climatic change in global distribution of tropical cyclones. Proc Nat Acad Sci 117(20):10706–19714. https://doi.org/10.1073/pnas.1922500117

Nelson RB (1999) An introduction to copulas. Springer, New York

Requema AI, Mediero L, Garrote L (2013) A bivariate return period based on copulas for hydrologic dam design: accounting for reservoir routing in risk estimation. Hydrol Earth Syst Sci 17:3023–3038. https://doi.org/10.5193/hess-17-3023-2013

Roberts MJ et al (2020) Projected future changes in tropical cyclone using the CMIP6 HighResMIP multimodel ensemble. Geophys Res Lett 47:e2020GL088772. https://doi.org/10.1029/2020GL088662

Sackl B, Bergmann H (1987) A bivariate flood model and its application. In: Singh VP (ed) Hydrologic frequency modeling. Dreidel, Dordrecht, pp 571–582

Salvadori G, Durante F, De Michele C, Bernardi M, Petrella L (2016) A multivariate copula-based framework for dealing with hazard scenarios and failure probabilities. Water Resour Res 52: 3701–3721

Vickery PJ, Lavelle FM (2014) The effects of warm Atlantic Ocean sea surface temperatures on the ASCE 7-10 design wind speeds. Proceedings of the conference on advances in hurricane engineering, Applied Technology Council, Structural Engineering Institute, Reston, VA, pp 13–22

Vickery PJ, Skerlj P, Twisdale L (2000) Simulation of hurricane risk in the US using empirical track model. J Struct Eng:1222–1237. https://doi.org/10.1061/(ASCE)0733-9445(2000)126:10 (1222)

Wahl T, Jain S, Bender J, Meyers SD, Luther ME (2015) Increasing risk of compound flooding from storm surge and rainfall to major U.S. cities. Nature. Climate Change 5(12):1093–1097

Xi D, Lin N, Smith J (2020) Evaluation of a physics-based tropical cyclone rainfall model for risk assessment. J Hydrometeorol 21:2197–2217. https://doi.org/10.1175//JHM-D-0035.1

Yue S (2001) A bivariate gamma distribution for use in multivariate flood frequency analysis. Hyrdrol Processes 15:1033–1045

Yue S, Ouarda TBMK, Bobeé B, Legendre P, Bruneau P (2001) The Gumbel mixed model for flood frequency analysis. J Hydrol 246:1–18

Zhang L, Singh VP (2006) Bivariate flood frequency analysis using the copula method. J Hydrol Eng-ASCE 11:150–164

Zhang L, Singh VP (2007) Trivariate flood frequency analysis using the Gumbel-Hougaard copula. Hydrol Eng-ASCE 12:431–439

Open Access This chapter is licensed under the terms of the Creative Commons Attribution 4.0 International License (http://creativecommons.org/licenses/by/4.0/), which permits use, sharing, adaptation, distribution and reproduction in any medium or format, as long as you give appropriate credit to the original author(s) and the source, provide a link to the Creative Commons license and indicate if changes were made.

The images or other third party material in this chapter are included in the chapter's Creative Commons license, unless indicated otherwise in a credit line to the material. If material is not included in the chapter's Creative Commons license and your intended use is not permitted by statutory regulation or exceeds the permitted use, you will need to obtain permission directly from the copyright holder.

Chapter 6
Current and Future Tropical Cyclone Wind Risk in the Small Island Developing States

Nadia Bloemendaal and E. E. Koks

Abstract Tropical cyclones (TCs) are amongst the costliest and deadliest natural hazards and can cause widespread havoc in tropical coastal areas. Small Island Developing States (SIDS) are particularly vulnerable to TCs, as they generally have limited financial resources to overcome past impacts and mitigate future risk. However, risk assessments for SIDS are scarce due to limited meteorological, exposure, and vulnerability data. In this study, we combine recent research advances in these three disciplines to estimate TC wind risk under past (1980–2017) and near-future (2015–2050) climate conditions. Our results show that TC risk strongly differs per region, with 91% of all risk constituted in the North Atlantic. The highest risk estimates are found for the Dominican Republic and Puerto Rico, with present-climate expected annual damages (EAD) of 1.51 billion and 1.25 billion USD, respectively. This study provides valuable insights in TC risk and its spatial distribution, and can serve as input for future studies on TC risk mitigation in the SIDS.

Keywords Small Island Developing State · Modelling · Risk · Damages · Tropical cyclones

---

Supplementary Information The online version contains supplementary material available at [https://doi.org/10.1007/978-3-031-08568-0_6].

N. Bloemendaal (✉)
Institute for Environmental Studies, Vrije Universiteit, Amsterdam, The Netherlands

Allianz SE, Munich, Germany
e-mail: nadia.bloemendaal@vu.nl

E. E. Koks
Institute for Environmental Studies, Vrije Universiteit, Amsterdam, The Netherlands

Oxford Programme for Sustainable Infrastructure Systems, Environmental Change Institute, University of Oxford, Oxford, UK
e-mail: elco.koks@vu.nl

6.1 Introduction

Tropical cyclones (TCs) are among the deadliest and costliest natural hazards, wreaking widespread havoc when they make landfall. Small Island Developing States (SIDS) in particular have been substantially affected by TCs. The 2017 hurricanes Irma and Maria, for example, caused catastrophic damage across multiple SIDS in the Caribbean, including Antigua and Barbuda, Saint Martin, the British and U.S. Virgin Islands, Puerto Rico, and Dominica (Cangialosi et al. 2018; Pasch et al. 2019). The latter island suffered damage totals exceeding 930 million USD, amounting to approximately 226% of the country's total Gross Domestic Product (GDP) (Government of the Commonwealth of Dominica 2017). Over the past 5 years, Fiji was hit by four major TCs, each causing major damage across the country. Almost 15% of Fiji's total population was displaced due to TC Winston (2016), and damages were estimated at 1.38 billion USD, which is 31% of the country's GDP (World Bank 2016).

The aforementioned historical events all demonstrate that SIDS are particularly vulnerable to, and can be disproportionately affected by, TCs. Over the last few years, authorities have been increasingly aware of this disproportionality (UNESCO 2019) and that these impacts are most likely to deteriorate even further towards the future (UNFCCC 2005). Under future climate change, TC intensity, storm surge inundation, and rainfall are projected to increase globally, thereby enhancing the associated risks (Knutson et al. 2020). SIDS, unfortunately, often face limited institutional capacity, scarce financial resources, and high vulnerability. They therefore struggle to overcome past impacts and mitigate future TC risk (United Nations 2021). This is driven by the fact that many SIDS are characterized by their dependency on a single economic sector providing employment and economic growth, such as tourism (Pathak et al. 2021). When such a sector is (substantially) disrupted after a TC event, it can take years for this sector, and thereby (the majority of) the economy of the SIDS, to overcome such impacts. It is therefore crucial to quantify their respective TC risk, to support the design of appropriate risk mitigation strategies, and thereby enhancing the resilience of these SIDS.

However, despite their exposure and vulnerability to TCs, (global) risk assessments focused on SIDS are scarce, and instead are more often performed on a regional scale or for particular TC events (e.g., Webb 2020; Stephenson and Jones 2017; Commonwealth of Australia 2013). A uniform study, assessing the economic risk for all SIDS prone to TCs, is currently still lacking. This gap can be attributed to two complicating factors: (1) scarce observational records over longer temporal scales on both hazard and impact/damage totals, and (2) the relatively small size of most SIDS, making it difficult to identify and study them in meteorological datasets, as the spatial resolutions of such datasets are often lower than the entire size of an island. Recent scientific advances, however, overcome these limitations. Global-scale synthetic TC models, for instance, overcome the limited spatial and temporal hazard information imposed by historical TC data (Bloemendaal et al. 2020b; Emanuel et al. 2006; Lee et al. 2018). These models take information, commonly spanning a 30–40 year time scale, from either historical data (Bloemendaal et al. 2020b) or global climate models (Lee et al. 2018; Emanuel et al. 2006) and resample this information to an equivalent of a few 1000 years under the same climate conditions. The resulting dataset contains a wealth of information on all theoretically possible TC events for all TC-prone locations on earth, including information on low-probability, high-impact events that might not have occurred in the historical records. At the same time, increasing efforts to collect high-resolution information on assets and economic activity at a global scale also allow the analysis of previously considered data-scarce areas, particularly in developing countries such as the SIDS (Koks et al. 2019).

In this study, we aim to improve our understanding of the current TC-induced windstorm risk in the SIDS. To do so, we make use of publicly available models and datasets. We combine state-of-the-art TC hazard modelling for the current and future climate (Bloemendaal et al. 2020b) with high-resolution exposure information obtained from both OpenStreetMap and globally-consistent gridded exposure (Eberenz et al. 2020) and vulnerability information (Eberenz et al. 2021).

The remainder of this chapter is organized as follows. Section 6.2 briefly describes the SIDS that are studied, followed by Sect. 6.3 in which the methodology is explained in detail. Section 6.4 presents the results of our analysis and puts them in a broader perspective. Finally, Sect. 6.5 provides some concluding remarks and ways forward.

6.2 Small Island Developing States

The United Nations (UN) defines SIDS as "a distinct group of 38 UN Member States and 20 Non-UN Associate Members that face unique social, economic and environmental vulnerabilities" (United Nations 2021). Of these 58 SIDS, six are found within 5°N/S of the Equator, a region generally unfavorable for TC development due to a too weak Coriolis force. In addition, Bahrein, Cabo Verde, and Guinea-Bissau are located in regions generally not considered to be prone to TCs; as such, we exclude these SIDS as well. Lastly, we exclude French Polynesia from our analysis due to absence in the exposure dataset (see Sect. 6.3). This leaves 48 SIDS included in this study, which are geographically mapped in Fig. 6.1.

6.3 Methods

This study estimates the current and future TC windstorm risk in the 48 SIDS presented in Sect. 6.2. To do so, we take a traditional catastrophe loss modeling approach (Kron 2005; Koks et al. 2019), which is commonly applied in catastrophe modeling by academia and the (re)insurance industry. More specifically, we define risk (in terms of USD) as a function of hazard – the probability of the TC event;

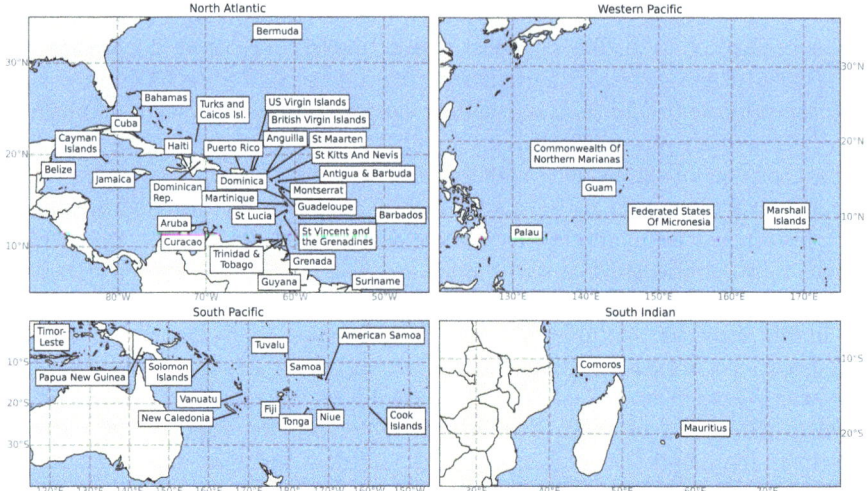

Fig. 6.1 Overview of the Small Island Developing States that are part of this study

exposure – the value of assets subject to the TC; and vulnerability – the capacity of a society to deal with the event (Koks et al. 2015b). In this study, both the exposure and vulnerability are kept constant throughout time. More specifically, only the hazard-component is changing under climate change. This means that TC risk results solely portray the effect of climate change on TC wind speeds, as socio-economic changes are not included.

6.3.1 Hazard

6.3.1.1 Baseline Climate Conditions

At the basis of the TC hazard modelling component lies the Synthetic Tropical cyclOne geneRation Model (STORM) (Bloemendaal et al. 2020b). STORM is a fully statistical model taking information on TC track, characteristics, and environmental variables (mean sea-level pressure and sea-surface temperature; SST) as input variables. These variables are resampled in STORM to extend the temporal scale of the input dataset to 10,000 years of TC activity under the same climate conditions. The resulting STORM synthetic TC dataset contains data on the position of the eye of the TC (longitude/latitude), 10-minute 10-m average maximum sustained wind speed (in m/s), minimum pressure (in hPa), and the radius to maximum winds (in km). The STORM dataset as presented in Bloemendaal et al. (2020b) resembles the climate conditions over 1980–2017, and was generated using historical TC statistics from the International Best-Track Archive for Climate

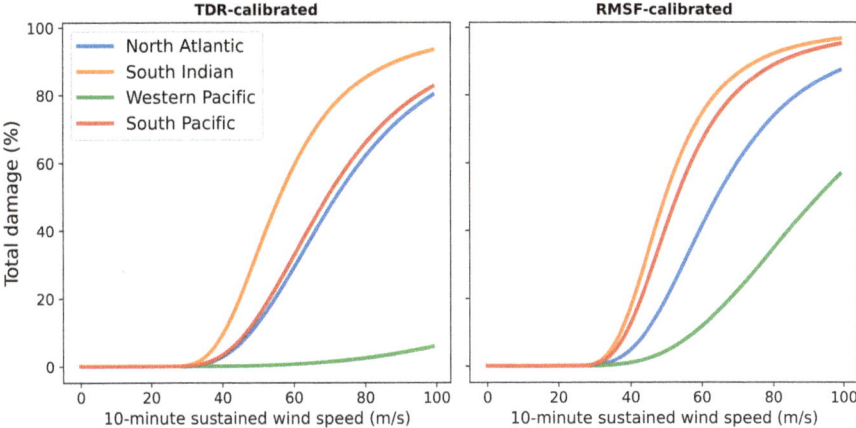

Fig. 6.2 Vulnerability curves used in this study. The left-panel presents the four vulnerability curves (one for each basin) which are calibrated using the Total Damage Ratio (TDR). The right-panel presents the four vulnerability curves (one for each basin) which are calibrated using the Root-mean-squared fraction (RMSF)

Stewardship (IBTrACS; Knapp et al. 2010). We direct readers to Bloemendaal et al. (2020b) for detailed information on STORM and the STORM baseline dataset.

Please note that STORM simulates TCs with wind speeds exceeding 18 m/s (Tropical Storm-classification on the Saffir-Simpson Hurricane Wind Scale; Simpson and Saffir 1974). SIDS can also be affected by Tropical Depressions and post-TCs; however, their accompanying wind speeds are typically too low to cause (substantial) damage to housing and infrastructure (see also Fig. 6.2 and Sect. 6.3.3).

In Bloemendaal et al. (2020a), these synthetic tracks were translated to a 2D-wind field by applying the refined 2D-parametric wind field model of Holland (1980) (Lin and Chavas 2012). In this parametric approach, asymmetry in the wind field arises from the background flow. Please note that in extratropical regions, however, asymmetry may also arise through enhanced wind shear from large-scale background flows or nearby troughs, which is not modeled here (Bloemendaal et al. 2020a; Ritchie and Elsberry 2001). Wind speed RPs were empirically derived from this set of events and at every 10-km grid cell in a basin. Here, we use this STORM RP dataset as the baseline climate RP dataset, with RPs up to 1000-year. Please refer to Bloemendaal et al. (2020a) for validation and access to the RP datasets.

6.3.1.2 Future Climate Conditions

To create a future climate synthetic TC dataset, we use TC information from four global climate models (GCMs): CMCC-CM2-VHR4 (Scoccimarro et al. 2017), CNRM-CM6-1-HR (Voldoire 2019), EC-Earth3P-HR (EC-Earth Consortium 2018), HadGEM3-GC31-HM (Roberts 2017). These GCMs are part of the

HighResMIP multi-model ensemble (Roberts et al. 2020), a model intercomparison study focused on simulating climate runs at higher spatial resolution to better capture small-scale processes such as tropical cyclones. HighResMIP follows the Coupled Model Intercomparison Project Phase 6 (CMIP6) modelling protocol (Eyring et al. 2016). We use the high-resolution coupled ocean-atmosphere GCM runs for the periods 1979–2014 (resembling the GCM baseline climate conditions) and for 2015–2050 (the GCM future climate conditions). These future climate conditions are based on the high-emission Representative Concentration Pathway 8.5 (RCP8.5) with the Shared Socioeconomic Pathway 5 (SSP5) scenario (SSP585) (O'Neill et al. 2016; Van Vuuren et al. 2011). The SSP585 scenario assumes no climate change mitigation strategies are implemented, and average global surface temperatures will increase by 3.3 °C–5.7 °C in 2100 compared to the late nineteenth century (IPCC 2021). While the plausibility of this scenario is under debate given the recent developments in the energy sector, we cannot rule out this scenario and its corresponding climate features altogether. As such, the future-climate results presented in this study should be perceived as a high-end view of future risk.

TCs, and in particular TC intensity, are generally poorly captured in GCMs (Roberts et al. 2020), mostly due to the relatively coarse spatial resolution (coarser than $0.25° \times 0.25°$), which does not adequately resolve smaller-scale TC features such as strong wind and pressure gradients (Murakami and Sugi 2010). As STORM resamples the TC statistics found in the input dataset to an equivalent of 10,000 years, this means that this poor TC representation will be propagated through STORM, and TC intensity will therefore also be underestimated in the STORM +GCM datasets. To overcome this, we use a novel methodology based on the delta approach. Here, we provide a summary of this method; the details of this method and an extensive data analysis can be found in Bloemendaal et al. (2022). In summary, we extract information on changes in the STORM input variables from the baseline and future-climate conditions of the four aforementioned GCMs. This way, we omit the first-order model bias of GCMs, such as an underestimation in TC intensity or an erroneous genesis frequency). Next, we project these changes onto the observed TC statistics (TC genesis frequency and location, track, and intensity) from IBTrACS, which was used for the generation of the baseline STORM dataset. This way, we create a future-climate version of these observed TC statistics. These future-climate statistics then form the input for STORM, hereby creating four synthetic TC datasets for each of the four GCMs.

Lastly, we point out that STORM is run on the basin-scale, meaning that TCs are cut off once they reach the basin boundaries (see Fig. 1 in Bloemendaal et al. 2020b for an overview of the basin boundaries). While almost all of the SIDS used in this study lie well within one basin boundary, Timor-Leste is located just east of the South Indian/South Pacific border in STORM. This means that in STORM, Timor-Leste is solely affected by TCs originating in the South Indian basin, while in reality, the country is also affected by TCs forming in the South Pacific. This can potentially affect the probability and intensity of TC wind hazard estimates for this country.

6.3.2 Exposure

6.3.2.1 OpenStreetMap

To gain insights into the type of assets that are exposed to baseline and future TCs, we make use of publicly available information provided by OpenStreetMap (OSM; www.openstreetmap.org). OSM is a global geospatial database covering a wide range of objects that can be spatially represented on a map. This ranges from administrative boundaries and land uses to roads and buildings. These elements are represented as polygons (e.g., land uses or buildings), linestrings (e.g., roads and electricity lines), and points (e.g. telecommunication towers). Supplementary Table 1 provides an overview of the number of assets per SIDS that are extracted from OSM. All assets were extracted from OSM on July 19, 2021. The geolocations (latitudinal/longitudinal) will be overlaid with the STORM RP maps to assess their exposure to TC wind speeds. As no future projections for the built-up area and infrastructure development exist at this point, we use the same database for both the current and future-climate analysis (i.e., exposure and vulnerability are kept constant).

It should be noted that objects in OSM are 'tagged' (named) and georeferenced by its users using satellite imagery. Due to its dependency on the activity of its users, completeness substantially varies between regions. While very few global numbers exist on actual completeness of buildings (or all assets combined), Barrington-Leigh and Millard-Ball (2017) estimated that in 2016, over 80% of the world's roads were included in OSM. For SIDS in particular, Goldblatt et al. (2020) assessed the completeness of OSM for Haiti, Dominica, and St. Lucia. They found a clear relationship between the population density and the completeness of OSM. More specifically, urban areas tend to be better mapped compared to rural areas. Here we make the implicit assumption that the relative amount of mapped assets between high and low-density built-up areas is similar and that the OSM represents the spatial distribution of assets across the SIDS correctly. To circumvent gaps in the absolute number of mapped assets in OSM, we will present the OSM-derived results in terms of percentage of assets affected by Category 3 TC wind speeds on the Saffir-Simpson Hurricane Wind Scale (Simpson and Saffir 1974). These wind speeds correspond to 10-minute wind speeds equal or greater than 43.4 m/s. We use this Category 3 threshold, as this sets the lower wind speed threshold from which major to catastrophic damage to well-built structures occurs (NOAA 2021).

Due to the lack of detailed information about the characteristics of each asset (e.g., construction materials of a building, the quality of a road, the design standards of electricity assets) and the limited availability of vulnerability curves for these individual assets, we only use the asset data extracted from OSM to better understand exposure (and how exposure may change towards the future as a result of climate change). We do not try to estimate asset-level damage and risk, as dealing with the uncertainty and unavailability of the required information to do so is out-of-scope for this study.

6.3.2.2 LitPop

As previously mentioned, the OSM dataset contains information on the location of assets but does not provide information on asset values. The "lit population" (LitPop) database does contain asset value information, aggregated at 1 km × 1 km resolution (Eberenz et al. 2020). Due to this 1 km × 1 km aggregation, we cannot use LitPop to assess an individual asset's exposure. Hence, we do not aim to connect the asset values from LitPop with the OpenStreetMap asset information.

The LitPop database is a globally consistent high-resolution exposure dataset describing the combined value of physical assets (e.g., buildings, infrastructure) within each 1 km × 1 km grid cell. The data are produced using a combination of gridded night light intensity and population data to disaggregate physical asset stock values (capital) proportionally within each country (Eberenz et al. 2020). The LitPop dataset is available for 224 countries; all SIDS but one (French Polynesia) incorporated in this analysis are included in the LitPop database. The LitPop values (in USD) used in this study are relative to the base year 2014. We direct readers to Eberenz et al. (2020) for more information on the LitPop database.

6.3.3 Vulnerability

To estimate the potential damage for each SIDS, we use the regionally calibrated vulnerability curves developed by Eberenz et al. (2021). These vulnerability curves were derived for nine regions; for all SIDS, we use the vulnerability curve corresponding to their respective region. Following Eberenz et al. (2021), we apply a sigmoidal vulnerability function satisfying two constraints: (i) a minimum threshold for the occurrence of damage with an upper bound of 100% direct damage (Emanuel 2011); (ii) a high power-law function for the slope, describing an increase in damage with increasing wind speeds (Pielke 2007). Figure 6.2 presents the four regionally calibrated curves for the four basins that are included in this study.

To account for the uncertainty in the estimation of these vulnerability curves, we use two vulnerability curves per basin: (i) the curve calibrated through using the root-mean-squared fraction (RMSF) and (ii) the curve calibrated through using the total damage ratio (TDR). The RMSF is defined as the relative deviation between modeled and reported damage for all matched events in a region, whereas the TDR is defined as the sum of simulated damages divided by the sum of normalized reported damages (Eberenz et al. 2021). Following the recommendation by Eberenz et al. (2021), the risk estimates reported in this study will be based on the TDR calibration. The risk estimates using the RMSF calibrated curves are provided in **Supplementary Materials**.

We point out that Eberenz et al. (2021) calibrated their vulnerability curves using 1-minute average sustained wind speeds. As the STORM wind speeds are given as 10-minute average sustained wind speeds, we convert the STORM wind speeds to their 1-minute equivalent using a wind speed conversion factor of 0.88 (Harper et al. 2008). Furthermore, we note that the Eberenz et al. (2021) curves are calibrated on total damages: such TC damages are composed of damages caused by wind speeds, storm surge, and precipitation. As a result, aside from wind speed-inflicted damages, the damages calculated in the current study also implicitly contain damages caused by storm surge and precipitation. For the future climate study, however, we use the same Eberenz et al. (2021) vulnerability curves, as equivalent future vulnerability curves have not been developed yet. This means that the future damages listed in this study do not account for changes in the construction quality of buildings and infrastructure, or whether adaptation measures have been implemented. In addition, as the curves constitute a relationship between damage and TC hazards, the latter represented as wind speed, any future-climate changes in sea-level rise (Shepard et al. 2012) or changes in precipitation (Knutson et al. 2020) are not reflected in the damage estimates.

6.3.4 Risk

We express TC risk in terms of Expected Annual Damages (EAD, in USD). This EAD is calculated as the integrated value of expected damages over all RPs. To do so, a trapezoid approach is taken to assess the EAD, see Eq. 6.1:

$$EAD = \sum_i \frac{(P_i - P_{i-1}) \cdot (D_{P_{i+1}} - D_{P_i})}{2}$$

$$P_i = \left\{ \frac{1}{p} | p \in [10, 20, 30, \ldots, 100, 200, 300, \ldots, 1000] \right\} \quad (6.1)$$

Here, P_i is the exceedance probability of a TC wind speed (given as the inverse of the RP), and D_{P_i} the damage related to that event (in USD) (Koks et al. 2015a). As indicated in Eq. 6.1, the RPs are given at 10-year interval levels between the 10-year and 100-year RP level, and at 100-year interval levels between the 100-year and 1000-year RP. This is done to comply with the RP levels as used in the STORM RP datasets.

6.4 Results and Discussion

6.4.1 Future Trends in Tropical Cyclone Hazard

From the STORM baseline and future climate RP datasets, we extract the wind speed RPs (up to 1000-year) in the capital cities of each of the 48 SIDS (Fig. 6.3). First of

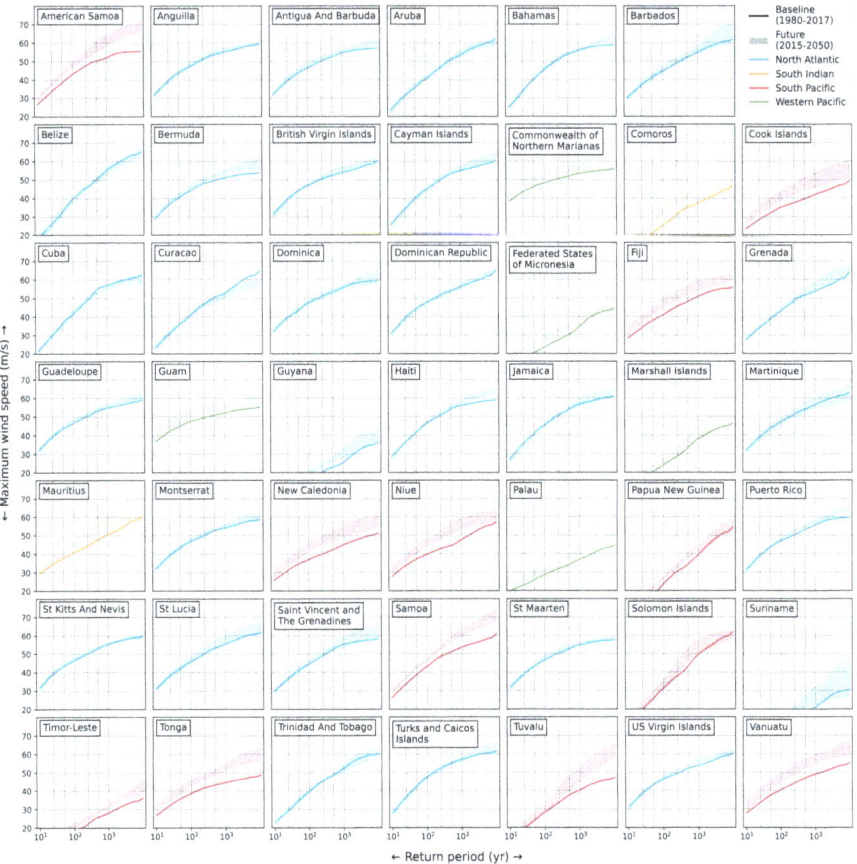

Fig. 6.3 Empirically derived return periods of maximum 10-minute 10-m average wind speeds in the capital city for each of the 48 SIDS. The solid line represents the STORM baseline climate conditions (corresponding to the average climate conditions over 1980–2017). The shaded areas represent the range of return periods of the four STORM future climate datasets (average climate conditions over 2015–2050). Colors indicate the respective basin: North Atlantic – blue, South Indian – yellow, South Pacific – red, and Western Pacific – green

all, we note that the changes in RP curves between the baseline and future climate conditions differ substantially per SIDS and basin. In the North Atlantic (the blue curves in Fig. 6.3), our results show that for the majority of SIDS, a minor change in RPs up until the 1000-year RP is projected under future climate change compared to the baseline scenario. In this basin, the average change in wind speed at the 1000-year RP amounts to 1.3 ± 1.6 m/s across the SIDS. This relatively minor difference is primarily driven by the number of intense (Category 3 or higher) TCs entering the Gulf of Mexico and Caribbean region staying approximately the same in the future-climate STORM synthetic datasets, despite a decrease in overall TC frequency (Bloemendaal et al., 2022). These findings are consistent with other literature (e.g., Bruyère et al. 2017).

Contrary to the North Atlantic, for the other basins, we project higher wind speeds at the 1000-year RP for the 2015–2050 period compared to the baseline. The largest difference between the baseline and future-climate curves is found for the South Pacific (red curves in Fig. 6.3), amounting to an average of 7.0 ± 2.0 m/s across the SIDS. Figure 6.3 illustrates the projected substantial increase in TC intensity, visualized through the future-climate RP curves (shaded areas in Fig. 6.3) all being substantially higher compared to their baseline counterparts. This change is likely caused by the combination of two factors. First, the STORM future-climate synthetic datasets show a slight increase in TC frequency in the South Pacific compared to its baseline-climate counterpart (Bloemendaal et al., 2022). Second, the SST fields that serve as input for STORM (see Sect. 6.3), play a direct role in the STORM TC intensity modelling. In STORM, the SST fields are used to calculate the Maximum Potential Intensity (MPI; Bister and Emanuel 1998). This MPI provides a theoretical upper bound of TC intensity in a given location. As the future-climate (2015–2050) STORM simulations make use of increasing SSTs present in the four input GCMs (see Sect. 6.3), this means modeled future climate conditions are more favorable for (further) TC intensification compared to the STORM baseline climate conditions. The combination of these two components (frequency and intensity) thus leads to increased probabilities of intense TC occurrences, and hence of such systems to affect the SIDS in this basin.

A few SIDS are noteworthy to discuss individually. Guyana and Suriname, for instance, are generally affected by lower TC wind speeds compared to the other North Atlantic SIDS, while their RP curves also show larger uncertainties towards the future. These two countries are located near the southern North Atlantic basin boundary, relatively close to the Equator compared to the other SIDS in the North Atlantic. As TCs in STORM are modeled to deflect away from the Equator, TC peak intensities are generally being reached northward of Guyana and Suriname (this also explains the difference in wind speeds amongst the different RP curves in the North Atlantic). However, it should be noted that a difference exists in TC genesis frequency between $5°$-$10°N$ between the STORM baseline and future-climate datasets. This region is of particular interest here, as Guyana and Suriname are generally hit by TCs forming in this latitudinal region. In the baseline STORM dataset, the probability of TC formation in this region amounts to $0.93 \pm 1.42\%$ of all TC formations, whereas this increases to $4.6 \pm 5.0\%$ of all TC formations in the future-climate STORM datasets (aggregated over all separate GCM runs). Note that in STORM, the probability of the TC genesis in a given location is weighted by the monthly TC genesis locations as extracted from the input dataset (either historical data or GCM data). This weighted number can never be smaller than zero. However, the large standard deviation hints at a large signal-to-noise ratio, implying that we cannot label this difference as a strong signal instigated by the effects of climate change, which is also reflected by the larger uncertainty in future-climate RPs shown in the Guyana and Suriname subplots.

6.4.2 Future Trends in Exposure

Figure 6.4 presents the RP curves of the fraction of buildings affected by Category 3 (43.4 m/s) wind speeds per SIDS in the baseline and future climate analyses. A similar figure for fraction of infrastructure can be found in the **Supplementary Materials** (Supplementary Fig. 2). By expressing the number of buildings as a fraction of the total in a SIDS, we circumvent possible data scarcity in OSM, as well as enable homogenized comparison across the SIDS (see Sect. 6.3). We point out to the reader that changes in exposure are solely driven by changes in TC wind speeds under climate change, as no future projections for the OSM database currently exist. For most SIDS, the RP curve shows a strikingly steep increase, indicative of the RP at which the capital city (most often the location with the

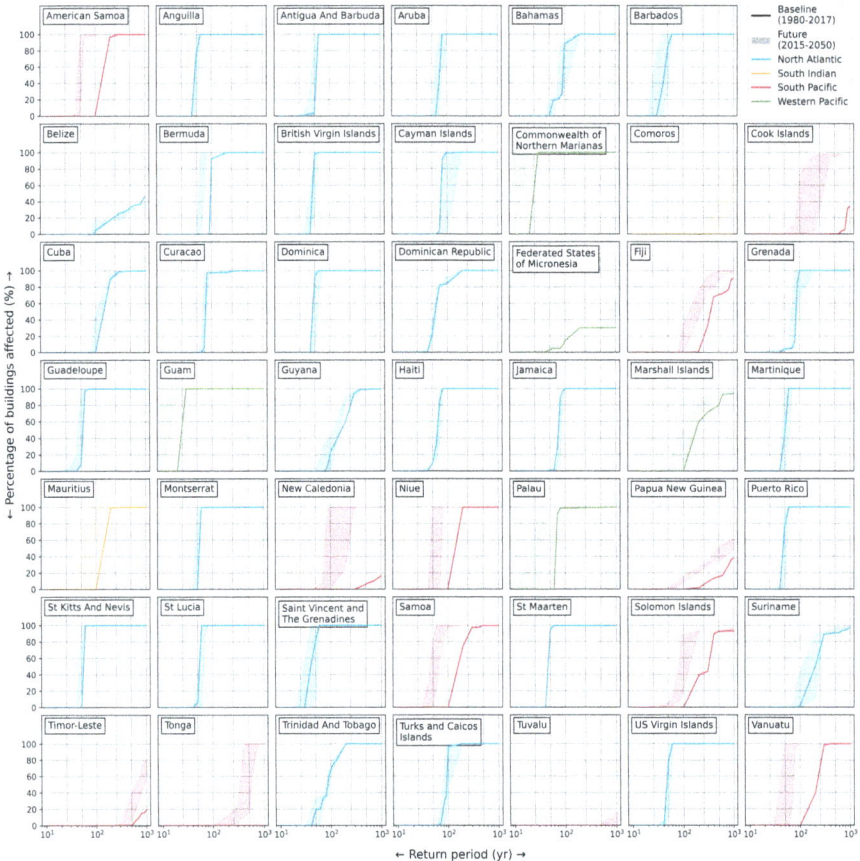

Fig. 6.4 Similar to Fig. 6.3, but now shown are empirically derived return periods of the percentage of buildings affected by Category 3 (43.4 m/s) wind speeds or higher in each of the 48 SIDS. In both climate scenarios, we use the same building data from OpenStreetMap; changes in the return period curve are therefore entirely driven by changes in wind speed hazard

highest density of assets) is hit by Category 3 wind speeds (Fig. 6.4). However, we find substantial differences in the exact RP at which this happens. These differences are discussed in the following paragraphs.

First of all, we observe a shift in RPs at which almost all buildings are affected under the 2015–2050 climate conditions compared to the 1980–2017 period. For all SIDS located in the North Atlantic (except Bermuda), the baseline climate RP curves lie within the range of future climate RP curves. This indicates a (very) minor change in the probabilities of a certain fraction of buildings being affected under climate change. This minor change, in turn, corresponds to a minor change in Category 3 wind speed RPs in the North Atlantic region (see Fig. 6.3 and Supplementary Fig. 1). On the contrary, in the other basins our results indicate a substantial shift towards higher probabilities of buildings being affected (lower RPs) under climate change. The most notable shifts in RP curves are found for the SIDS located in the South Pacific: all SIDS show a substantial increase in affected buildings under climate change. This is driven by a similar substantial increase in Category 3 wind speed probabilities (decrease in RP) in these regions (see Fig. 6.3 and Supplementary Fig. 1). Particularly SIDS located poleward of 20°S are projected to face a more-than-tenfold increase in Category 3 probability; this includes the Cook Islands, Tonga, and New Caledonia. For the latter island country, no buildings are affected at the 300-year RP, but this number is projected to increase to 99–100% in the future climate conditions. In the Comoros (in the South Indian), buildings are not affected by Category 3 wind speeds (with a RP below 1000-year) in the baseline climate (see also Fig. 6.3). Our results, however, project that under climate change, between 20% and 60% of all buildings will be affected at the 700-year RP. This range increases to 43–85% at the 1000-year RP.

In addition, we note the apparent difference in the shape of the baseline and future-climate (2015–2050) RP curves across the SIDS: SIDS such as Montserrat and Guam show a sudden, steep increase from 0% to 100%, while other SIDS like Guyana or the Solomon Islands show a more gradual increase. The explanation for this difference is two-fold: one hazard-related and one exposure-related. First, on the hazard side, the decay of TC wind speeds after landfall plays a prominent role. In STORM, TC decay occurs when a TC is over land for more than 9 hours (Bloemendaal et al. 2020b). This implies that for SIDS where a TC generally takes less than 9 hours to pass (e.g., the SIDS that are part of the Windward Islands, or other small islands), no wind decay is modelled. For mainland countries such as Belize, a modeled TC will typically be over land for more than 9 hours, and as such this TC will be decaying after landfall. As a result, Category 3 wind speeds in the inland part of this country will have a RP greater than 1000-year. In the case of Belize, where built-up areas are spatially distributed across the country, this means that not all assets will be exposed to Category 3 wind speeds and hence a maximum of 47% of the total number of assets will be affected by these wind speeds at the 1000-year RP in the baseline climate; In the future climate, this portion amounts to 32–40%. For Guyana and Suriname, however, most assets are found along the coastline, meaning that these have a higher probability of being affected by these wind speeds. This translates to 95% of all assets being exposed to Category 3 wind speeds at the 314-year and 857-year RP, respectively. Another illustrative example where this effect of inland decay comes into play is for Haiti and the Dominican Republic, which are both located on the island of Hispaniola. The majority of their assets (68% and 65%, respectively) are located on the southern side of the country (south of 19°N), which is the direction from which TCs typically approach Hispaniola. However, in Haiti, 95% of the buildings are affected at the 76-year RP in the baseline climate, while in the Dominican Republic, 83% of all buildings are exposed at this RP. This difference is driven by the fact that, in Haiti, the other large cities are also located along the coastlines (western and northern), while in the Dominican Republic, the second largest city of the country, Santiago de los Caballeros, is located further northward and inland. Compared to the coastal cities, the assets in Santiago de los Caballeros are less exposed to Category 3 force winds as the modeled TCs moving over the Dominican Republic will generally be decaying in STORM.

Aside from the hazard component, differences in the spatial distribution of built-up areas play an important role in clarifying the different characteristics of the RP curves. For SIDS where almost all assets are located at one part of the island (e.g., small island nations such as Saint Lucia, Montserrat, or St. Maarten), these assets will face an (approximately) equal probability of Category 3 wind speed exposure. As a result, the RP curve will show a sharp increase at the RP where this Category 3 threshold is met. Such devastating impacts with near-total destruction in small island nations was for instance seen in the aftermath of Hurricane Irma (2017)'s passage over Barbuda. Barbuda was directly hit by the Category 5 TC, and it was estimated that 95% of all structures were damaged or destroyed, leaving the island uninhabitable for the first time in 300 years (Cangialosi et al. 2018).

On the other hand, for SIDS where assets are spread over a larger area or multiple islands, we find clear spatial differences in Category 3 wind speed RP across the country. This means that not all assets share the same probability of being affected by such intense wind speeds. Examples of such SIDS include the Solomon Islands and the Bahamas. The latter archipelago was directly hit by Hurricane Dorian's Category 5 wind speeds in 2019. The second largest city of the country, Freeport, was catastrophically impacted by the TC (Avila et al. 2020). Meanwhile, the capital city of Nassau, located 200 km southeast of Freeport, suffered little damage and served as shelter for those impacted by the TC (IDMC 2020).

6.4.3 Future Trends in Damage

By combining TC wind speed RP maps with asset values from LitPop (see Sect. 6.3), we estimate damages at RPs ranging from 10-year to 1000-year (see Fig. 6.5). We note that the data from LitPop was derived for the baseline climate conditions and does not contain future projections; as such, differences in damage RPs are entirely driven by changes in TC wind speeds under climate change.

6 Current and Future Tropical Cyclone Wind Risk in the Small... 135

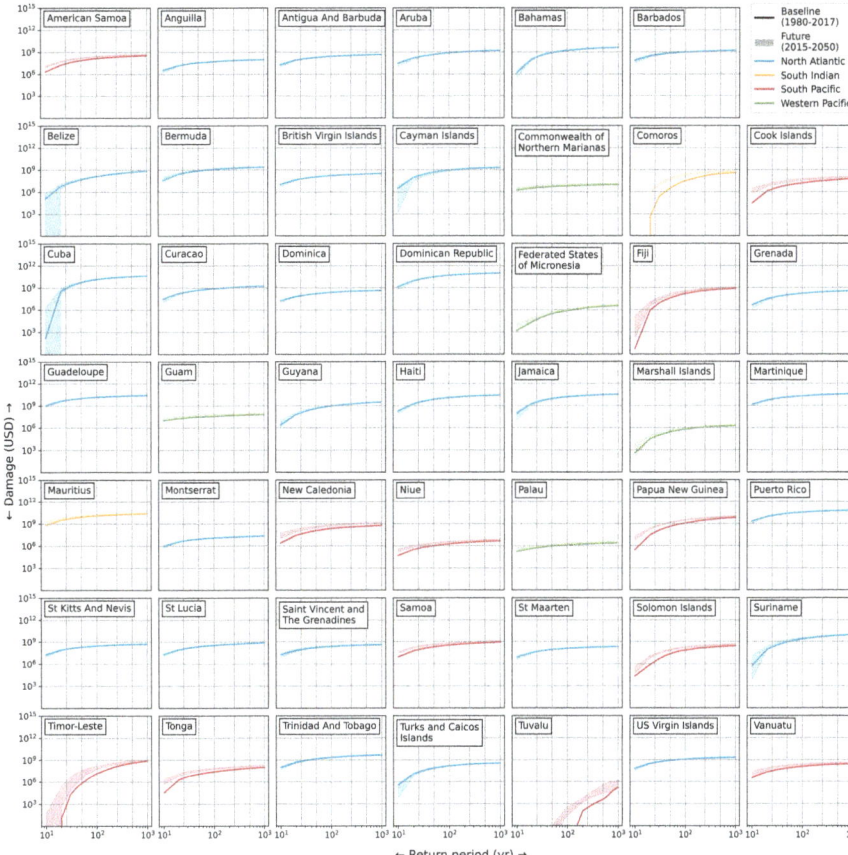

Fig. 6.5 Similar to Figs. 6.3 and 6.4, but now shown are empirically derived return periods of total damages to assets (in USD, plotted on a logarithmic scale) in each of the 48 SIDS. Damages are calculated as the aggregated damage over all physical assets in the Litpop database, see Sect. 6.3

The graphs in Fig. 6.5 illustrate the combined effects of the wind speed RP curves (Fig. 6.3) and the applied vulnerability curve (Fig. 6.2; left panel). Unlike the other basins, the vulnerability curve for the Western Pacific does not show a steep increase towards higher wind speeds. This means that the maximum damage in this basin does not exceed 6% of the total GDP, regardless of a future increase in wind speeds. As a result, the highest total damage at the 1000-year RP in the Western Pacific is found for Guam, totaling to 60.8 million USD in the baseline scenario, and estimates for the 2015–2050 period ranging from 72.9 million to 1.01 billion USD. The other basins, however, do show a steep vulnerability curve, with maximum total damages of up to 80%, 83%, and 94% for the North Atlantic, South Indian, and South Pacific, respectively. In these three basins, highest damages at the 1000-year RP are found for the Dominican Republic (85.5 billion USD), Mauritius (23.2 billion USD), and Papua New Guinea (6.39 billion USD) in the baseline climate. The largest relative change in damage at the 1000-year RP is found for Tuvalu, amounting to approximately 419% compared to the baseline climate.

6.4.4 Future Trends in Risk

Based on our results, the EAD for the baseline situation, for all SIDS combined, is estimated at 6.2 billion USD, or 1.2% of their aggregated GDP. For the 2015–2050 time period, this combined EAD increases by 2.4% (−8.1–16.2%, depending on the global climate model) compared to the baseline situation. This increase, however, varies significantly between the different basins. While we find a decrease of −3.8% (−12.1–8.1%) for the North Atlantic basin, we find an increase of almost 153.6% (104.8–201.0%) for the South Pacific. For the South Indian and Western Pacific basins, the increases are approximately 58.0% (25.5–89.4%) and 64.5% (37.6–117.6%), respectively. As the North Atlantic basin constitutes 91.6% of the total EAD, it is clear that SIDS located in this basin are the main contributors to these changes.

Taking a closer look at the individual SIDS (Fig. 6.6), we find that the top 10 SIDS with highest EAD contains eight SIDS located in the North Atlantic,

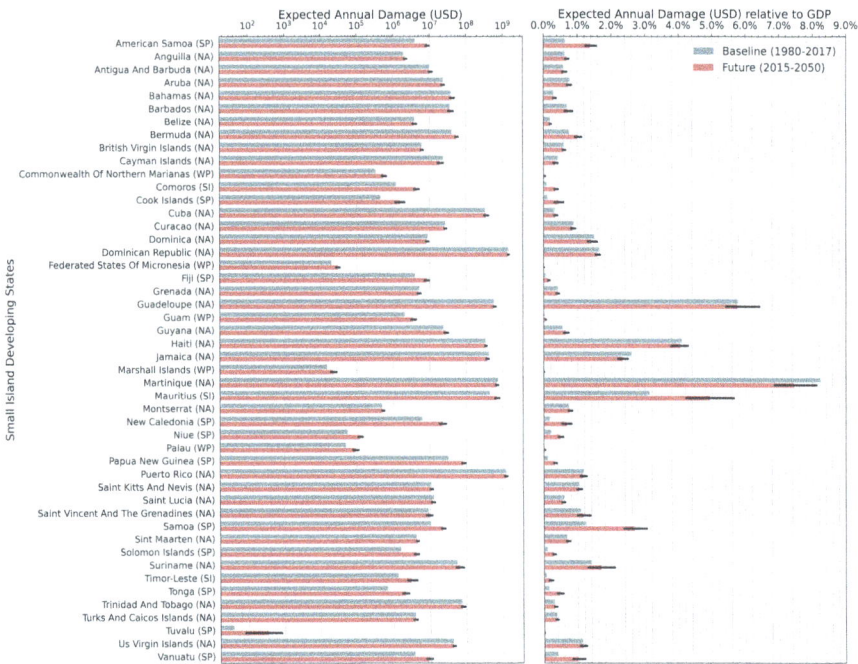

Fig. 6.6 Expected Annual Damage (EAD, in USD) for the baseline and future climate conditions using the TDR-calibration method. The error bar for the future situation represents the spread of EAD across the four separate STORM future-climate datasets (see Sect. 6.3)

complying with the earlier finding that this basin constitutes most TC risk. Our results show that the Dominican Republic (1.47 billion USD), Puerto Rico (1.25 billion USD), and Martinique (763 million USD) experience the largest risk. For these three SIDS, these EADs slightly decrease under future climate change, with average values across the four STORM global climate model datasets amounting to 1.41 billion, 1.22 billion, and 691 million USD, respectively.

We point out here that the asset value exposed to certain TC wind speeds is the key driver of EAD. This becomes visible when looking at the EAD for Guadeloupe, Dominica, and Martinique. Despite their RP curves in Figs. 6.3, 6.4 and 6.5 being approximately similar, there are substantial differences in the EAD estimates (Fig. 6.6). This difference in EAD is driven by the fact that Dominica has a total asset value of 895 million USD in LitPop, whereas this value amounts to 61 billion and 78 billion USD for Guadeloupe and Martinique, respectively.

Validation of risk estimates is an essential part of any risk assessment. As pointed out in the introduction, few studies were conducted to assess the TC risk to SIDS and were mostly executed on a case study basis. A direct validation of our results using these other studies, however, is difficult. This is because there are numerous components in the entire model chain (composed of the hazard, exposure, and vulnerability modeling) that can each impose a substantial influence on the EAD estimates, and accounting for each of them lies beyond the scope of this research. Instead, we compare our results against EADs derived in another, uniformly executed study for a selection of SIDS, and briefly list some reasons for differences in EAD estimates.

The Commonwealth of Australia (2013) presents a series of reports on TC risk for Pacific island nations. These reports also list wind speed estimates at the 100-year RP for the different island nations; in Bloemendaal et al. (2020a), these estimates were compared against their counterparts from the STORM baseline climate dataset. Results showed that for eight out of 12 locations, STORM wind speeds were within 5 m/s of those in the reports. In terms of EAD, we find that for the Cook Islands, Timor-Leste, Papua New Guinea, and Samoa, our estimates for the baseline climate are within 50% of their listed values. For Fiji and Niue, their EAD estimates are approximately a factor 20 higher than ours. This can mostly be attributed to the difference in total exposure (asset) value. For Niue, this value is a factor 20 higher in the Commonwealth of Australia case studies; for Fiji, this difference amounts to a factor 10 higher compared to our results. Furthermore, our damage estimates predominantly consist of wind damage to physical assets (buildings and infrastructure), whereas the case study reports also assess flood damage from precipitation and consider damages to agriculture.

6.4.5 *Limitations and Directions for Future Research*

In the previous sections, we have demonstrated and discussed the different hazard, exposure, and risk datasets for the SIDS. These datasets, however, impose some limitations, which we will briefly reflect upon here, as well as provide some directions for future research.

First, TC wind speeds after landfall are modeled using an empirical inland decay function (Kaplan and Demaria 1995; Bloemendaal et al. 2020b) and translated to a 2D wind field using the Holland parametric wind field model (Holland 1980; Lin and Chavas 2012; Bloemendaal et al. 2020a). The decay function was calibrated based on US landfalling events; as such, this function may perform less in other regions. Furthermore, this decay function is activated in STORM once the TC eye will be over land for more than 9 hours. For TCs crossing land in shorter time periods, no decay is assumed. In reality, the inland surface winds will start to decay prior to landfall in response to enhanced surface friction caused by the land mass (Done et al. 2020; Nanaji Rao et al. 2021). In addition, the 2D-parametric wind field model does not take terrain effects into account, meaning that wind speeds on the leeward side of e.g., a mountain ridge, might be overestimated compared to reality. These limitations can be overcome by applying wind field models that also account for terrain effects, such as the Done et al. (2020) model. Such models, however, also require information on variables that are currently not simulated by most synthetic models (including STORM). In addition, they are computationally more intensive, meaning that application to large synthetic datasets is challenging. Nonetheless, future developments in both the field of synthetic modeling and parametric wind field modeling can aim to overcome these limitations, thereby improving wind speed estimates from synthetic models.

In addition, while we solely focused on wind risk in this chapter, TCs can also cause substantial damage through flooding, triggered by storm surges, waves, and rainfall. These hazards, however, are not simulated in the current version of STORM. It is possible to model storm surges and waves by forcing a hydrodynamical model with synthetic TCs from STORM (as was for instance done for the baseline climate in Dullaart et al. 2021), but currently no STORM module exists that allows the modeling of TC rainfall. We plan to develop such a module in future work.

Next, while OpenStreetMap is currently the most widely used and consistent database of geospatial objects available globally, it should be noted that consistency does not equal completeness (Koks and Haer 2020). Unfortunately, very few databases are available to identify how well or how bad certain places are mapped, and comparison can often only be done case-specific. For the SIDS in the South Pacific, we can compare our building estimates with estimates from island specific case studies (Commonwealth of Australia 2013). Comparison estimates show variations between 10% of the buildings in OSM versus the respective case study (e.g., Timor-Leste and Papua New Guinea) up to 135% (Tuvalu) and 168% (Niue) of the buildings in OSM versus the case study analysis. Other islands, such as Vanuatu or Samoa, show almost identical numbers of buildings between OSM and their case study analyses. While this has no implications for our risk estimates, as the LitPop database is used to estimate those values, it could mean that the RP at which 95% of the buildings are exposed to Category 3 wind speeds may shift when more buildings are included. However, we do believe this shift will be limited as our results present relative exposure values. While it is hard to underpin the following claim with numbers, visual exploration of OSM data does indicate that high-density areas do include more buildings compared to low-density areas on each of the SIDS. This implies that our general observations and conclusions with regards to exposure will still hold, even when more buildings are included.

6.5 Concluding Remarks

This study presented in this chapter is the first to provide a uniform TC risk assessment across all SIDS prone to TCs using publicly available data. To calculate this risk, we have combined state-of-the-art TC hazard modelling for the current and near-future (2015–2050) climate (represented through four global climate models), combined with high-resolution exposure information obtained from both OpenStreetMap (OSM) and globally consistent gridded exposure and vulnerability information. The analysis is performed for 48 SIDS, subdivided across four basins (North Atlantic, South Pacific, Western Pacific, and South Indian). We presented RP curves for maximum wind speeds, affected assets (buildings and infrastructure), and damages, both for the baseline and future climate. Our results showed that the largest changes in RPs across these three elements were all found in the South Pacific basin; smallest changes were observed for SIDS located in the North Atlantic.

The aggregated expected annual damage (EAD) for the baseline situation and for all SIDS combined, is estimated at 6.2 billion USD, or 1.2% of their combined GDP, which increases by 2.4% (−8.1–16.2%) towards the future. The results show a clear change towards the future for all risk components, but with distinct differences between the four basins. The North Atlantic basin experiences overall the highest absolute risk, constituting 91.6% of all EAD across the SIDS, but also shows a slight decrease in this EAD towards the future, amounting to −3.8% (−12.1–8.1%). The South Pacific, on the other hand, experiences an increase of approximately 153.6% (104.8–201.0%) compared to the baseline climate. The risk in the Western Pacific and South Indian basins increases by 64.5% and 58.0%, respectively. In terms of absolute EAD, the Dominican Republic experiences the highest absolute risk, totaling to 1.47 billion USD in the baseline climate. Even though this estimate decreases under climate change to 1.41 billion USD, it is still the highest risk across all SIDS.

References

Avila L, Stewart SR, Berg R, Hagen A (2020) Tropical cyclone report: hurricane Dorian, 24 August–7 September 2019. National Hurricane Center

Barrington-Leigh C, Millard-Ball A (2017) The world's user-generated road map is more than 80% complete. PLoS One 12:e0180698

Bister M, Emanuel KA (1998) Dissipative heating and hurricane intensity. Meteorog Atmos Phys 65:233–240

Bloemendaal N, de Moel H, Muis S, Haigh ID, Aerts JCJH (2020a) Estimation of global tropical cyclone wind speed probabilities using the STORM dataset. Sci Data 7:377

Bloemendaal N, Haigh ID, de Moel H, Muis S, Haarsma RJ, Aerts JCJH (2020b) Generation of a global synthetic tropical cyclone hazard dataset using STORM. Sci Data 7:40

Bloemendaal N, De Moel H, Martinez AB, Muis S, Haigh ID, Van Der Wiel K, Haarsma RJ, Ward PJ, Roberts MJ, Dullaart JCM, Aerts JCJH (2022) A globally consistent local-scale assessment of future tropical cyclone risk. Sci Adv 8

Bruyère C, Rasmussen R, Gutmann E, Done J, Tye M, Jaye A, Prein A, Mooney P, Ge M, Fredrick S (2017) Impact of climate change on Gulf of Mexico hurricanes. NCAR tech note. NCAR/TN535

Cangialosi JP, Latto AS, Berg R (2018) Tropical cyclone report: hurricane Irma, 30 August–12 September 2017. National Hurricane Center

Commonwealth of Australia (2013) Current and future tropical cyclone risk in the South Pacific: South Pacific regional risk assessment. Australian Government

Done JM, Ge M, Holland GJ, Dima-West I, Phibbs S, Saville GR, Wang Y (2020) Modelling global tropical cyclone wind footprints. Nat Hazards Earth Syst Sci 20:567–580

Dullaart JCM, Muis S, Bloemendaal N, Chertova MV, Couasnon A, Aerts JCJH (2021) Accounting for tropical cyclones more than doubles the global population exposed to low-probability coastal flooding. Communications Earth & Environment 2:135

Eberenz S, Stocker D, Röösli T, Bresch DN (2020) Asset exposure data for global physical risk assessment. Earth Syst Sci Data 12:817–833

Eberenz S, Lüthi S, Bresch DN (2021) Regional tropical cyclone impact functions for globally consistent risk assessments. Nat Hazards Earth Syst Sci 21:393–415

EC-Earth Consortium (2018) EC-Earth-Consortium EC-Earth3P-HR model output prepared for CMIP6 HighResMIP. Earth System Grid Federation

Emanuel K (2011) Global warming effects on U.S. hurricane damage. Weather, Climate, and Society 3:261–268

Emanuel K, Ravela S, Vivant E, Risi C (2006) A statistical deterministic approach to hurricane risk assessment. Bull Am Meteor Soc 87:299–314

Eyring V, Bony S, Meehl GA, Senior CA, Stevens B, Stouffer RJ, Taylor KE (2016) Overview of the Coupled Model Intercomparison Project Phase 6 (CMIP6) experimental design and organization. Geosci Model Dev 9:1937–1958

Goldblatt R, Jones N, Mannix J (2020) Assessing OpenStreetMap ompleteness for management of natural disaster by means of remote sensing: a case study of three small island states (Haiti, Dominica and St. Lucia). Remote Sens 12:118

Government of the Commonwealth of Dominica (2017) Post-disaster needs assessment, Hurricane Maria, 18 Sept 2017

Harper BA, Kepert JD, Ginger JD (2008) Guidelines for converting between various wind averaging periods in tropical cyclone conditions. World Meteorological Organization

Holland GJ (1980) An analytic model of the wind and pressure profiles in hurricanes. Mon Weather Rev 108:1212–1218

IDMC (2020) Displacement in paradise, hurricane Dorian slams the Bahamas

IPCC (2021) Climate change 2021: the physical science basis. In: Masson-Delmotte V, Zhai P, Pirani A, Connors SL, Péan C, Berger S, Caud N, Chen Y, Goldfarb L, Gomis MI, Huang M, Leitzell K, Lonnoy E, Matthews JBR, Maycock TK, Waterfield T, Yelekçi O, Yu R, Zhou B (eds) Contribution of working Group I to the sixth assessment report of the intergovernmental panel on climate change

Kaplan J, Demaria M (1995) A simple empirical model for predicting the decay of tropical cyclone winds after landfall. J Appl Meteorol 34:2499–2512

Knapp KR, Kruk MC, Levinson DH, Diamond HJ, Neumann CJ (2010) The international best track archive for climate stewardship (IBTrACS) unifying tropical cyclone data. Bull Am Meteor Soc 91:363–376

Knutson T, Camargo SJ, Chan JC, Emanuel K, Ho C-H, Kossin J, Mohapatra M, Satoh M, Sugi M, Walsh K (2020) Tropical cyclones and climate change assessment: part II: projected response to anthropogenic warming. Bull Am Meteorol Soc 101:E303–E322

Koks EE, Haer T (2020) A high-resolution wind damage model for Europe. Sci Rep 10:6866

Koks EE, Bočkarjova M, de Moel H, Aerts JCJH (2015a) Integrated direct and indirect flood risk modeling: development and sensitivity analysis. Risk Anal 35:882–900

Koks EE, Jongman B, Husby TG, Botzen WJW (2015b) Combining hazard, exposure and social vulnerability to provide lessons for flood risk management. Environ Sci Pol 47:42–52

Koks EE, Rozenberg J, Zorn C, Tariverdi M, Vousdoukas M, Fraser SA, Hall JW, Hallegatte S (2019) A global multi-hazard risk analysis of road and railway infrastructure assets. Nat Commun 10:2677

Kron W (2005) Flood Risk = Hazard • Values • Vulnerability. Water Int 30:58–68

Lee C-Y, Tippett MK, Sobel AH, Camargo SJ (2018) An environmentally forced tropical cyclone hazard model. JAMES 10:223–241

Lin N, Chavas D (2012) On hurricane parametric wind and applications in storm surge modeling. J Geophys Res - Atmos 117:n/a

Murakami H, Sugi M (2010) Effect of model resolution on tropical cyclone climate projections. SOLA 6:73–76

Nanaji Rao N, Yesubabu V, Srinivas CV, Naresh Krishna V, Langodan S (2021) Impact of surface roughness parameterizations on tropical cyclone simulations over the Bay of Bengal using WRF-OML model. Atmos Res 105779

NOAA (2021) Saffir-Simpson hurricane wind scale [Online]. Accessed 28 July 2021

O'Neill BC, Tebaldi C, van Vuuren DP, Eyring V, Friedlingstein P, Hurtt G, Knutti R, Kriegler E, Lamarque JF, Lowe J, Meehl GA, Moss R, Riahi K, Sanderson BM (2016) The scenario model intercomparison project (ScenarioMIP) for CMIP6. Geosci Model Dev 9:3461–3482

Pasch RJ, Penny AB, Berg R (2019) Tropical cycone report: hurricane Maria, 16–30 September 2017. National Hurricane Center

Pathak A, Van Beynen PE, Akiwumi FA, Lindeman KC (2021) Impacts of climate change on the tourism sector of a small island developing state: a case study for the Bahamas. Environmental Development 37:100556–100556

Pielke RA Jr (2007) Future economic damage from tropical cyclones: sensitivities to societal and climate changes. Phil Trans R Soc A 365:2717–2729

Ritchie EA, Elsberry RL (2001) Simulations of the transformation stage of the extratropical transition of tropical cyclones. Mon Weather Rev 129:1462–1480

Roberts M (2017) MOHC HadGEM3-GC31-HM model output prepared for CMIP6 HighResMIP. Earth System Grid Federation

Roberts MJ, Camp J, Seddon J, Vidale PL, Hodges K, Vanniere B, Mecking J, Haarsma R, Bellucci A, Scoccimarro E, Caron L-P, Chauvin F, Terray L, Valcke S, Moine M-P, Putrasahan D, Roberts C, Senan R, Zarzycki C, Ullrich P (2020) Impact of model resolution on tropical cyclone simulation using the HighResMIP–PRIMAVERA multimodel ensemble. J Clim 33:2557–2583

Scoccimarro E, Bellucci A, Peano D (2017) CMCC CMCC-CM2-VHR4 model output prepared for CMIP6 HighResMIP. Earth System Grid Federation

Shepard CC, Agostini VN, Gilmer B, Allen T, Stone J, Brooks W, Beck MW (2012) Assessing future risk: quantifying the effects of sea level rise on storm surge risk for the southern shores of Long Island, New York. Nat Hazards 60:727–745

Simpson RH, Saffir H (1974) The hurricane disaster-potential scale. Weatherwise 27:169–186

Stephenson T, Jones JJ (2017) Impacts of climate change on extreme events in the coastal and marine environments of Caribbean Small Island Developing States (SIDS)

UNESCO (2019) Advocating for Small Island Developing States on the frontlines of climate change [Online]. Available: https://en.unesco.org/news/advocating-small-island-developing-states-frontlines-climate-change. Accessed 18 July 2021

UNFCCC (2005) Climate change, Small Island Developing States. Climate Change Secretariat

United Nations (2021) About small island developing states [Online]. Available: https://www.un.org/ohrlls/content/about-small-island-developing-states. Accessed 18 July 2021

Van Vuuren DP, Edmonds J, Kainuma M, Riahi K, Thomson A, Hibbard K, Hurtt GC, Kram T, Krey V, Lamarque J-F, Masui T, Meinshausen M, Nakicenovic N, Smith SJ, Rose SK (2011) The representative concentration pathways: an overview. Clim Chang 109:5

Voldoire A (2019) CNRM-CERFACS CNRM-CM6-1-HR model output prepared for CMIP6 HighResMIP. Earth System Grid Federation

Webb J (2020) What difference does disaster risk reduction make? Insights from Vanuatu and tropical cyclone Pam. Reg Environ Chang 20:20

World Bank (2016) World Bank commits $50 million to support Fiji's long-term Cyclone Winston recovery. https://www.worldbank.org/en/news/press-release/2016/06/30/world-bank-commits-50m-to-support-fijis-long-term-cyclone-winston-recovery. Accessed 18 July 2021

Open Access This chapter is licensed under the terms of the Creative Commons Attribution 4.0 International License (http://creativecommons.org/licenses/by/4.0/), which permits use, sharing, adaptation, distribution and reproduction in any medium or format, as long as you give appropriate credit to the original author(s) and the source, provide a link to the Creative Commons license and indicate if changes were made.

The images or other third party material in this chapter are included in the chapter's Creative Commons license, unless indicated otherwise in a credit line to the material. If material is not included in the chapter's Creative Commons license and your intended use is not permitted by statutory regulation or exceeds the permitted use, you will need to obtain permission directly from the copyright holder.

Chapter 7
Development of a Simple, Open-Source Hurricane Wind Risk Model for Bermuda with a Sensitivity Test on Decadal Variability

Pinelopi Loizou, Mark Guishard, Kevin Mayall, Pier Luigi Vidale, Kevin I. Hodges, and Silke Dierer

Abstract A hurricane-catastrophe model was developed for assessing risk associated with hurricane winds for Bermuda by combining observational knowledge with property value and exposure information. The sensitivity of hurricane wind risk to decadal variability of events was tested. The historical record of hurricanes passing within 185 km of Bermuda was created using IBTrACS. A representative exposure dataset of property values was developed by obtaining recent governmental Annual Rental Value data, while Miller et al. (Weather Forecast 28:159–174, 2013) provided a vulnerability relationship between increasing winds and damage. With a probabilistic approach, new events for 10,000 years were simulated for three different scenarios using (1) the complete record of annual TC counts; (2) two high-frequency periods and; (3) two low-frequency periods. Exceedance probability curves were constructed from event loss tables, focusing on aggregating annual losses from damaging events. Expected losses of low-frequency scenarios were less than losses of high-frequency scenarios or when the whole historical record was

P. Loizou (✉)
Department of Meteorology, University of Reading, Reading, UK
e-mail: pinelopi.loizou@pgr.reading.ac.uk

M. Guishard
Bermuda Weather Service, Bermuda Airport Authority, St. George's, Bermuda
e-mail: mguishard@airportauthority.bm

K. Mayall
Locus Ltd, Hamilton, Bermuda
e-mail: kmayall@locus.bm

P. L. Vidale · K. I. Hodges
National Centre for Atmospheric Science, Department of Meteorology, University of Reading, Reading, UK
e-mail: p.l.vidale@reading.ac.uk; k.i.hodges@reading.ac.uk

S. Dierer
Axis Capital, Zurich, Switzerland
e-mail: silke.dierer@axiscapital.com

© The Author(s) 2022
J. M. Collins, J. M. Done (eds.), *Hurricane Risk in a Changing Climate*, Hurricane Risk 2, https://doi.org/10.1007/978-3-031-08568-0_7

used. This framework suffers from uncertainties due to different assumptions and biases within IBTrACS. Small data sizes limit our ability to conduct a formal model validation and results should be interpreted in this context. In the future, sensitivity tests on the different components of the model will be performed.

Keywords Hurricanes · Catastrophe modelling · Decadal variability

7.1 Introduction

Tropical cyclones (TCs) belong to the category of weather systems which bring severe damage and destruction across many regions of the planet in respect to rain, winds, and storm surge. Studies by McCarthy et al. (2015) and Goldenberg et al. (2001) have shown that TC activity around the globe undergoes important variability through the decades. Insurance and re-insurance companies can be particularly impacted by TCs, especially in countries that are more likely to see a TC making landfall.

Table 7.1 provides information about the ten costliest Atlantic hurricanes (NOAA 2020a; U.S. Bureau of Labor Statistics n.d.; Kishore et al. 2018). It can be seen that four of them occurred during the last Atlantic hurricane seasons (2017 and 2018) while nine of them occurred during the past 20 years. The columns indicate: name; year; maximum achieved intensity; total numbers of fatalities; total cost in billions of US dollars unadjusted for inflation; and adjusted total cost for 2017 in billions of US dollars. Deaths and damage costs refer to the total numbers of fatalities (direct and indirect) and damages across all the affected areas and countries. It should be highlighted that when thinking about damage and impact from a hazard, it is useful to use a metric of the affected area's wealth, for example the Gross Domestic Product (GDP). Hurricane Maria (2017) can be seen as a notable example: even though the total damage caused by the hurricane was around $91.6 bn, the impact on Dominica was way more significant than the impact on the United States. The damage after adjusting for inflation was 244% of Dominica's 2017 GDP. In addition, the

Table 7.1 Top 10 costliest Atlantic Hurricanes (as of 2019)

Name	Year	Category	Deaths	Cost (in bn)	Cost 2017 (in bn)
Katrina	2005	5	1200	$125	$164.9
Harvey	2017	4	68	$125	$129.5
Maria	2017	5	Estimates up to >8500	$91.6	≥$94.9
Irma	2017	5	47	$77.2	$66.5
Sandy	2012	3	233	$68.7	$76.3
Ike	2008	4	103	$38	$43.3
Wilma	2005	5	23	$27.4	$34.4
Andrew	1992	5	26	$27.3	$47.6
Ivan	2004	5	92	$26.1	$33.9
Michael	2018	5	74	$25.1	$25.1 (US)

7 Development of a Simple, Open-Source Hurricane Wind Risk Model for...

uncertainty and large range of fatalities (particularly in Puerto Rico) caused by Hurricane Maria can be attributed to the fact that the assessment of deaths was difficult to perform and that many people died because of delays (or inability) in receiving medical care (Kishore et al. 2018).

Insurance and re-insurance companies often use catastrophe models to quantify the risk associated with hurricanes. A catastrophe model is used for assessing financial impacts of catastrophes, for estimating physical damages of properties, and assigning probabilities to the range of potential outcomes (RMS 2020). There are three main components: hazard (in this case, information about tropical cyclones), exposure (information about properties), and vulnerability (information about the damage a property can get). The goal of catastrophe modelling is to combine the three main components for the estimation of financial loss from hazards.

The aim of the study is to combine what is known from the historical hurricane record with information about property values, exposure, and vulnerability to develop a hurricane catastrophe risk model to assess the risk for Bermuda. It is worth noting that the intent of this study is not to rigorously reproduce the methodology of traditional catastrophe model development used by insurers and re-insurers. However, we use the conceptual process as a guide to develop our hurricane wind risk model. Figure 7.1 presents time series of annual numbers of tropical cyclones. The red line indicates the time series for the whole North Atlantic basin, while the black bars present the number of storms that came within 185 km (or 100 nm) of Bermuda. The historical record of hurricanes is created by using the International

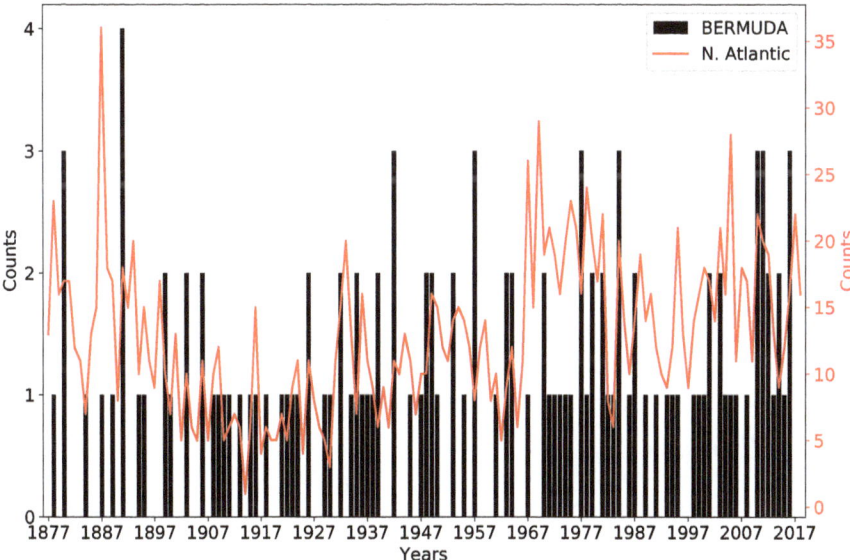

Fig. 7.1 Time series of annual numbers of tropical cyclones. Red line shows the time series for the whole North Atlantic basin. The black bars show the time series of storms that came within 185 km of Bermuda

Best-Track Archive for Climate Stewardship (IBTrACS) (Knapp et al. 2010). Recent Annual Rental Value data, taken from the Bermuda Government (2019), are used for the development of a representative dataset of property values for each of the 36 electoral constituencies in Bermuda. Miller et al. (2013) have performed damage analysis for Hurricane Fabian (2003) that shows the estimation of damage functions incorporating effects of topography. The study concluded that when topographic effects are taken into consideration for the near-surface wind speeds, there is a correlation between increasing damage and elevation.

7.2 Methodology

7.2.1 Data

For the purposes of this study, IBTrACS was used for obtaining a historical record for storms that have impacted Bermuda. In addition, Annual Rental Value data from the Bermuda Government are used for developing a representative dataset of property values for each of the 36 electoral constituencies in Bermuda.

7.2.1.1 Best-Track Dataset (Observations)

IBTrACS is a combination of the best track data taken from different agencies such as the Regional Specialized Meteorological Centers (RSMCs), the Tropical Cyclone Warning Centers (TCWCs), as well as other national agencies. The IBTrACS-ALL (v03r03) dataset, which includes data taken from all agencies, is used for this study. Full details can be found in Knapp et al. (2010). Data are available from 1877 until 2018. The agencies provide information about the best estimated position of each storm in terms of longitude and latitude in addition to reporting wind speed and mean sea level pressure (MSLP) values. The different agencies use different wind-averaging periods, and the values are reported in knots. The North Atlantic data are derived from the Hurricane Databases (HURDAT2) and are provided at 6-hour intervals. The wind speeds are 1-min sustained winds at 10 m, and they have been converted from knots to meters per second (multiplied by 1.94).

7.2.2 Exposure

7.2.2.1 Annual Rental Value Data

The Government of Bermuda's Land Valuation Department collects information about locations, types of property, size of living accommodation, size of any ancillary accommodation, amenities, and characteristics (Land Valuation

Table 7.2 Examples for the ARV data

2009 ARV	PV	Description	Address (fictional)	Parish
$15.600	780.000	APARTMENT	5 HARRY STREET HM01	CITY OF HAMILTON
$21.600	1.080.000	SHOP	8 HARRY STREET HM01	CITY OF HAMILTON
$13.800	690.000	APARTMENT	2 RONALD ROAD HM38	DEVONSHIRE
$33.600	1.680.000	HOUSE	8 FRED LANE MA12	SANDYS
$40.800	2.040.000	HOUSE	11 FLER LANE GE14	ST. GEORGE'S

Department 2019). They provide the Land Valuation List which includes location, type, and annual rental value (ARV) data. The ARV data used in this study are from 2009, but accessed in 2019, since more recent data were unavailable. A few representative examples of the ARV data are shown in Table 7.2. In order to protect the householders' personal information, the addresses displayed on the table are anonymized. The annual rental value is converted to estimated actual property value (PV) by multiplying by a factor of 50. In operational catastrophe models developed for re/insurance applications, building parameters such as construction type and number of stores are often used as second-order modifiers. In the case of the current analysis, secondary modifiers such as property type and location are available in the ARV dataset and could be used in future to refine and enhance this modelling framework.

7.2.3 Bermuda's Historical Record of Hurricanes

The first step of the process was to obtain a historical record of hurricanes that have either made landfall or that have been in close proximity to Bermuda. Therefore, by using the complete record for IBTrACS (1877–2018), for every year, for every storm, every track point which came within 185 km of Bermuda (32.39°N, 64.68°W) along with the wind speed information is kept for further analysis. The choice of 185 km is based on the threat parameter used by the Bermuda Weather Service (NOAA 2020a). The process is summarised on Fig. 7.2a. Figure 7.2b presents all the points that were kept for further analysis. Bermuda is indicated with a black cross.

For each point that is kept, the distance from Bermuda is calculated by using the Haversine formula given by:

$$d = r * c \qquad (7.1)$$

where $r = 6371$ km is the Earth's radius and c is given by:

$$c = 2 * \arctan\left(\frac{\sqrt{a}}{\sqrt{1-a}}\right) \qquad (7.2)$$

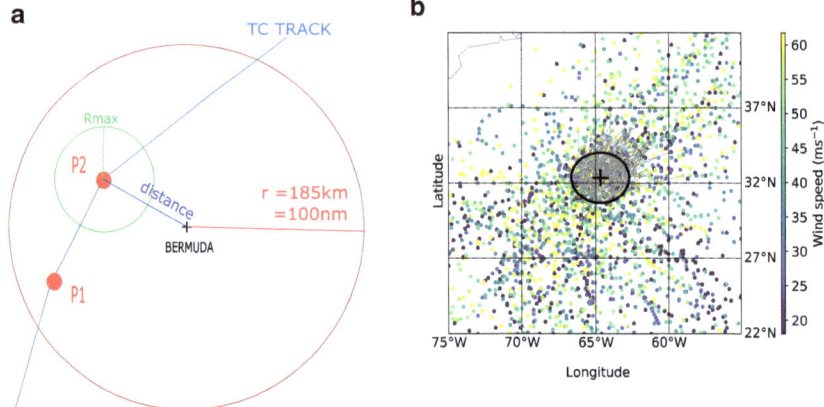

Fig. 7.2 (**a**) Schematic of the methodology for creating the dataset of storms for Bermuda. (**b**) Tropical cyclones that passed within 185 km from Bermuda (black cross)

where

$$a = \sin^2\left(\frac{\phi_2 - \phi_1}{2}\right) + \cos(\phi_1) * \cos(\phi_2) * \sin^2\left(\frac{\lambda_2 - \lambda_1}{2}\right) \quad (7.3)$$

with ϕ_1 and λ_1 the latitude and longitude coordinates of the storm track point in radians and ϕ_2 and λ_2 the latitude and longitude coordinates of Bermuda in radians. For each point, a radius of maximum wind (r_{max}) was chosen based on Eq. 7.4:

$$r_{max} = \begin{cases} 200 \text{ km}, v \leq 17 \text{ ms}^{-1} \\ 125 \text{ km}, 17 < v \leq 32 \text{ ms}^{-1} \\ 95 \text{ km}, 32 < v \leq 42 \text{ ms}^{-1} \\ 50 \text{ km}, 42 < v \leq 49 \text{ ms}^{-1} \\ 30 \text{ km}, 49 < v \leq 58 \text{ ms}^{-1} \\ 25 \text{ km}, 58 < v \leq 70 \text{ ms}^{-1} \\ 20 \text{ km}, v > 70 \text{ ms}^{-1} \end{cases} \quad (7.4)$$

where v is the intensity from the IBTrACS. The values for r_{max} were chosen empirically based on a collection of data from H*WIND (NOAA 2020b) which included tropical cyclones that affected Bermuda during the period 2006–2014.

Then for each point that was saved, the intensity of the storm at Bermuda is calculated by:

7 Development of a Simple, Open-Source Hurricane Wind Risk Model for...

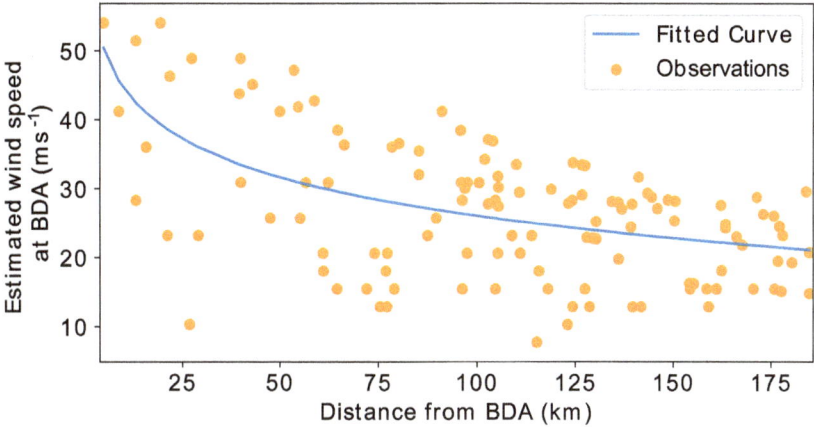

Fig. 7.3 Plot of estimated wind speeds at Bermuda against the distance from Bermuda (orange line). The blue line indicates a fitted logarithmic curve

$$v_{BDA} = \begin{cases} v * \sqrt{\frac{r_{max}}{d}}, & d > r_{max} \\ v, & d \leq r_{max} \end{cases} \quad (7.5)$$

where d is given by the aforementioned Haversine formula. The first part of Eq. (7.5) is a variation of the Rankine Vortex (Holland et al. 2010). Afterwards, for each year, for each storm, the point with the highest estimated intensity in Bermuda is retained. Eventually, all the points of the highest estimated intensities for all the storms that passed within 185 km of Bermuda are obtained. By sorting the data according to distance and fitting a logarithmic curve, a relationship between the distance of a storm from Bermuda and its estimated wind speed in Bermuda is obtained. The relationship is presented in Fig. 7.3 and it is described by:

$$f(x) = 63.1 - 8.05 * \ln(x), \quad (7.6)$$

where x is the distance (d) in km and $f(x)$ is the wind speed in ms^{-1}.

A very important component of a catastrophe model is the relationship between wind and damage. According to Sealy and Strobl (2017) the appropriate way to simulate the relationship is by varying the damage of the property with the cubic power of the wind speed. For the purposes of this study, a damage index, f, proposed by Emanuel (2011) is used for the calculation of the proportion of damage as a function of wind speed, V:

$$f = \frac{u_i^3}{1 + u_i^3} \quad (7.7)$$

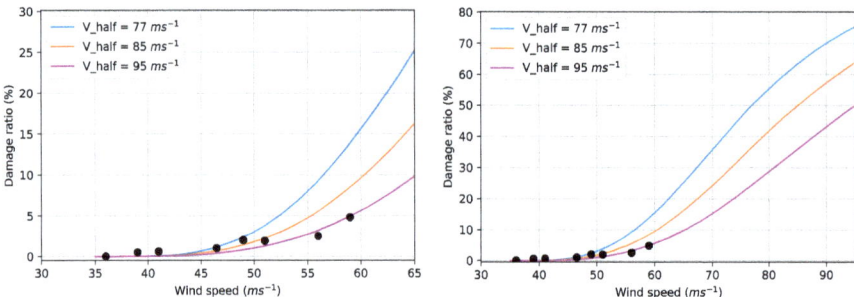

Fig. 7.4 Relationship between damage and wind speed. Black dots indicate an estimate of the Miller et al. (2013) figure 12 data. (**a**) Fitting the data with the different damage index curves by varying V_{half}; (**b**) projection of the different curves that were fitted to the data

where

$$u_i = \frac{MAX[(V_i - V_{thresh}), 0]}{V_{half} - V_{thresh}} \quad (7.8)$$

where V_i is the estimated wind speed in Bermuda (calculated by Eq. 7.6), V_{thresh} is the wind speed below which no damage occurs, and V_{half} is the value of wind speed at which half of the property is damaged. Different studies (Emanuel 2011; Elliott et al. 2015; Sealy and Strobl 2017) have used a threshold of around 25.7 ms^{-1} (50 kts) for V_{thresh}, while for V_{half} different values were chosen based on the nature of each study. For this study, in order to choose appropriate thresholds, the data by Miller et al. (2013) were used. They found a threshold of approximately 37.5 ms^{-1} for the occurrence of roof damage. Therefore, by using $V_{thresh} = 37.5$ ms^{-1} and by varying V_{half} to best fit the Miller et al. (2013) data (see Fig. 7.4), it was found that at $V_{half} = 95$ ms^{-1} half of the property was damaged.

7.2.4 Generating New Datasets

The next step of the study involved using the historical record of annual number of storms for Bermuda shown in Fig. 7.1 to generate new random events with their potential losses. The process of generating new datasets, as summarized in Fig. 7.5, begins by calculating the probability of a number of hurricanes occurring. Previous studies (Jagger et al. 2001; Klotzbach 2010; Emanuel 2011; Scherb et al. 2015; Sealy and Strobl 2017) have suggested using the Poisson distribution since it provides a simple method for computing the probability of hurricane occurrence. The Poisson distribution is given by:

$$P(X = k) = \frac{\lambda^k}{k!} e^{-\lambda} \quad (7.9)$$

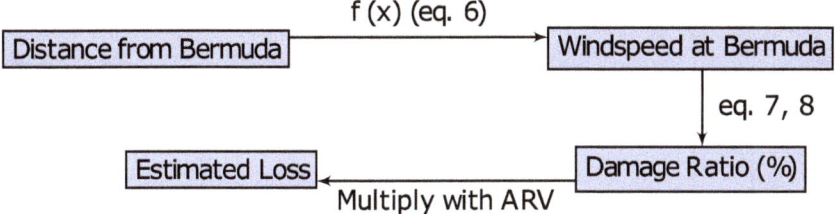

Fig. 7.5 Schematic of the process of generating the new datasets

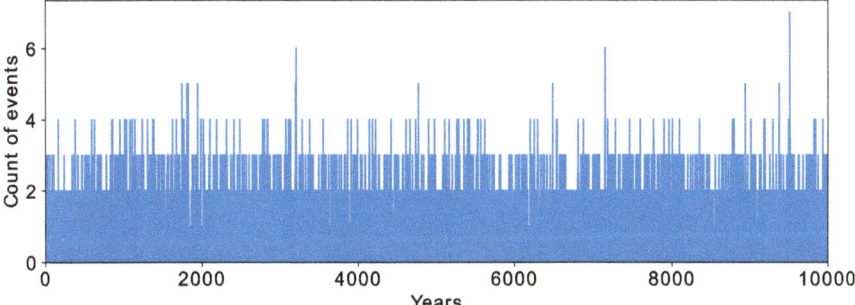

Fig. 7.6 Time series of simulated events for 10,000 years

where λ is taken as the average annual number of hurricanes ($\mu = 0.86$) for Bermuda from the historical record. Afterwards, random events (Fig. 7.6) for the period of 10,000 years were generated from the Poisson distribution. To each event, a randomly generated number for distance (in km) between 0 and 185 was assigned. Then, by using Eqs. 7.6, 7.7, and 7.8, a wind speed and a damage ratio value are calculated and assigned to each event. Eventually, by using the PV data, the potential loss for each event can be estimated and then the sum of all the losses in each year is calculated.

7.2.5 Incorporating Decadal Variability

Numerous studies have shown that on decadal time scales, TC activity in the North Atlantic can be influenced by the Atlantic Multidecadal Oscillation (AMO) through variations of sea surface temperatures (Goldenberg et al. 2001; McCarthy et al. 2015; Ting et al. 2019; Murakami et al. 2020; Mann et al. 2021; Hallam et al. 2021). The associated warm and cold phases of the AMO can last for 20–40 years and they can lead, either directly or via modulation of other modes, such as the El-Niño Southern Oscillation (ENSO), to more or less active hurricane seasons (Knight 2005; Zhang and Delworth 2006; Klotzbach and Gray 2008). Therefore, the final step of the study

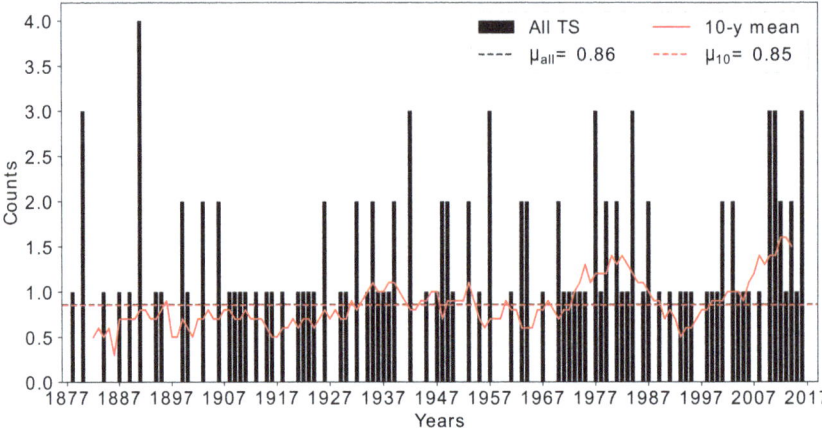

Fig. 7.7 Time series for Bermuda (black bars). The 10-year moving average is shown with the red solid line and its mean is shown with the dashed red line

was to test the model for different climate scenarios. To do that, different time periods with either increased or decreased TC activity within the time series were examined. The different periods were chosen based on the following steps:

1. Find the mean (μ_{all}) of the time series for the annual number of tropical storms in Bermuda.
2. Calculate the 10-year moving average of the time series (centred–red solid line in Fig. 7.7).
3. Find the mean of the 10-year moving average (μ_{10} – red dashed line in Fig. 7.7).
4. For high-frequency phases, take at least 10 consecutive years for which the 10-year moving average is greater than μ_{10} and the mean of the 10+ consecutive years to be greater than μ_{all}. Two high frequency phases were found: 1973–1989 and 1999–2014.
5. For low-frequency phases, take at least 10 consecutive years for which the 10-year moving average is less or equal than μ_{10} and the mean of the 10+ consecutive years to be less than μ_{all}. Two low-frequency phases were found: 1882–1895 and 1897–1930.

Then, for each phase, by using the mean of the phase, new events were randomly generated by following the process outlined in Sect. 7.2.4.

7.3 Results

The output of a catastrophe model is the loss amount from a catastrophic peril. This is given in the form of an Event Loss Table – the format and a data sample are shown in Table 7.3. By looking at the time series of simulated events on Fig. 7.6, the

Table 7.3 Example event loss table

Year	Event 1	Event 2	Event 3	Event 4	Event 5	Event 6	Event 7	Sum
0	14,512,670	–	–	–	–	–	–	14,512,670
1	0	–	–	–	–	–	–	0
2	0	0	–	–	–	–	–	0
3	0	0	0	10,238,454	–	–	–	10,238,454
4	0	0	0	0	–	–	–	0
5	45,319,873	–	–	–	–	–	–	45,319,873
6	0	8,749,651	45,319,873	–	–	–	–	54,069,523
7	–	–	–	–	–	–	–	0
8	13,099,546	14,512,670	–	–	–	–	–	27,612,216
9	16,005,999	0	–	–	–	–	–	16,005,999
10	0	10,238,454	–	–	–	–	–	10,238,454

maximum number of events in a single year in that scenario is seven individual events. Therefore, for Table 7.3 column 1 corresponds to the year number, columns 2–8 correspond to the losses from the individual events, and column 9 corresponds to the amount of loss in a year (the sum of losses from the individual events). It can be seen that there were years with no events (e.g., year 7), years with a single non-damaging event (e.g., year 1), years with a single damaging event (e.g., year 0), years with multiple non-damaging events (e.g., year 4), years with multiple damaging events (e.g., year 8), and years with both damaging and non-damaging events (e.g., year 3). From this tabulated output one can construct an Exceedance Probability (EP) curve. The EP curve describes the annual probability that an amount of loss will be exceeded. In constructing the EP curves, we focus on aggregating annual losses. If a more granular analysis were needed, effort would have been made to establish a specific identifier for each event. However, it is worth noting here that the model output has been constructed from an Aggregate Exceedance Probability (AEP) perspective and neglects further analysis of individual contributions to the annual losses (Occurrence Exceedance Probability - OEP). A future refinement would be to assess the variability of loss events on an annual basis via an OEP analysis.

Figure 7.8 presents the cumulative distribution function (CDF) of expected losses on all properties in all parishes of Bermuda when the whole time series of annual counts of tropical storms for Bermuda is used for simulating new events. The histogram (empirical results) indicates the actual losses from events that were intense enough to cause damage, meaning events that had an estimated wind speed at Bermuda greater than 37.5 ms^{-1} (based on Miller et al. 2013). The black dashed line indicates the theoretical CDF, meaning what one would expect to observe if there was an infinite number of damaging events. Non-damaging events were excluded from the analysis, but we present all the model output. For example, by

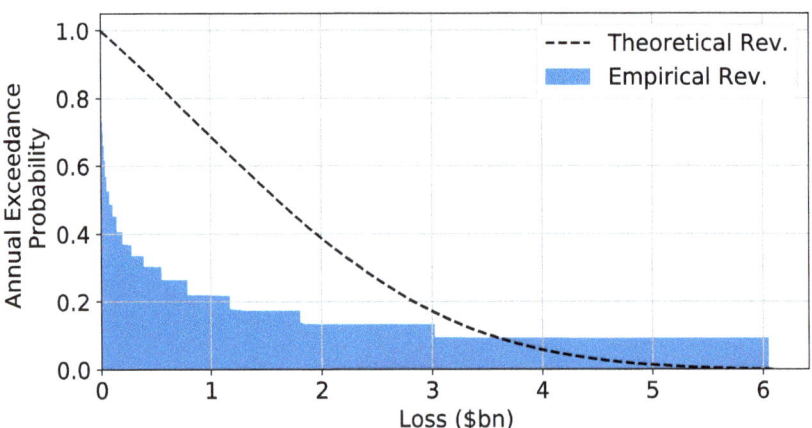

Fig. 7.8 Cumulative distribution function (CDF) of expected AEP losses on all properties, all parishes of Bermuda when the 1877–2018 record for Bermuda was used for simulating new events. The histogram indicates the empirical CDF of losses from damaging events

looking at the histogram from the empirical results, there is a 21.9% chance that during a year with at least one damaging event, losses will exceed $1bn.

Examination of the decadal variability of TCs revealed two high-frequency and two low-frequency phases. High-phases A and B correspond to the periods 1973–1989 (with mean $\mu = 1.18$) and 1999–2014 (with mean $\mu = 1.31$), respectively, during which the 10-year moving average was greater than the mean of the 10-year moving average. Low-phases A and B correspond to the periods 1882–1895 (with mean $\mu = 0.64$) and 1897–1930 (with mean $\mu = 0.68$), respectively, during which the 10-year moving average was less than or equal to the mean of the 10-year moving average.

For each one of the four phases, the mean was calculated and used as described in Sect. 7.2.4 to find the Poisson rate probability of number of events occurring, from which new events were randomly generated for each phase. Empirical and theoretical CDFs were plotted for each phase, as well as for the CDF shown in Fig. 7.8 (hereafter referred to as no-phase), and are shown in Fig. 7.9.

Results showed that losses from damaging events sampled from both high-frequency scenarios were larger than the losses from damaging events sampled from the two low-frequency scenarios and the no-phase scenario. In addition, the annual exceedance probabilities for low-phase B were smaller than the ones for low-phase A, while the probabilities for high-phase B were greater than the ones for high-phase A. It should be noted that, since the process of simulating new events is random, the output of the model will not always resemble the results presented here.

Furthermore, the number of simulated events is dependent on the average annual number of hurricanes. For this study, the means of the different phases ranged from 0.64 to 1.13 hurricanes per year. It is expected that a significantly higher annual

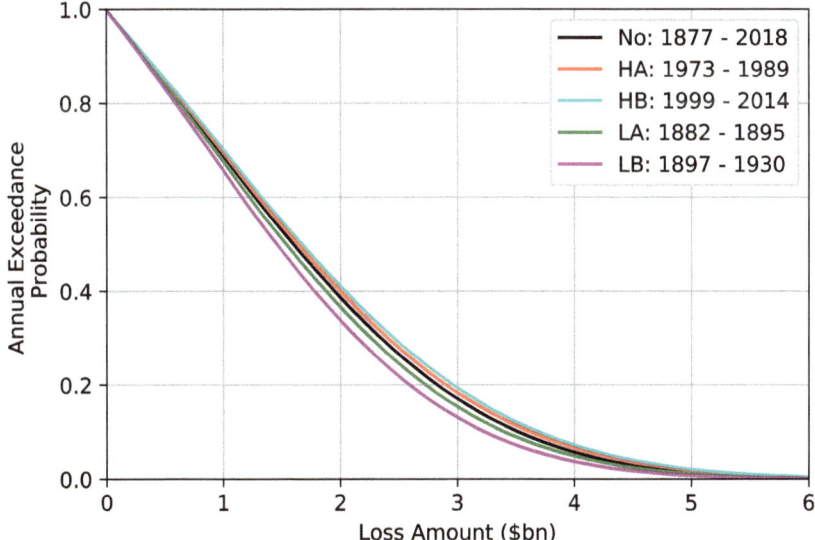

Fig. 7.9 EP curves for all five scenarios

average will result in a significantly increased number of simulated events, and thus larger losses. Lastly, it is important to highlight that, based on Eq. 7.6, the highest wind speed of a simulated event can be up to 63.1 ms^{-1}, while the lowest damaging wind speed is 37.5 ms^{-1}. It is certain that if the former value were higher or if the latter value were lower, the resulting EP curves would be very different, showing greater losses particularly in a high-frequency scenario. In future work, the sensitivity of the model on both the effect of the average annual rate of hurricanes and the estimated wind speed in Bermuda will be explored.

Information about return periods (RP) of catastrophic events can be obtained from EP curves. The return period (in years) corresponds to 1/EP. Table 7.4 presents examples of EP values and the corresponding RPs for all five scenarios shown in Fig. 7.9 for six different loss amounts. The loss amounts are shown in column 1, EPs and RPs for the no-phase scenario are in columns 2 and 3, for the two high-phase scenarios in columns 4–7 and for the two low-phase scenarios in columns 8–11. For example, in the no-phase scenario there is a 5.7% probability that an amount of $4bn will be exceeded in a year with at least one damaging event. This probability corresponds to a RP of around 17.5 years. The probability that the same amount will be exceeded rises for both high-frequency scenarios and dips for both low-frequency scenarios.

Table 7.5 presents examples of estimated losses for certain return periods of catastrophic events. The first and second columns indicate the return periods and

Table 7.4 Exceedance probability and return period values for different loss amounts

Loss ($bn)	No-Phase		High-A		High-B		Low-A		Low-B	
	EP (%)	RP (year)	EP (%)	RP (year)	EP (%)	RP (year)	EP (%)	RP (year)	EP (%)	RP (year)
0.50	84.8	1.2	85.2	1.2	85.5	1.2	84	1.2	83.1	1.2
1.00	68.9	1.5	69.3	1.4	70.1	1.4	67.4	1.5	65.7	1.5
2.00	38.7	2.6	40.0	2.5	41.2	2.4	36.7	2.7	33.7	3
3.00	17.1	5.8	18.3	5.5	19.4	5.2	15.4	6.5	13.1	7.6
4.00	5.7	17.8	6.6	15.3	7.2	13.9	4.9	20.6	3.7	27.2
5.00	1.3	75.9	1.8	54.5	2.1	47.4	1.2	86.6	0.7	138.5

Table 7.5 Loss amounts for different return period values

RP (years)	EP (%)	Loss ($bn)				
		No-Phase	High-A	High-B	Low-A	Low-B
5	20	2.8	2.9	3	2.7	2.6
10	10	3.5	3.6	3.7	3.4	3.2
25	4	4.3	4.4	4.5	4.1	3.9
50	2	4.7	4.9	5	4.6	4.4
75	1.3	5	5.2	5.3	4.9	4.7
100	1	5.2	5.4	5.5	5.1	4.8
200	0.5	5.5	5.8	6	5.5	5.2

corresponding exceedance probabilities, while expected losses for each scenario are shown in the remaining columns. For example, a once-in-200-years catastrophic event is expected to cause $5.5bn worth of damage across all parishes and all types of buildings in a no-phase scenario, compared to $6bn in a high-phase B scenario. These losses are halved for a catastrophic event with a return period of once-per-five-years.

Event Loss Tables and EP curves provide the ability to yield indicative return periods of threshold loss events (or changes in magnitude of losses for a given return period). This is very important for insurance and re-insurance companies since they are provided with necessary information that can help in the process of decision making.

7.4 Limitations and Future Work

This study serves as a simple catastrophe model for assessing annual hurricane wind risk in Bermuda, with a scientifically informed sensitivity test on long-term frequencies. It does not intend to reproduce traditional catastrophe modelling methodologies widely used by insurance and re-insurance companies, but to merely serve as a guide for the development of a hurricane wind risk model.

The process of building a catastrophe model entails various sources of uncertainties in all the different components.

Firstly, a key limitation of the study is the use of the observational record. Despite the fact that the record for the North Atlantic basin is considered to be the longest and most comprehensive record compared to other basins (Strachan et al. 2013), it suffers from homogeneity problems due to changes in operational procedures (Landsea 2007), whose most important source of uncertainty is the observational error (Tolwinski-Ward 2015). In addition, evaluation of the model with historical losses is problematic, as there are only a few very recent official reports from damaging storms affecting Bermuda that can be used for calibration purposes. So, not only are the basin-wide statistics a source of uncertainty, but the damaging impacts in Bermuda are also insufficient to affect a useful calibration of the model.

Secondly, different decisions made in the process of exploring the relationship between the distance of a storm from Bermuda and the estimated intensity in the country (see Sect. 7.2.3) is another source of uncertainty. These decisions include the arbitrary choices for r_{max}, the use of the variation of the Rankine vortex, the different conversions, and the curve fitting. In the future it would be really beneficial to explore different techniques for simulating hurricane intensity such as the ones outlined in Holland et al. (2010), Justus et al. (1978), and Jagger and Elsner (2006). In addition, wind asymmetries, the exclusion of which can have a negative impact on TC risk assessment (Pahwa 2007; Alvehag and Soder 2011), could be addressed following suggestions by studies such as Olfateh et al. (2017) and Chang et al. (2020).

Thirdly, it is important to highlight the lack of studies in Bermuda that explore the relationship between the intensity of a storm and the proportion of damage on properties. Since the vulnerability component of a catastrophe model is of great importance, there is a necessity for more studies like Miller et al. (2013) to be conducted in future catastrophic events, since they will provide an opportunity for updates and sensitivity tests on this model framework.

Lastly, the impact of decadal variability of TCs on potential losses has been examined only in terms on frequency. It will be of great interest to explore the impact in terms of intensity as well. The reasoning behind this comes from the fact that, particularly for the North Atlantic basin, it has been shown that even during low-activity hurricane seasons, very intense tropical cyclones can cause a lot of damage and destruction should they make landfall.

7.5 Discussion

We have developed a simple model for the assessment of hurricane wind risk. Despite the limitations of the study outlined above, this methodology may be useful for jurisdictions with limited availability of property exposure or vulnerability datasets. In the absence of a set of robust engineering studies or readily available property exposure data, assessments of the variability of risk can still be achieved, especially for small island jurisdictions. In our study, we utilized a real estate dataset and a published damage survey as the bases for development of exposure and vulnerability inputs, respectively. The hazard portion of this model is constructed by randomly generating multiple location-centric events that are constrained using the historical record. However, this approach is simple compared to the Monte Carlo simulations used to develop the stochastic storm track datasets in commercial catastrophe models (RMS 2019). Our method can quickly and easily be applied to assess the variability of wind hazard in different climate regimes, such as ENSO or the NAO, and can utilize other input such as historical anecdotal document archives (e.g. Chenoweth and Divine 2008), climate model simulations of future storm regimes (e.g. Wehner et al. 2015), or geological proxy datasets, such as those provided in Wallace et al. (2014). The simple nature of the model may also be of benefit in quick sensitivity tests of modelled losses to changes in hazard, vulnerability, or exposure. This may be especially useful for the purposes of teaching different aspects of risk and its estimation. The code underlying the model itself is written in Python, and it is accessible freely via Github here https://github.com/PinelopiLoizou/Risk_Model.

Acknowledgements Pinelopi Loizou acknowledges PhD studentship funding from the SCE-NARIO NERC Doctoral Training Partnership NPIF grant NE/R008868/1, the contribution by the UK Associates of BIOS in support of the 12-week internship at the Bermuda Institute of Ocean Sciences, during which this study was undertaken and CASE funding support from AXIS Capital. Mark Guishard acknowledges support via the Risk Prediction Initiative of the Bermuda Institute of Ocean Sciences.

References

Alvehag K, Soder L (2011) A reliability model for distribution systems incorporating seasonal variations in severe weather. IEEE Trans Power Deliv 26:910–919. https://doi.org/10.1109/TPWRD.2010.2090363

Chang D, Amin S, Emanuel K (2020) Modeling and parameter estimation of hurricane wind fields with asymmetry. J Appl Meteorol Climatol 59:687–705. https://doi.org/10.1175/JAMC-D-19-0126.1

Chenoweth M, Divine D (2008) A document-based 318-year record of tropical cyclones in the Lesser Antilles, 1690-2007. Geochem Geophys Geosyst 9. https://doi.org/10.1029/2008GC002066

Elliott RJR, Strobl E, Sun P (2015) The local impact of typhoons on economic activity in China: a view from outer space. J Urban Econ 88:50–66. https://doi.org/10.1016/j.jue.2015.05.001

Emanuel K (2011) Global warming effects on U.S. hurricane damage. Weather Clim Soc 3:261–268. https://doi.org/10.1175/WCAS-D-11-00007.1

Goldenberg SB, Landsea CW, Mestas-Nuñez AM, Gray WM (2001) The recent increase in Atlantic hurricane activity: causes and implications. Science (80-) 293:474–479. https://doi.org/10.1126/science.1060040

Hallam S, Guishard M, Josey SA, Hyder P, Hirschi J (2021) Increasing tropical cyclone intensity and potential intensity in the subtropical Atlantic around Bermuda from an ocean heat content perspective 1955–2019. Environ Res Lett 16:034052. https://doi.org/10.1088/1748-9326/abe493

Holland GJ, Belanger JI, Fritz A (2010) A revised model for radial profiles of hurricane winds. Mon Weather Rev 138:4393–4401. https://doi.org/10.1175/2010mwr3317.1

Jagger TH, Elsner JB (2006) Climatology models for extreme hurricane winds near the United States. J Clim 19:3220–3236. https://doi.org/10.1175/JCLI3913.1

Jagger T, Elsner JB, Niu X (2001) A dynamic probability model of hurricane winds in coastal counties of the United States. J Appl Meteorol 40:853–863. https://doi.org/10.1175/1520-0450(2001)040<0853:ADPMOH>2.0.CO;2

Justus CG, Hargraves WR, Mikhail A, Graber D (1978) Methods for estimating wind speed frequency distributions. J Appl Meteorol 17(3):350–353

Kishore N et al (2018) Mortality in Puerto Rico after Hurricane Maria. N Engl J Med 379:162–170. https://doi.org/10.1056/NEJMsa1803972

Klotzbach P (2010) 3A.4 Caribbean/Central American hurricane landfall probabilities

Klotzbach PJ, Gray WM (2008) Multidecadal variability in North Atlantic tropical cyclone activity. J Clim 21:3929–3935. https://doi.org/10.1175/2008JCLI2162.1

Knapp KR, Levinson DH, Kruk MC, Howard JH, Kossin JP (2010) The International Best Track Archive for Climate Stewardship (IBTrACS) project: overview of methods and Indian ocean statistics. Indian Ocean Trop Cyclones Clim Chang 215–221. https://doi.org/10.1007/978-90-481-3109-9_26

Knight JR (2005) A signature of persistent natural thermohaline circulation cycles in observed climate. Geophys Res Lett 32:L20708. https://doi.org/10.1029/2005GL024233

Land Valuation Department (2019) Bermuda government - land valuation department. https://www.landvaluation.bm/. Accessed 1 July 2019

Landsea C (2007) Counting Atlantic tropical cyclones back to 1900. EOS Trans Am Geophys Union 88:197–202. https://doi.org/10.1029/2007EO180001

Mann ME, Steinman BA, Brouillette DJ, Miller SK (2021) Multidecadal climate oscillations during the past millennium driven by volcanic forcing. Science (80-) 371:1014–1019. https://doi.org/10.1126/science.abc5810

McCarthy GD, Haigh ID, Hirschi JJM, Grist JP, Smeed DA (2015) Ocean impact on decadal Atlantic climate variability revealed by sea-level observations. Nature 521:508–510. https://doi.org/10.1038/nature14491

Miller C, Gibbons M, Beatty K, Boissonnade A (2013) Topographic speed-up effects and observed roof damage on Bermuda following Hurricane Fabian (2003). Weather Forecast 28:159–174. https://doi.org/10.1175/waf-d-12-00050.1

Murakami H, Delworth TL, Cooke WF, Zhao M, Xiang B, Hsu P-C (2020) Detected climatic change in global distribution of tropical cyclones. Proc Natl Acad Sci 117:10706–10714. https://doi.org/10.1073/pnas.1922500117

NOAA (2020a) Costliest U. S. tropical cyclones. NOAA Tech Memo NWS NHC-6:1–5

NOAA (2020b) H*WIND-Real-time hurricane analysis project. https://storm.aoml.noaa.gov/hwind/. Accessed 10 July 2019

Olfateh M, Callaghan DP, Nielsen P, Baldock TE (2017) Tropical cyclone wind field asymmetry-development and evaluation of a new parametric model. J Geophys Res Ocean 122:458–469. https://doi.org/10.1002/2016JC012237

Pahwa A (2007) Modeling weather-related failures of overhead distribution lines. In: 2007 IEEE Power Engineering Society general meeting, vol. 21 of, IEEE, 1–1

RMS (2019) About us https://www.rms.com/catastrophe-modeling

RMS (2020). Understanding catastrophe. https://www.rms.com/catastrophe-modeling. Accessed 26 Jan 2021

Scherb A, Garrè L, Straub D (2015) Probabilistic risk assessment of infrastructure networks subjected to hurricanes. 12th Int Conf Appl Stat Probab Civ Eng ICASP 2015:1–9

Sealy KS, Strobl E (2017) A hurricane loss risk assessment of coastal properties in the caribbean: evidence from the Bahamas. Ocean Coast Manag 149:42–51. https://doi.org/10.1016/j.ocecoaman.2017.09.013

Strachan J, Vidale PL, Hodges K, Roberts M, Demory ME (2013) Investigating global tropical cyclone activity with a hierarchy of AGCMs: the role of model resolution. J Clim. https://doi.org/10.1175/JCLI-D-12-00012.1

Ting M, Kossin JP, Camargo SJ, Li C (2019) Past and future hurricane intensity change along the U.S. East Coast. Sci Rep 9:7795. https://doi.org/10.1038/s41598-019-44252-w

Tolwinski-Ward SE (2015) Uncertainty quantification for a climatology of the frequency and spatial distribution of North Atlantic tropical cyclone landfalls. J Adv Model Earth Syst 7:305–319. https://doi.org/10.1002/2014MS000407

U.S. Bureau of Labor Statistics (n.d.) CPI inflation calculator. https://data.bls.gov/cgi-bin/cpicalc.pl. Accessed 7 July 2019

Wallace DJ, Woodruff JD, Anderson JB, Donnelly JP (2014) Palaeohurricane reconstructions from sedimentary archives along the Gulf of Mexico, Caribbean Sea and western North Atlantic Ocean margins. Geol Soc London Spec Publ 388:481–501. https://doi.org/10.1144/SP388.12

Wehner M, Prabhat KAR, Stone D, Collins WD, Bacmeister J (2015) Resolution dependence of future tropical cyclone projections of CAM5.1 in the U.S. CLIVAR hurricane working group idealized configurations. J Clim 28:3905–3925. https://doi.org/10.1175/jcli-d-14-00311.1

Zhang R, Delworth TL (2006) Impact of Atlantic multidecadal oscillations on India/Sahel rainfall and Atlantic hurricanes. Geophys Res Lett 33:L17712. https://doi.org/10.1029/2006GL026267

Open Access This chapter is licensed under the terms of the Creative Commons Attribution 4.0 International License (http://creativecommons.org/licenses/by/4.0/), which permits use, sharing, adaptation, distribution and reproduction in any medium or format, as long as you give appropriate credit to the original author(s) and the source, provide a link to the Creative Commons license and indicate if changes were made.

The images or other third party material in this chapter are included in the chapter's Creative Commons license, unless indicated otherwise in a credit line to the material. If material is not included in the chapter's Creative Commons license and your intended use is not permitted by statutory regulation or exceeds the permitted use, you will need to obtain permission directly from the copyright holder.

Chapter 8
Climate Change Impacts to Hurricane-Induced Wind and Storm Surge Losses for Three Major Metropolitan Regions in the U.S.

Peter J. Sousounis, Roger Grenier, Jonathan Schneyer, and Dan Raizman

Abstract Climate change is expected to have increasingly significant impacts on U.S. hurricane activity through this century (Hayhoe et al., Our changing climate. In: Reidmiller DR, Avery CW, Easterling DR, Kunkel KE, Lewis KLM, Maycock TK, Stewart BC (eds) Impacts, risks, and adaptation in the United States: fourth national climate assessment, volume II. U.S. Global Change Research Program, Washington, DC, pp 72:144. https://doi.org/10.7930/NCA4.2018.CH, 2018). A key concern for private insurers is how the relative contributions to loss from wind and water may change because damage from flood is not typically covered in the residential market. This study addresses the concern by considering how climate change by 2050 under an extreme climate scenario may impact hurricane frequency and damage. Using a stochastic catalog of 100,000 years of possible events that can occur in today's climate, and available information on how hurricane frequency and intensity may change, multiple catalogs of events are created to reflect future hurricane activity. Climate change impacts on precipitation rate are not accounted for here, although sea level rise is included to understand how much worse storm surge may become. Relative changes to wind loss and coastal flood loss are examined for three economically significant and hurricane prone urban locations: Houston-Galveston, Miami, and New York. Results show that relative changes in wind loss may pale in comparison to relative changes in storm surge loss. Houston shows large increases in relative contribution of surge to total loss because the contribution is currently small, New York shows the least significant increases because contributions are currently large, and Miami is in the middle.

P. J. Sousounis (✉) · R. Grenier · D. Raizman
AIR Worldwide, Boston, MA, USA
e-mail: psousounis@air-worldwide.com; rgrenier@air-worldwide.com; draizman@air-worldwide.com

J. Schneyer
Willis Re, Boston, MA, USA
e-mail: jonathan.schneyer@rsmas.miami.edu

Keywords Hurricane · Climate change · Storm surge · Sea level rise

8.1 Introduction

Climate change is expected to have significant impacts on hurricanes and the damage they cause by the end of this century (Hayhoe et al. 2018). Even by mid-century, the windspeeds and storm surge inundation heights and extents may be greater than what the U.S. coastline is exposed to at present (Villarini and Vecchi 2013; Little et al. 2015). Moreover, the relative changes in damage to property from wind, storm surge, and precipitation-induced flooding may not be equal. Increases in storm surge damage from rising seas, along with increasing damage further inland due to precipitation-induced inland flooding from the Clausius Clapeyron effect (Liu et al. 2019), may outpace increases in wind damage. Arctic amplification is contributing significantly to melting of the Greenland ice sheet (Hofer et al. 2020) and global warming in general is heating and expanding ocean water. Sea levels have risen globally on average some 10 cm in the last 50 years (Frederikse et al. 2020), but along some portions of the U.S. coastline, sea levels have experienced two to three times that equivalent rate of rise in the last 15 years. (e.g., Ocean City, MD has experienced 6 mm/year of sea level rise according to data from NOAA 2021). Other factors, like a slowdown in the Atlantic Meridional Overturning Circulation (Caesar et al. 2021), which is contributing to sea level rise (SLR), may also be related to climate change.

The physics and thermodynamics of melting ice contributing to SLR may be relatively straightforward compared to how climate change may affect hurricane intensity. Increasing sea surface temperatures (SSTs) certainly contribute to an increase in the potential intensity of a storm but how outflow temperature, vertical wind shear, and distribution of moisture (to name a few) may change are less well understood and may counter some of the potential intensity increases from increased SSTs. How the relative risks may change in the future are especially important considering that damage from (coastal) flood for residential property is not typically insured by private carriers and is underinsured in general, contributing to a "protection gap" where losses are borne by individuals, businesses, and taxpayers.

The topic for this study is directly motivated by increasing concerns from the insurance industry regarding how climate change may impact hurricane losses. Companies that develop catastrophe models for use by the insurance industry are being asked quite frequently if not only do the models that they provide account for climate change for the short time horizons that insurance companies typically focus on but whether catastrophe models can provide a view of the risk on longer time horizons.

A previous similar study (Grenier et al. 2020) focused on total loss from all hurricane-related sub-perils combined. In this study, we consider how climate change by mid-century may impact hurricane wind and coastal flood risk separately. Specific attention is paid to how the relative breakdown of wind losses vs. surge losses may change. The relative breakdown is a concern from an insurance standpoint because residential flood damage is not a risk that is typically insured by commercial insurance companies. Coverage is available through the National Flood Insurance Program (NFIP), although take-up is low (NAIC 2017), especially outside of designated special flood hazard areas. Although risk from inland flood is also likely to increase because of higher atmospheric moisture content, stronger horizontal moisture convergence associated with stronger hurricanes (Liu et al. 2019), and possibly slower moving storms (Hall and Kossin 2019), we do not explicitly evaluate the impact from a change in that risk in this paper.

The basic methodology that is used leverages the AIR Worldwide Hurricane Model for the United States (hereafter AIR Hurricane Model). This is a proprietary model, although some good information is publicly available from the State Board of Administration of Florida (AIR 2021) and from Grenier et al. (2020). The AIR Hurricane Model provides two different 100,000-year catalogs of North Atlantic hurricane activity to represent the risk under (1) all sea surface temperature (SST) conditions as well as (2) under anomalously warm conditions. The latter catalog is referred to as the Warm SST catalog or the WSST catalog, and it was built based on years when the Atlantic Multidecadal Oscillation (AMO) Index has been positive. The warm AMO phase is the one that currently exists, and which has existed since the mid-1990s. More information on the warm SST catalog is available from AIR (2008).

In principle, the large catalog of events also contains events that could occur in a future climate, albeit with a different frequency. The current climate catalogs can therefore be sub-sampled according to a climate change target to create ones that reflect future climate risk. A side benefit of the sub-sampling approach is that because the new catalog consists of different combinations of existing events, the losses by event and other hazard characteristics of the event are already known. Sea level rise is also accounted for in this study in terms of its impacts on storm surge, although resampling cannot create the desired result, so new storm surge footprints for existing events have to be created and the losses to property calculated separately.

Climate change information consistent with an extreme climate scenario is used to define the climate change target. The Representative Concentration Pathway 8.5 (RCP 8.5) scenario is the most extreme of the family of four introduced by the Intergovernmental Panel on Climate Change (IPCC), but several reasons make it an appropriate choice for guidance in this study. One reason is that because none of the RCP scenarios are forecasts, there cannot be any probability associated with any of them, and in that sense, they are all plausible. A second reason is that as an extreme climate scenario it is a good test of the tail of the climate risk distribution. A third reason is that some recent publications (Schwalm 2020) have suggested we have been most closely tracking along the RCP 8.5 scenario in terms of emissions over the last 15 years and that it is a likely scenario for the next 30 years.

Section 8.2 describes the specific methodology for creating the climate change conditioned catalogs from both a wind and storm surge standpoint. Section 8.3 presents hazard and loss results to show the impact of climate change from several different perspectives on wind and storm surge loss. Section 8.4 provides a discussion as well as some next steps to take from a research perspective.

8.2 Methodology

The overall approach involves creating a set of stochastic catalogs of hurricane events to account for the impacts of climate change including SLR. Two different methodologies are employed to account for the climate change impacts on wind and storm surge. Climate change impacts to precipitation are not accounted for explicitly except to the extent that stronger storms tend to have higher precipitation rates (Lonfat et al. 2004).

8.2.1 Accounting for Climate Change Impacts to Wind and Storm Surge

To address wind, the basic approach involves sub-sampling the existing AIR Worldwide U.S. Hurricane 100,000-year WSST catalog of events that reflects the risk in the current climate. The sub-sampling is done in such a way so that the new catalog reflects the potential risk from future climate change. The end result is referred to as a climate change conditioned catalog. A significant benefit of the sub-sampling approach is that the events comprising the climate change catalogs already exist and already have losses computed. Thus, once the climate change catalog is created, the impact on losses is known instantly. This benefit allows many different versions of such catalogs to be created and evaluated very quickly. Each event in the catalog is defined in terms of a maximum windspeed footprint, a maximum storm surge inundation height footprint, and precipitation-induced inland flood depths for on- and off-plain locations.

A prerequisite for creating a climate change conditioned catalog is defining a climate change target to guide the sub-sampling. This target is typically a set of criteria used for deciding whether a randomly drawn event from the parent catalog should be kept or not. The target is typically informed by available information either from peer-reviewed literature, in-house analyses, or otherwise expert judgement. In some instances, to some degree, the target can have some subjectivity associated with it given the fact that the science of climate change and its impacts on complicated weather phenomena is incomplete. In that sense, the target can effectively define a climate change scenario – e.g., how would losses change if the following were to occur...

The sub-sampling approach has been used in the past by AIR on a variety of AIR client and industry sponsored projects (e.g., Robinson et al. 2017). The methodology works well, especially when the parent catalog that is being sampled contains a very large inventory of events and when the climate change target does not require events of intensity or landfall location that are not contained in the parent. The more extreme the climate change target is, the less representative the subsampled catalog may be in terms of reflecting expected intensities and frequencies because of the emissions scenario in combination with the time horizon (e.g., RCP 8.5 for 2090).

However, by virtue of using a 100,000-year catalog that represents the current climate to create 10,000-year climate change catalogs that represent the future climate, it is likely that many "new" events that are not in the current climate catalog will appear in the future climate one. The following sub-sections lay the groundwork for creating the climate change target used in this study and how the sub-sampling was actually performed.

8.2.2 Accounting for Changes in Storm Activity

The climate change target for storm activity was created by considering much of the available literature. A recent article by Knutson et al. (2020), and the supplemental material contained therein, was particularly useful for identifying many studies and the wide range of results of how climate change may impact both weak and strong storms. In searching for relevant studies, one challenge was that few, if any, of the studies really show the climate change impacts on an extreme climate scenario (e.g., RCP 8.5) for mid-century. Most studies focus on either an RCP 8.5 or an RCP 4.5 scenario for late century. However, given the known sensitivity of tropical cyclone activity on sea surface temperatures (Evans 1993), it is reasonable to consider how future activity may change according to global temperature increases rather than to RCP scenarios for select time-horizons. To that end, we note that projected increases in global temperature for 2050 under an RCP 8.5 climate scenario are very similar to those for late century under an RCP 4.5 scenario, which is approximately two degrees Celsius (IPCC 2013). Using this equivalence thus allows for some studies to be relevant for defining our climate change target. We present a subset below.

Camargo (2013) examined output from eight different CMIP5 models and found no statistically significant difference between current and future end-of-century climate for RCP 4.5, although the study noted that the coarse resolution likely was a factor, especially in reproducing current tropical cyclone climatology. Knutson et al. (2013) used high resolution numerical downscaling of CMIP5 output to show that category 4/5 storm frequency would increase over the North Atlantic by 39% by the end of the century under an RCP 4.5 climate scenario. The study also showed an overall decrease in storm frequency of 28% that would occur primarily because of decreases in weaker storms. The methodology was applied globally in a later study (Knutson et al. 2015) to show similar results for other basins. Bacmeister et al. (2018) used the Community Atmospheric Model with a horizontal resolution of 28 km and found that under RCP 4.5 for late century, overall tropical cyclone activity decreases over the North Atlantic, but that Category 4/5 storm activity doubles. Roberts et al. (2020) used the CMIP6 HighResMIP Multi-model Ensemble to examine changes in tropical cyclone activity assuming an RCP 8.5 scenario valid for early to mid twenty-first century (2020–2050) relative to mid to late twentieth century (1950–1980) and found no significant changes in tropical cyclone activity. Emanuel (2021) downscaled CMIP6 model output using CHIPs (Emanuel et al. 2004) and showed that major hurricane frequency would increase by 26% and overall hurricane frequency would increase by 17% for a doubling of atmospheric CO_2, which, according to the IPCC (2013), would occur by mid-century under an RCP 8.5 scenario. Although the result was a global average, the study noted even more significant changes for the North Atlantic.

The considerable spread of results for changes in the frequency of strong hurricanes and overall is summarized in Knutson et al. (2020). We therefore consider the study by Knutson et al. (2013) as a moderate result and as a guide to define changes in the frequencies of strong hurricanes. Table 8.4 from that study was useful for guidance to increase the frequency of major category storms. But it was not followed category by category. Changes in the frequency of weaker storms, even qualitatively in terms of increases or decreases, are less certain as we have noted in our brief review above. To add to the uncertainty, a study by Lee et al. (2020) notes that the overall number of storms projected for the future depends critically on which moisture variable is used in the Genesis Potential Index. Because of the uncertainty, especially about whether weak storms will increase or decrease, we choose for this study to leave the frequencies of category 0–2 storms unchanged. Previous studies have shown that U.S. hurricane losses are dominated by damage from major hurricanes (e.g., Pielke et al. 2008). For stronger, category 3–5 storms, we use the information from Knutson et al. (2013) as well as from the other abovementioned studies as a guide, and define increases of 15%, 25%, and 35% for category 3, 4, and 5 storms respectively. Thus, while the defined frequency target does not necessarily reflect the results from any one study or even in terms of the consensus, it is certainly within the interquartile spread of uncertainty for results that correspond to an RCP 4.5 late-century climate scenario as shown by Knutson et al. (2020), which we note is equivalent to that of an RCP 8.5 mid-century climate scenario. It is also worth noting that by virtue of increasing the frequency of category 3, 4, and 5 storms that the average intensity is implicitly increased. The results of the frequency adjustments by Saffir Simpson category are shown in Fig. 8.1.

Other possible impacts on hurricane activity were considered as part of the target definition, but ultimately excluded. Despite some recent studies that have shown a decrease in forward speed (e.g., Kossin 2018), especially post-landfall, and a poleward migration of the latitude of lifetime maximum intensity (Kossin et al. 2014), Knutson et al. (2019) indicated there is low confidence that they are the result of climate change, and Knutson et al. (2020) expressed low confidence that such trends would continue in the future.

Despite the end result being guided by relevant peer-reviewed literature, and because we will not be considering changes in other storm characteristics like storm size or rainfall rates (as we describe shortly), it is more appropriate to interpret our climate change target as one representing an extreme mid-century climate change scenario rather than an RCP 8.5 climate change scenario.

Sub-sampling was conducted from a landfall perspective, which implicitly makes the assumption that changes in basin activity are the same at landfall. Ting et al. (2019) do suggest that the relatively high vertical wind shear along the U.S. coast that has existed in the past during positive phases of the Atlantic Multidecadal Oscillation, which helps reduce intensities and acts as a protective barrier, may be

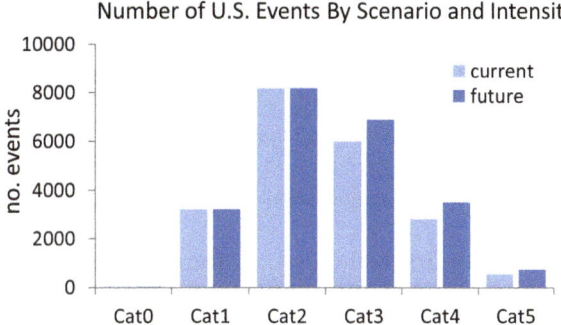

Fig. 8.1 Climate change (future) target frequencies for events by Saffir Simpson Category (by central pressure) at landfall in the U.S. Frequencies for current climate shown for perspective

less effective because of climate change. Thus, the landfall percent increases of intense storms could be larger than those over the basin, although quantifying the result at this point has high uncertainty. The entire catalog of events was partitioned into different Saffir-Simpson categories so that Bin ι contained only storms that made landfall as a category ι storm ($\iota = 0, 1, \ldots 5$). Adjusted frequency targets were created for each category, and then events were drawn at random until the frequency targets for all categories were met.

Many catalogs were created based on the climate change target. The primary reason is because sub-sampling by its very nature begins with a random seed, and because the target is not specified uniquely, a 30% increase in category 4 storms (for example) can be achieved in many different ways. Each catalog that is created meets the target but will differ slightly in other ways. The corresponding industry loss will also be slightly different. And, for any one catalog, the increase of Category 3–5 storms may differ slightly in certain regions from that of the specified target, but for several catalogs together, the target is essentially preserved. We show the sensitivity of this in the next section. Generating one thousand different catalogs provides enough samples so that the spread in losses can be accurately obtained.

8.2.3 Accounting for Sea Level Rise Impacts to Storm Surge

In order to account for the impacts of sea level rise on storm surge, new storm surge footprints for each existing hurricane event in the 100,000-year catalog had to be created regardless of storm intensity. Sub-sampling the existing storm surge footprints was therefore not an option, and a more complex solution and procedure had to be developed and implemented. Losses for these new storm surge footprints were calculated using the loss module of the AIR Hurricane Model.

Storm surge is currently modeled in the AIR Hurricane Model using a form of the National Oceanic and Atmospheric Administration (NOAA) Sea, Lake, and Overland Surges from Hurricanes (SLOSH) Model (Jelenianski et al. 1992). The AIR SLOSH model accounts for hurricane parameters, coastal geography, coastline features, tidal rivers, and flood defenses, but uses AIR Hurricane Modeled winds from a stochastic catalog of events. Tidal effects are included by computing a tidal height for each event, considering the simulated landfall date and time, the landfall location, and other adjustments based on the local geography and seasonality. The raw surge elevation output is then post-processed using a high-resolution 30 m digital terrain model to calculate storm surge depths at a 30 m resolution. Damage to property and contents is computed using that information along with AIR-developed damage functions that are appropriate for the building and contents within.

The AIR 100,000-year U.S. Hurricane catalogs each contain over 200,000 storm surge events. Resimulating these for different sea level rise scenarios is computationally expensive and time consuming. An alternate technique was therefore developed that allows the current climate storm surge footprint to be spread across a future climate land-seascape with the prescribed amount of SLR. The technique begins by increasing the storm surge heights at all land cells that are wet and all water cells by the prescribed amount of SLR. A series of sweeps is then performed to determine which additional cells will get wet. With no underlying friction present, a dry cell is wetted to an average surge height based on the average of the surge heights from the surrounding wet cells. If the average surge height is higher than the elevation of the grid cell in question, then the cell remains wet with the average surge height elevation. Without friction, the inland extent of the new surge footprint is limited by the distribution of terrain height. The methodology is explained in more detail in McInnes et al. (2013). This strategy allows for new storm surge events to appear in the catalog to the extent the base (e.g., without SLR) storm surge footprint contained some footprint just offshore that could be extended inland.

In this study, to improve upon this terrain-height-limited footprint, friction is introduced to allow a more realistic inland penetration of storm surge. The friction is parameterized according to the land-use of the underlying surface. A single value is used here for all cells that favors urban areas near the coast that would typically exhibit the most damage from storm surge. The value is chosen based on a limited set of calibration runs with SLOSH. We note that the approach used here does not capture the change in dynamics that sea level rise may have on storm surge, which on very local scales can be significant. These effects have been studied by numerous authors (Lin et al. 2012; Zhang et al. 2013; Bilskie et al. 2016), and although nonlinearities do exist as a function of bathymetry, land slope and friction, storm characteristics, etc., the general finding is that the impact of SLR produces a linear addition to the height of the water. Once the footprints are adjusted for SLR, the output has to be reformatted in order for the files to run within the loss estimation module in AIR's Touchstone® Software. This requires almost as much processing time as the SLR adjustment phase. Thus, only three regions of the U.S. are processed for this study: Galveston-Houston, Miami, and New York City.

8.2.4 Regional Sea Level Rise Projections

Regional projections of SLR are obtained from Sweet et al. (2017). This source provides very detailed information for a variety of scenarios that incorporate different RCP information and more. Additionally, there is an accompanying user-interactive web site that allows scenarios to be obtained for a large number of tide gage stations along the U.S. coastline. Typically, six different scenarios are shown for every station and are labeled as Low, Intermediate-Low, Intermediate, Intermediate-High, High, and Extreme. Figure 8.2 shows the global mean sea level rise version and the correspondences to the RCP scenarios.

Because of our focus on mid-century, the values in 2050 are relevant. Additionally, to capture some of the uncertainty of sea level rise within the RCP 8.5 scenario, we choose the Intermediate-Low and Intermediate-High scenarios, which are the two that most closely flank the 5–95% certainty boxes in Fig. 8.2. Increases were determined from 2010, which is the vintage of the bathymetry and elevation data in the AIR hurricane model. An example of how the values was determined for the Intermediate-High scenario for New York is shown in the bottom panel of Fig. 8.3. The values for both SLR scenarios for each region are shown in Table 8.1.

These SLR amounts were used in the adjustment process for all storm surge events that affected each of the three regions (The regions are shown in Figs. 8.10, 8.15, and 8.20). For example, for all events affecting Galveston, 14.2 inches and 25.6 inches were used in the storm surge adjustment process described above.

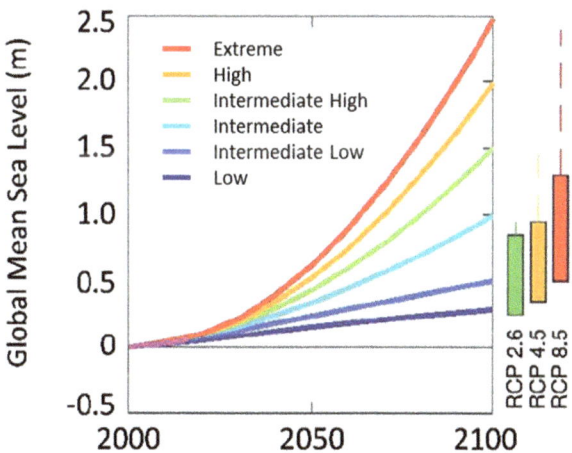

Fig. 8.2 Six representative Global Mean Sea Level rise scenarios for 2100 (six colored lines) relative to historical reconstructions from 1800–2015 and central 90% conditional probability ranges (colored boxes) of RCP-based Global Mean Sea Level Rise projections. Central 90% probability ranges augmented (dashed lines) by difference between median Antarctic contribution of Kopp et al. (2014) probabilistic GMSL/RSL study and the median Antarctic projections of DeConto and Pollard (2016), which have not yet been incorporated into a probabilistic assessment of future Global Mean Sea Level Rise scenarios. (Adapted from Sweet et al. 2017)

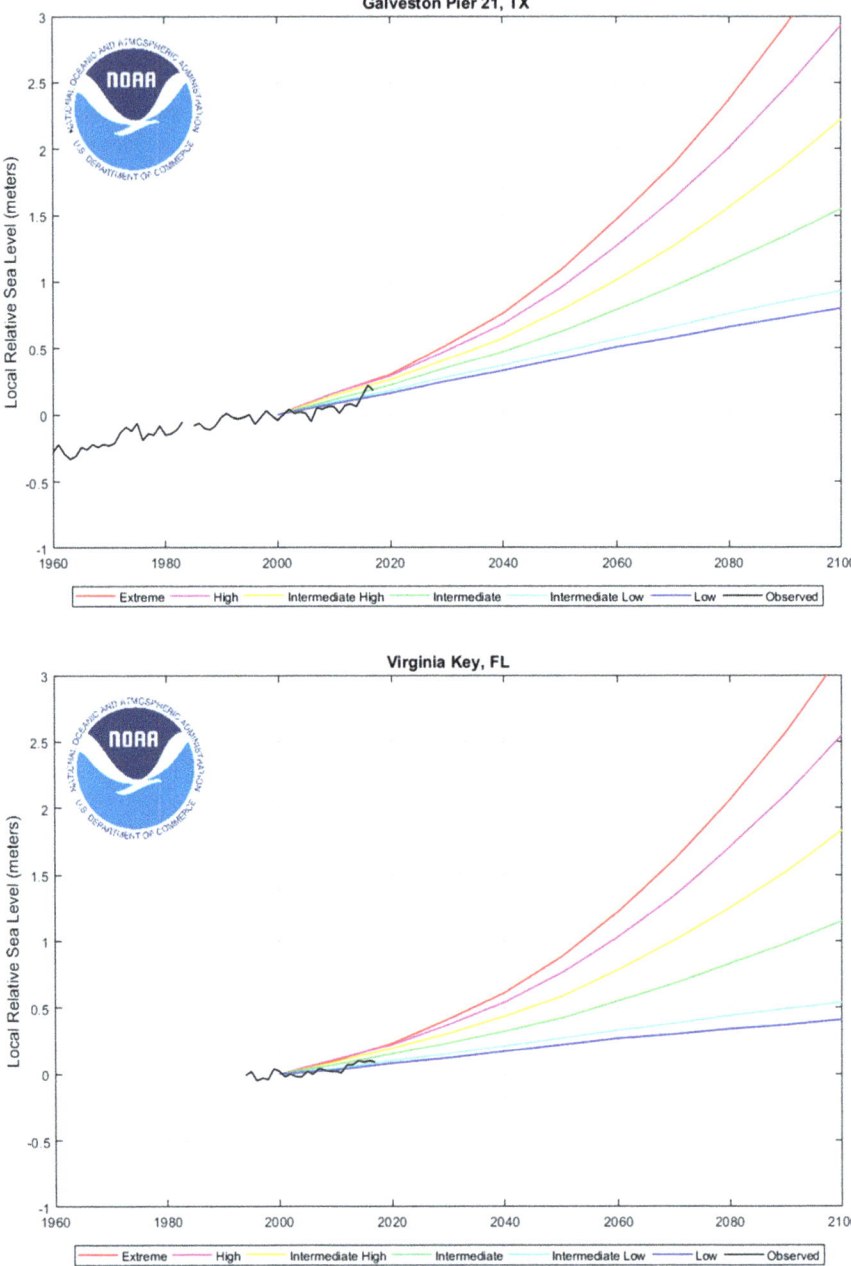

Fig. 8.3 NOAA SLR scenarios for locations used in the study. Annotations in bottom panel indicate considerations for adjusting the SLR value needed for 2050 for one NOAA scenario (yellow, intermediate high) because of the vintage (2010) of the storm surge output from the SLOSH model. More details provided in the text. (Plots obtained from NOAA website: Sea Level Trends – NOAA Tides & Currents)

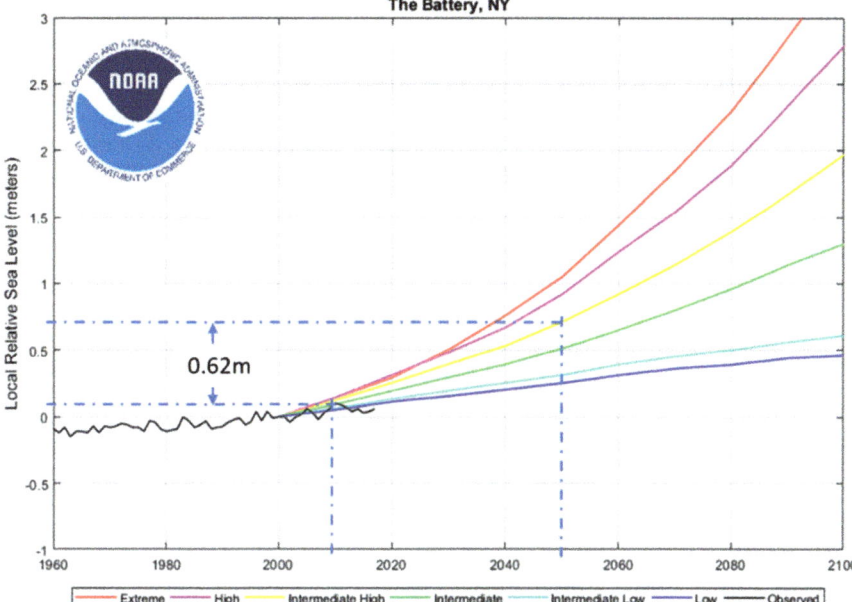

Fig. 8.3 (continued)

Table 8.1 Sea level rise values calculated for use in the three regions of study for the two SLR scenarios chosen

Region (buoy)	Intermediate low (inches)	Intermediate high (inches)
New York, NY (The Battery)	8.3	24.4
Miami, FL (Virginia Key)	7.5	20.1
Galveston, TX (Galveston)	14.2	25.6

All stochastic events that generated storm surge along the coastlines and even just offshore of each of the regions, including a 50-nautical-mile wide buffer on either side from the first 50,000 years in the AIR 100,000-year WSST catalog, were identified and evaluated for impacts from sea level rise. This resulted in 8527 events for the New York region, 16,643 events for the Miami region, and 20,173 events for the Galveston region. The number of events that actually cause loss in the counties shown in the three regions depended on the SLR scenario.

8.3 Results

We describe here some hazard and loss results at the national and county levels before focusing in more detail on the three urban regions. More detail for the national and county level results is presented in Grenier et al. (2020).

8.3.1 Regional Distribution of Landfall Activity

The subsampling of the AIR U.S. hurricane catalog to create a frequency/intensity distribution reflective of our mid-century extreme climate scenario yields a 20% net increase in major hurricanes making landfall. Because there was no further constraint on activity – e.g., regional changes – the 20% increase occurred more or less uniformly over the entire coastline affected by hurricane activity, but only when multiple 10,000-year catalogs of activity are considered. Figure 8.4 shows the

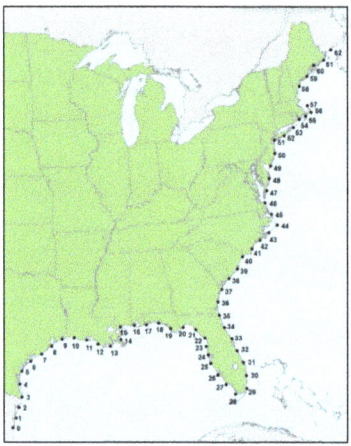

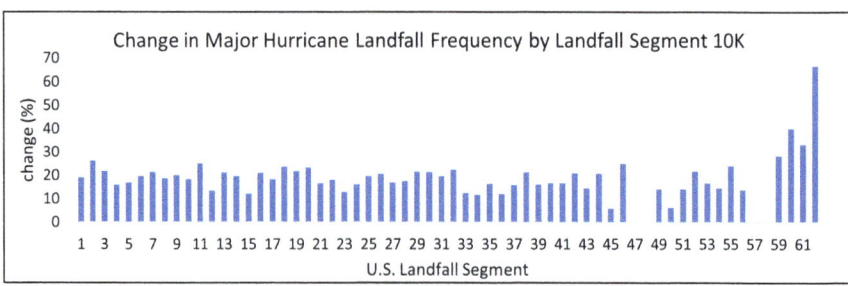

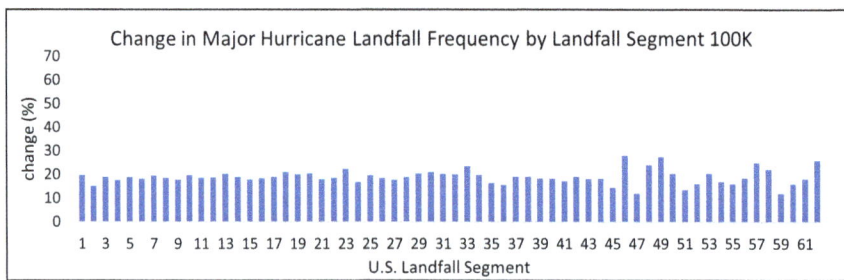

Fig. 8.4 Landfall segments used in the AIR U.S. Hurricane Model (upper panel) and percent changes in landfall frequency of major hurricanes by segment from a 10 K climate change catalog (middle panel) and from a 100 K climate change catalog (lower panel) relative to current climate

geographical landfall distribution of major hurricanes from a single 10,000-year (10 K) catalog and from the aggregate activity of 10 such catalogs together (e.g., effectively 100,000 years of activity). Clearly, there are more accentuated regional differences indicated by the 10 K view, but it is important to note that they are spurious and that there is no physical reason to explain it. The variations by coastline segment are a result of the openly defined target used to create the catalog.[1] Even with the 100 K catalog, the distribution by landfall segment is not perfectly uniform. Segment 47, for example, shows an increase of only 12%, although it is shouldered on either side by 28% and 24% increases. As a larger region, segments 45–49 together yield an average increase of ~21%. The non-uniformity for this region, even from a 100 K catalog, is related to the relatively low landfall rate, which is a result of the NW-SE oriented coastline at a latitude where storms are typically recurving away (e.g., NE-ward) from the coast.

8.3.2 National Loss Results

Impacts on loss from each of the climate change catalogs were evaluated at national, state, and county levels. One big advantage of the sub-sampling strategy is that because the events that go into the sub-sampled catalogs have already been run through the AIR Hurricane Model loss module, the information about where and how much loss, as well as further granularity of loss to particular lines of business (e.g., residential property, commercial property, automobiles, etc.), is already known. That is a great advantage because processing even one 10 K catalog to generate losses is computationally expensive. The loss estimates here consider the direct damage to exposures, such as residential, commercial, and industrial properties, and automobiles, etc., and the temporary loss of use of those exposures caused by that damage. The AIR Industry Exposure Database has been developed using data obtained from a variety of data sources, including private vendors, government reports and databases, and remotely sensed information. It includes the primary building material (wood, steel, concrete), use type (residential, commercial, industrial), number of stories, year built, etc., and replacement cost, all of which impact building vulnerability, damage, and loss. The database reflects all insurable property on a 90 m grid for the entire U.S. It does not include non-modeled exposures such as public infrastructure, marine, or cargo, or indirect sources of loss, such as lost wages or economic productivity. More information on the AIR industry exposure database is available from Hayes and Rowe (2008).

Loss metrics typically involve an average annual loss, an average occurrence loss, and return period losses, either from an occurrence or aggregate standpoint and from an insurable or insured standpoint. Average Annual Loss (AAL) is the average loss

---

[1] It is important to note that variation from one 10 K catalog to another in this respect is not an indication of the lack of convergence in the 10 K catalogs that are provided to clients because of other calibration measures that are implemented.

across all years in a stochastic catalog, or the expected loss per year averaged over many years. Average Occurrence Loss (AOL) is the average of the largest annual single event losses. Return period losses represent the magnitudes of loss at different exceedance probabilities. A 10,000-year loss in the 10,000-year catalog is the largest loss in the catalog. The 5000-year loss is the second largest and so on. Equivalently, a 5000-year loss can be referred to as having a 5000-year return period or a 0.02% exceedance probability. Again, this can be either from an aggregate (all event losses in a year summed) or from an occurrence (single largest loss in a year) standpoint. Insurable losses are based on the fraction of damage/loss relative to the total replacement value. Insured losses account for deductibles, limits, etc. that characterize the exposure in question. The AIR Industry Exposure Database uses location-averaged information for deductibles, etc. Clients who use the AIR software to compute their own losses can enter more specific information about buildings, contents, and policies.

The impact to losses for the entire U.S. is summarized in Fig. 8.5. Specifically, the distributions in AAL, 100-year return period aggregate loss, and 250-year return period aggregate loss from the 1000 different 10 K climate change catalogs are shown relative to the distribution of losses from the 1000 different 10 K catalogs that were sub-sampled to represent the current climate. To a large degree, the 20% fingerprint from changes in frequency of major hurricanes (storm activity) is evident in the AAL. The median loss change for the AAL is 19%. The agreement is more than coincidental because the result demonstrates that the bulk of the damage to U.S. property occurs by far and away from major hurricanes. The slightly larger spread in losses for the future climate is a reflection again of the open-ness of the target used for sub-sampling, the heterogeneity of the exposure and its vulnerability across the U.S., and the regional landfall frequency, particularly of major hurricanes, even for the current climate. It is important to note that the spread in losses is the result of the sampling variability associated with achieving the climate change target, not a reflection of the scientific uncertainty associated with the climate change impact. To that end, it is notable that the spread in loss results for the two return periods (both show median changes of ~15%) is notably larger than that for the AAL for both the current climate and the future climate. Again, this reflects the openness of the target and the heterogeneity of the exposure distribution across the U.S.

The percent changes by sub-peril (not shown) are also comparable given that the sub-sampling target did not specify any constraints on precipitation or storm surge. The 20% increase in flood AAL is more related to the net increase in storm activity and is much less sensitive to stronger hurricanes generating more precipitation. For precipitation and inland flood, it is reasonable to expect that because of the Clausius-Clapeyron effect, and all else equal, that precipitation and possibly flood would increase by about another 7% (given the assumed additional one-degree Celsius increase that would occur by 2050 under RCP 8.5). Storm surge would also increase more across the entire U.S. hurricane-affected coastline because of sea level rise, an aspect we evaluate later in this section in more detail for select locations.

It is important to note that the loss results in Fig. 8.5 reflect the direct damage to the full database of exposure (i.e., insurable), not just the insured portion. The insured portion would be significantly lower particularly for the surge and flooding components, which are significantly underinsured.

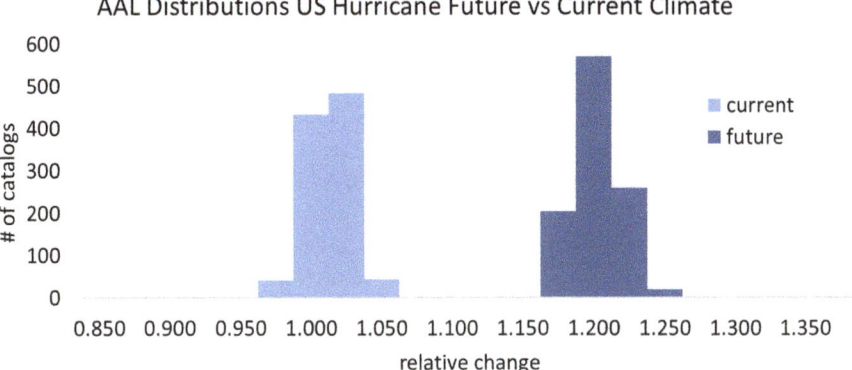
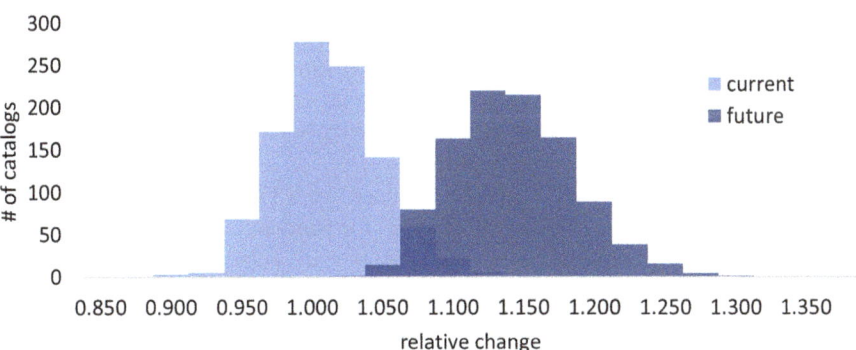
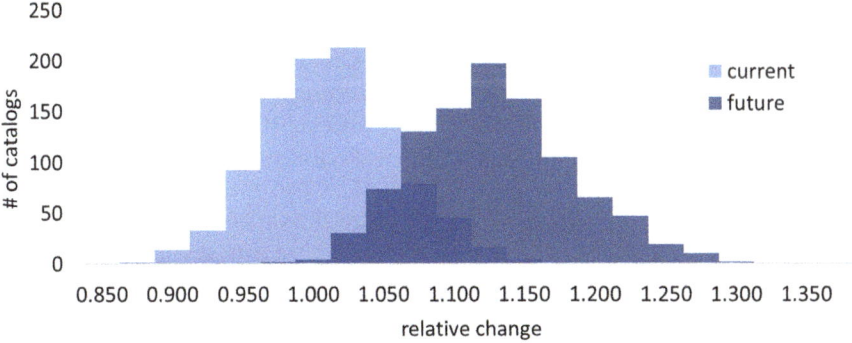

Fig. 8.5 Distribution of aggregate insurable loss changes for future climate catalogs relative to current climate. Spread is result of each catalog yielding a slightly different loss. Current climate catalogs are generated in a way similar to future ones to provide a better basis for comparison. Losses normalized by mean value of current climate loss

8.3.3 County Level Loss Changes

Additional insight into the geographical changes in hurricane risk for the climate change scenario as we have defined it is gained by considering loss results by county. Figure 8.6 presents county level detail of the AAL for current and future climate and for the change in AAL for all sub-perils combined. At first glance, there appears to be little difference in the distribution of current and future climate losses, and to a large degree this is likely attributable to the absence of any regional differences in landfall activity as shown in Fig. 8.5 (bottom panel). However, a view of the percentage difference plot in Fig. 8.6 does show coherent structure. The spread of percentages is admittedly small, as might be expected, but given the numbers of events involved in the calculations, the differences are statistically significant. One curious feature is that the highest percent changes are not exactly right along the coastline everywhere as might be expected. In fact, the only coastal area where changes are highest is in Florida across the eastern two thirds of the peninsula. A discontinuous band of highest percentage loss increase exists 100–500 km inland stretching from east Texas, eastward to Alabama then northeastward to southern North Carolina. Lower percentage changes exist farther northeast, with the smallest percentage increases over the central Appalachian region.

Some insight to this pattern can be obtained from Fig. 8.7, which shows the relative changes by sub-peril. Percent changes for wind and inland flood losses are even farther inland, while those for storm surge are adjacent to the coast. The distribution of sub-peril relative loss changes for wind and inland flood are likely a combined result of distribution of the hazard changes, damage function dependence on wind speed and flood depth respectively, and distribution of exposure. The second point is worth explaining more. Damage functions for windspeed typically are nonlinear, given that the force and power of the wind increase as the square and cube of the windspeed respectively, but even more so because at some high windspeed value, building fixtures, including parts of roofs, overhangs, exterior lamps, etc., tear off and become projectiles, which can cause even more damage by breaking windows of other buildings, thus breaching other building envelopes so that wind and water can enter those structures and cause even more damage (the AIR damage functions account for such effects). The fact that the highest percent increases in wind loss are slightly inland may reflect the nonlinearity of wind damage to wind speed – specifically, it is the zone where the change in storm activity creates the greatest impact from a fractional building damage perspective.

For flood (both precipitation-induced and coastal), the water typically has to reach a certain depth for a structure to be damaged. The farther inland location of maximum increase in precipitation-induced flood damage in Fig. 8.7 (middle panel) may be the result of more storms generating flood water that affects a greater number of properties annually on average, plus the fact that heavy precipitation typically extends much farther inland even as tropical cyclone winds decay below property-damaging strength. The distribution of changes in storm surge loss is easier to interpret, with a near continuous band of change along the coast and with some

8 Climate Change Impacts to Hurricane-Induced Wind and Storm Surge... 177

Fig. 8.6 Average annual loss (all sub-perils) for current climate (upper), future climate (middle), and percent change (lower). Note: changes for all sub-perils are from change in hurricane frequencies by Saffir Simpson Category only

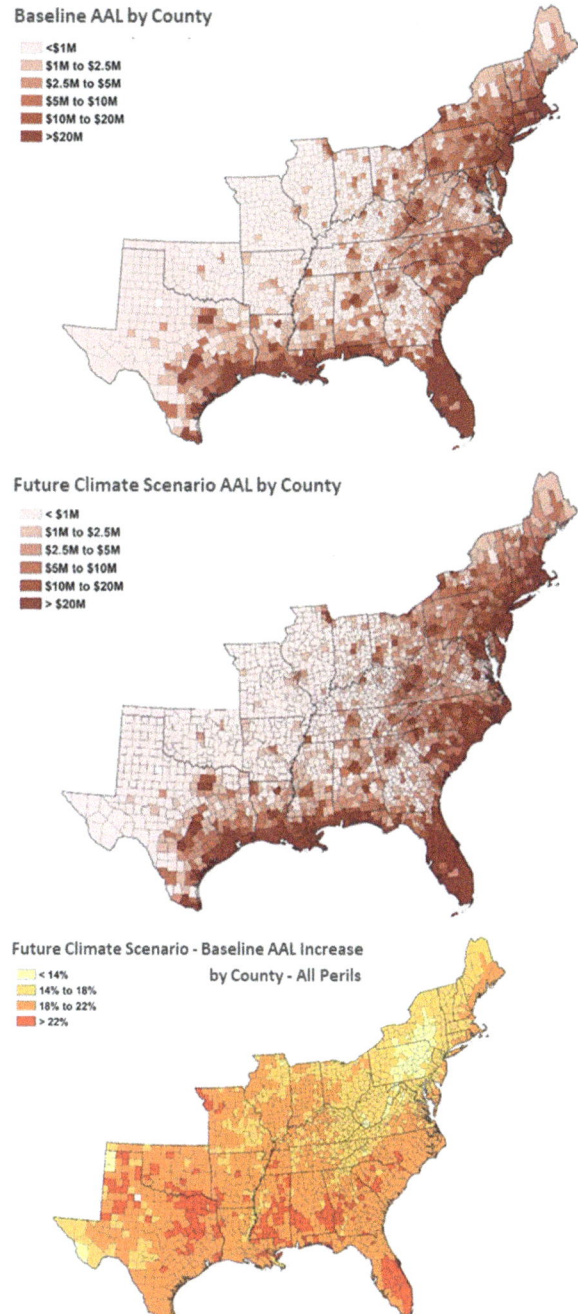

Fig. 8.7 Relative changes to AAL from sub-peril losses as shown. Note: changes for all sub-perils are from change in hurricane frequencies by Saffir Simpson Category only

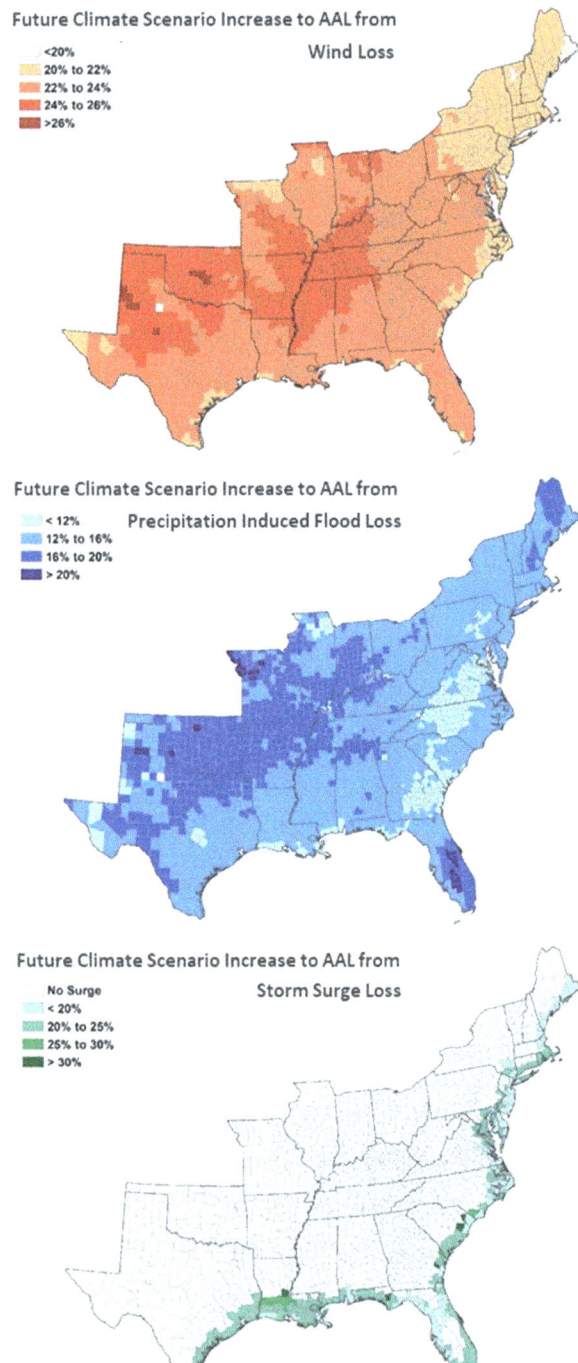

innermost locations exhibiting the highest change. This last feature is likely a result of areas experiencing storm surge damage from additional strong storm activity in the future climate that do not in the current climate.

The slightly lower percent increases in loss over the mid-Atlantic region extending northward appear to be driven more from wind loss rather than from precipitation-induced flood loss. Even though this region still experiences a 20% increase in major hurricanes, it is almost certainly the case that on the whole the storm intensities are weaker than those impacting the southern U.S. and that storms decay faster inland (because of mountainous terrain), and thus likely that fewer properties experience damage consistent with a 20% increase in major storm activity. This explanation is consistent with the fact that increases in storm surge loss for the U.S. coastline from Virginia northward are basically the same as for locations farther south, especially for coastlines oriented perpendicular to storms approaching from a southern direction.

8.3.4 Detailed Loss Analyses for Selected Urban Locations

The results in the previous sub-section showed potential climate change impacts to storm surge loss from increases in storm frequency (i.e., Fig. 8.7 bottom panel). Impacts from SLR were not included for that analysis because the methodology described in Sect. 8.2 to adjust storm surge footprints for SLR is still computationally involved – especially because hundreds of thousands of events need to be processed. Thus, for this study, we focus on three regions where relative impacts to storm surge loss from changes in storm activity and sea level rise are compared, and storm surge loss relative to wind loss is also compared. The three regions are Houston, TX; Miami, FL, and New York City, NY, and the counties included are shown in Fig. 8.8. Despite spanning less than ten percent of the coastline impacted by hurricane activity, these three regions alone account for approximately one-third of the AAL from U.S. hurricane activity.

For some additional background information, we show in Fig. 8.9 the contribution of storm surge loss to wind and surge AAL by Saffir Simpson category from the AIR Hurricane Model. The result for the entire coastline (see Fig. 8.4 for segments) is shown (i.e., segments 1–63), as well as for several subregions of coastline including the Gulf Coast (segments 1–17), Florida Coast (segments 18–35), southeast U.S. coast (segments 36–44), and the northeast U.S. coast (segments 45–63). In all cases, the relative contribution decreases with increasing hurricane intensity. Although there is little information in the literature on the topic, this result may be understood in part by realizing that the wind damage functions are very nonlinear in terms of fractional damage with respect to wind speed, and that storm surge damage functions tend to be more linear with respect to water depth (Sealya and Strobl 2017; USACE 2020). Additionally, even though storm surge height and depth depend non-linearly on wind speed (Harris 1957), wind damage will typically extend farther inland and do more damage with increasing storm strength than storm surge damage

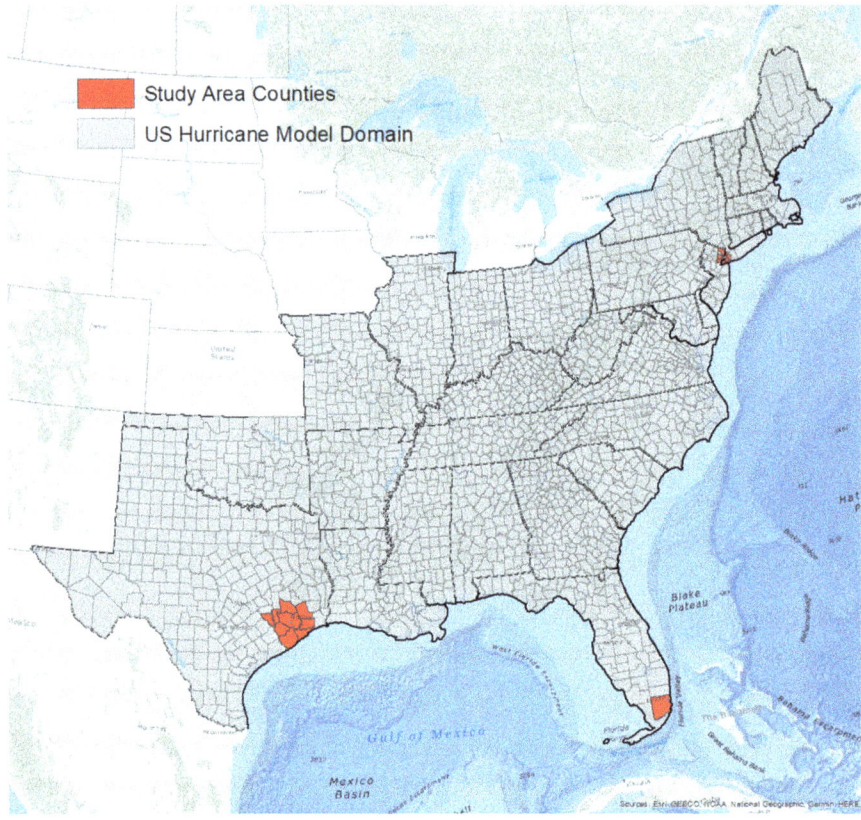

Fig. 8.8 Locations and extents of the three study regions for detailed SLR and storm surge analysis: Metro NYC, Miami-Dade, and Houston/Galveston (regions range from one to nine counties). Shaded states indicate domain of AIR Hurricane Model

can, especially in steep coastal terrain. Specific behavior of wind and surge damage functions depends on a multitude of factors, including building material, building height, flood mitigation, etc., and hence the different behaviors in Fig. 8.9 for the different coastline sections (e.g., convex vs. concave) cannot be completely understood without a more detailed assessment.

8.3.5 Galveston-Houston Return Period Wind and Storm Surge Results

This region is typically impacted by hurricanes. Although Hurricane Harvey in 2017 made landfall well south of Galveston, seventeen storms of category strength one or higher have made landfall within 50 nmi of Galveston since 1900. This frequency is

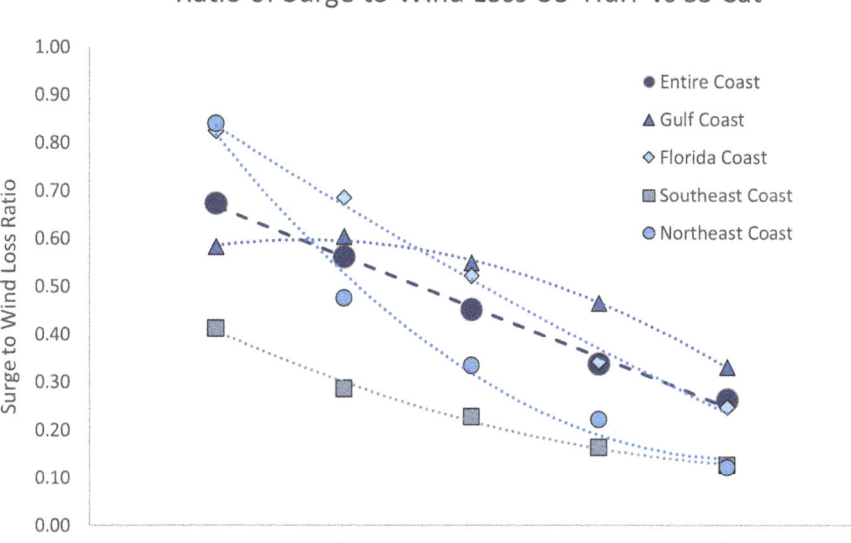

Fig. 8.9 Ratio of storm surge loss to windspeed loss by Saffir Simspn category in the AIR Hurricane Model. Region definitions in terms of landfall segments explained in text. Best-fit curves using second-degee polynomials included for all segments except for Entire Coast which has best fit line

accurately reflected in the current climate of the AIR Hurricane Model. As noted for this study, sub-sampling increases the frequency of major category strength storms by ~20%, and as we have seen, because of those storm changes, wind and surge losses increase by a little over 20%, and inland flood losses increase by a little less than 20%. However, SLR will increase storm surge loss relative to wind loss, and so it is worth looking at how the relative contribution of storm surge to wind and surge loss changes because of SLR. Figure 8.10 shows the contributions for current sea levels by county for reference.

The coastally adjacent counties (Brazoria, Galveston, and Chambers) currently experience relatively high percentages of loss from storm surge. Even Harris County, where downtown Houston is located, shows a 5% contribution because of a short stretch of coastline along the northern lobe of Galveston Bay and because of an adjacent deeper inlet that gets flooded. The inlet is in fact the reason why Liberty County also experiences some coastal flooding. Select return period storm surge inundation maps from the AIR (SLOSH based) storm surge model for the current climate are shown in Fig. 8.11 and provide some additional perspective on areas prone to coastal flooding. Plotted resolution for all return period storm surge maps is roughly 250 m, and each grid cell inundation height (e.g., storm surge height above

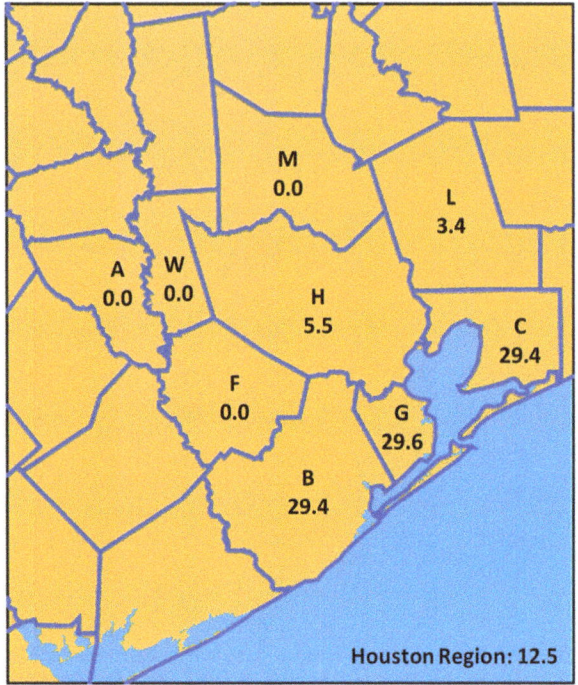

Fig. 8.10 Percent contribution (numbers) of storm surge to wind+surge AAL by county (indicated by first letter) for Austin, Brazoria, Chambers, Fort Bend, Galveston, Harris, Liberty, Montgomery, and Waller Counties for current climate. Nine county average for Houston Region is 12.5%

mean sea level) represents the return period value from all storm surge events impacting that grid cell.

As noted previously, the higher frequency of major hurricanes in our future climate scenario increases the wind and storm surge losses by approximately 20%, and thus the percent contribution of surge to wind and surge loss at the county level is essentially no different than for the current climate result shown in Fig. 8.10, so it is not shown. Sea level rises differentiates the relative contribution from storm surge loss significantly and is explored in more detail. Table 8.2 summarizes the impacts of SLR and increases in storm frequency on storm surge losses. Both SLR scenarios result in significant increases in storm surge loss.

The Intermediate-Low and Intermediate-High SLR scenarios increase storm surge AAL by 41% and 84% respectively. The increases in AAL are primarily attributable to increases in depth of areas that already get inundated without SLR, although some increase in AAL is the result of additional areas getting inundated. The percent breakdown is not calculated here, but some insight can be obtained by examining the storm surge footprints. To that end, storm surge footprints for select RPs for both SLR scenarios, including increases in storm frequency, are shown relative to those for current sea level and frequency conditions (the base case) in Fig. 8.12 to illustrate the point. We note that from Fig. 8.11, for a given RP, a comparison of the inundation areas for the Intermediate-Low and Intermediate-High SLR scenarios shows very little change in area inundated. One has to look carefully

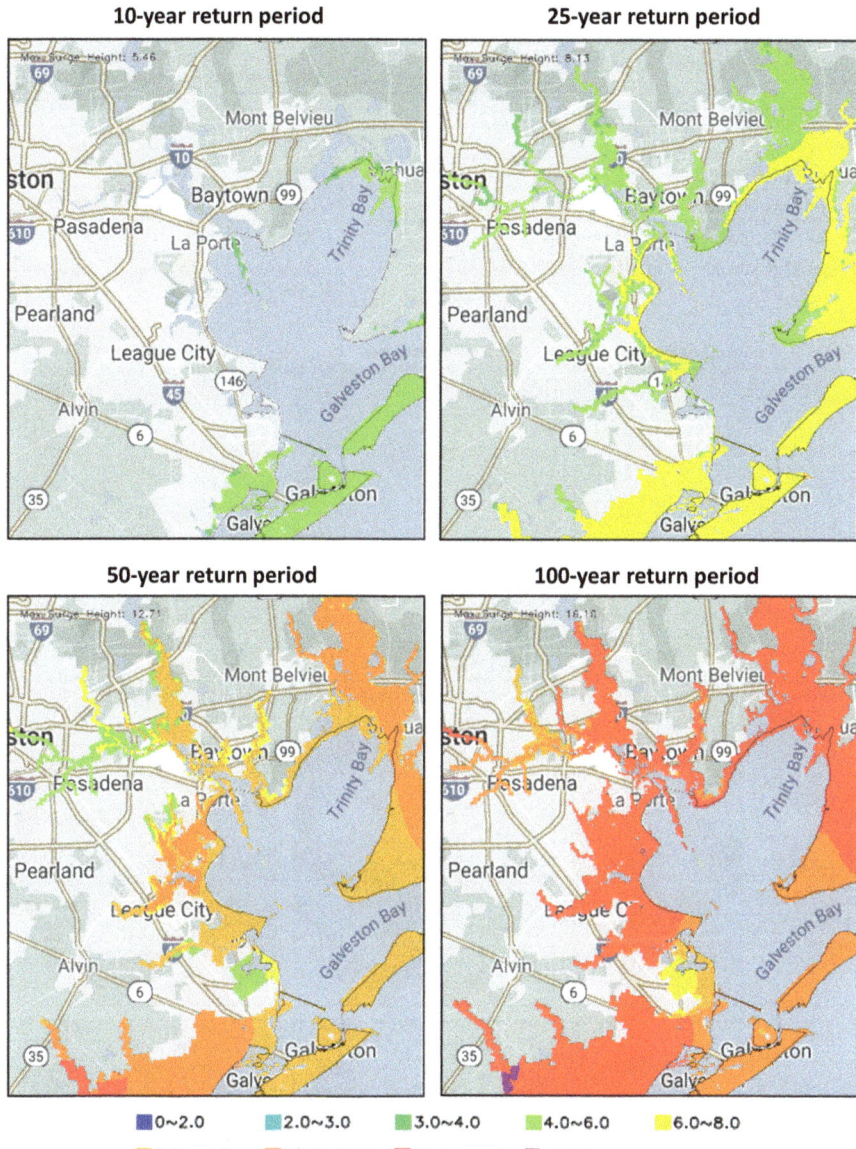

Fig. 8.11 Storm surge inundation height (feet) for Galveston-Houston Region for return periods and current sea levels. Plotted domain spans 0.60° lat × 0.60° lon. Downtown Houston is just outside western edge of plots

to identify signatures. For the 100-year RP, the differences are shown in Fig. 8.13. The contribution to the footprint size from just SLR and from SLR and increased storm frequency is shown. A comparison of the two plots shows that the additional

Table 8.2 Summary of impact of SLR and changes to storm frequency on storm surge loss for Galveston-Houston Region

Scenario	Pct change in surge AAL	Pct contr surge to wind+surge AAL
Current climate	–	12.5
Incr freq	23	12.5
Int-lo SLR scenario	41	16.8
Int-lo SLR scenario + incr freq	72	16.7
Int-hi SLR scenario	84	20.8
Int-hi SLR scenario + incr freq	123	20.6

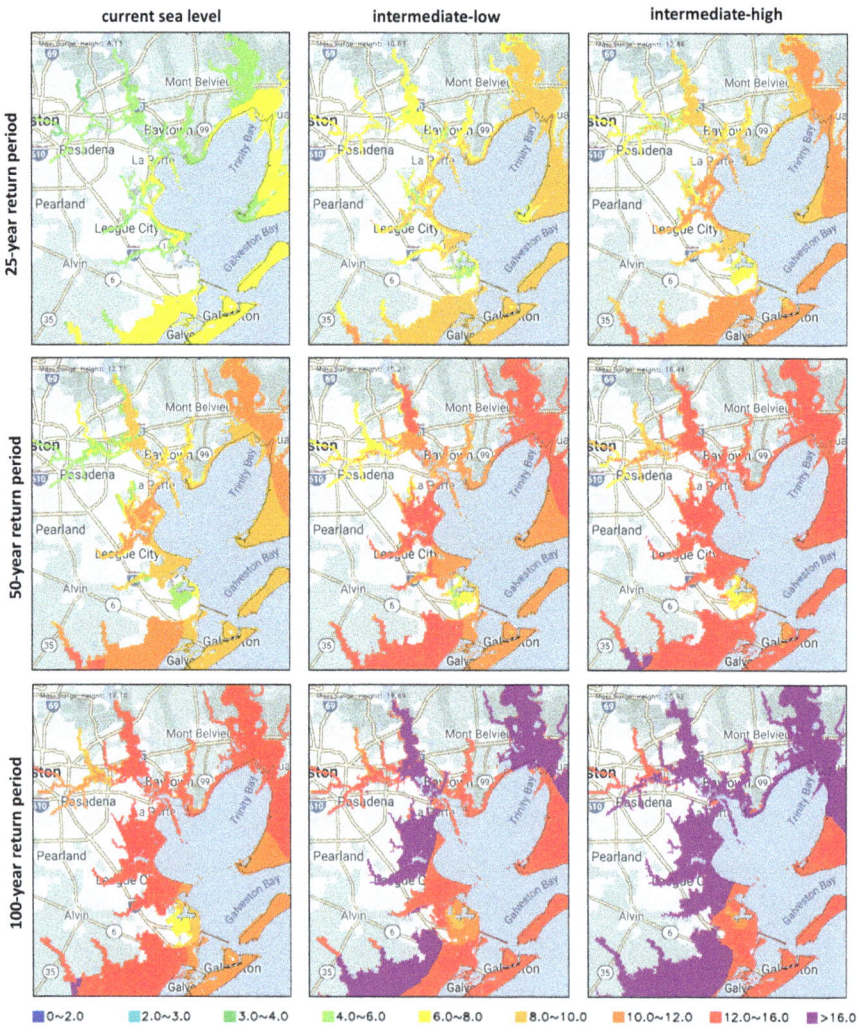

Fig. 8.12 Storm surge inundation heights (feet) for Galveston-Houston Region for return periods and SLR scenarios as shown. Plotted domain spans 0.60° lat × 0.60° lon. Downtown Houston is just outside western edge of plot

8 Climate Change Impacts to Hurricane-Induced Wind and Storm Surge...

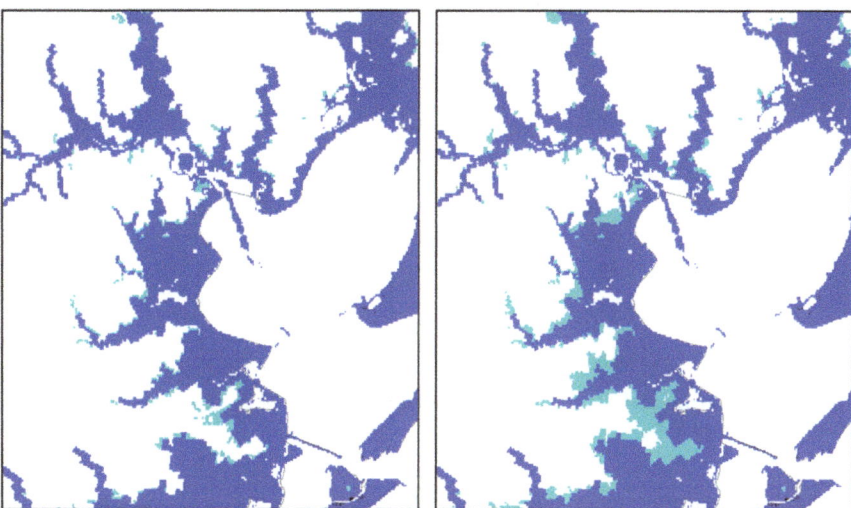

Fig. 8.13 Additional areas inundated (light blue) relative to current sea levels and storm frequency (dark blue) for Intermediate-Hi SLR scenario without frequency increases (left) and for Intermediate-Hi SLR scenario with frequency increases (right) for Galveston-Houston region for 100 year RP. Base map removed for clarity. Plotted domain spans 0.60° lat × 0.60° lon

Table 8.3 Relative contribution to storm surge footprint area for Galveston-Houston region for select return periods (top row in years) and for varous climate scenarios

Scenario	1000	500	250	100	50	25	10	5
Base	62,160	56,979	51,328	42,461	36,255	28,357	8337	0
Int lo w/o freq incr	1.1	0.9	0.7	2.6	2.8	4.4	62.9	–
Int lo w/ freq incr	4.0	3.7	3.2	6.5	7.1	10.4	105.4	–
Int hi w/o freq incr	1.4	1.1	1.0	**3.2**	3.3	5.0	70.9	–
Int hi w/ freq incr	5.1	6.1	6.1	**10.1**	10.8	17.9	145.6	–

Numbers for base indicate the numbr of inundated cells for current conditions. Bold numbers correspond to the changes shown in Fig. 8.14

impact of increased storm frequency combined with SLR is comparable to the impact from SLR alone. Table 8.3 shows a more complete summary in terms of the impacts to other RPs as well as for the Intermediate-Low SLR scenario. The contributions from increased storm frequency combined with SLR are obviously more significant at lower RP values, as whatever protective measures to prevent/inhibit coastal flooding are overcome. The numbers provide some insight into how losses increase because of changes in footprint size vs. increases in depth, but they do not tell the whole story.

In contrast to the small changes in footprint size, inundation heights increase on average by the amount of the SLR over large, already-inundated areas. Table 8.4 illustrates this more clearly. Maximum inundation heights are shown for different combinations of storm frequency and SLR scenarios. Especially for the higher return

Table 8.4 Maximum storm surge inundation heights (feet) for Galveston-Houston region for return periods (left column, years) and climate scenarios as shown

RP	Current SLR		Interm low		Interm high	
	w/o freq incr	w freq incr	w/o freq incr	w freq incr	w/o freq incr	w freq incr
5	0.00	4.32	6.03	5.50	6.03	6.83
10	5.46	5.95	7.59	7.13	7.59	8.53
25	8.13	9.18	10.26	10.63	10.26	12.86
50	12.71	13.54	14.90	15.21	14.90	16.44
100	16.10	17.51	18.38	18.69	18.38	20.92
250	22.61	24.16	24.74	25.34	24.74	27.08
500	26.97	28.57	29.10	29.75	29.10	33.85
1000	33.42	34.90	35.55	36.08	35.55	37.04

Current refers to current storm frequency, future refers to increased frequency for major category storms as descibed in the text

periods, augmentation of the maximum inundation height from a higher frequency of major storm activity is just as significant as the impact from SLR alone. For example, the 100-year RP maximum inundation height for the Intermediate-High SLR scenario is 2.28 feet higher than for the current SLR scenario without an increase in storm frequency. The inclusion of additional storm activity to the Intermediate-High SLR scenario adds another 2.54 feet to that maximum (although not necessarily in the same place).

A side note to the RP inundation footprints shown in Fig. 8.12 is the linear nature of the response (e.g., the maximum inundation height increases by the amount of SLR). A range of published literature shows the impact of SLR on storm surge in general to be somewhat nonlinear in either direction and dependent on several factors, including the magnitude of the SLR and coastal geometry to name but two (Lin et al. 2012; Zhang et al. 2013; Bilskie et al. 2016). That is, in some instances/locations, the inundation height because of SLR is greater than the sum of SLR and base storm surge while in other instances/locations, it is less than the sum.

Increases in storm frequency further increase the storm surge AALs by 72% and 123% for the Intermediate-Low and Intermediate-High SLR scenarios respectively as the SLR effectively amplifies the impact from frequency increase (e.g., 1.23 × 1.41 = ~ 1.72). The relative contributions of storm surge to the wind and surge AAL, however, changes very little from the SLR results without storm frequency increases, simply because the additional increases to surge loss are countered completely and even a little bit more by increases in wind loss. Recalling the behavior of storm surge loss relative to wind loss in Fig. 8.9, especially for the Gulf Coast Region, adding more major category strength storms does not change the relative contribution of storm surge loss to wind loss very much (there is a very slight decrease).

Impacts from SLR and increases in storm frequency to RP storm surge losses are comparable to or even greater than those for storm surge AAL – approaching a factor of two for the 25-year RP result. The RP results are shown in more detail in Grenier et al. (2020) and are not repeated here. The impacts on the percent contribution from surge for return period loss results are harder to assess given that a range of events needs to be evaluated and then the average contribution has to be calculated. The results at a county level are also difficult to extract from the software used to compute losses but can be inferred from the base sea level ones. For example, the counties with no storm surge contribution in Fig. 8.9 likely remain that way because they are land-locked, with surge contributions in other counties increasing approximately by one-third and two-thirds for the Intermediate-Low and Intermediate-High SLR scenarios. For the coastally adjacent counties of Brazoria, Chambers, and Galveston, which together contribute nearly one-third to the Houston Region AAL, the percent contributions could increase from roughly 30% to roughly 40% for the intermediate low SLR scenario.

Despite the significant contributions from storm surge to the Houston Region AAL, the bulk of the loss comes from wind (inland flood losses are significant but contribute only half of the wind total). From a mitigation standpoint, it is therefore beneficial to increase resilience to wind in addition to storm surge.

To that end it is worth considering how the RP windspeeds change for the region because of increases in storm frequency. Here, the region is defined by a 5-degree latitude-longitude box surrounding Galveston. Each of the 1000 sub-sampled catalogs for both the current and future climates will yield slightly different results within the box, so it is useful to examine the variability in the results from the sub-sampled catalogs, for both the current and future climates. Figure 8.14 shows how the distributions from the 1000 sub-sampled catalogs for the current and future climates differ for three different return periods. Each graph shows the number of sub-sampled catalogs (vertical axis) that yield a windspeed (horizontal axis) as the return period value. For example, the upper panel shows that for the current climate, the redrawn catalogs show that 2 catalogs have a 10-year return period windspeed value between 55 and 60 mph (not visible), 165 catalogs have a value between 60 and 65 mph, 778 catalogs have a value between 65 and 70 mph, and 55 catalogs have a value between 70 and 75 mph.

The 10-year RP windspeed distribution shows a nearly 5 mph shift from the current to the future climate while higher RP windspeed distributions show a lesser effect from current to future climate as indicated by the broader distributions, although for the 100-year RP, the most probable RP windspeed does shift by a 5-mph band. Additional RP windspeed analyses for the 5-degree × 5-degree region surrounding Galveston shows that major hurricane strength windspeed occurs for the current climate at around the 30-year RP. Because the future climate frequency is assumed to increase by ~20%, the return period for a major hurricane strength windspeed decreases to about 25 years (30/1.2). The information regarding how RP storm surge inundation heights may change because of SLR and storm activity (latter not shown here) and how RP windspeeds could change is useful for conducting more detailed cost-benefit analyses to determine optimal mitigation strategies including changes in building codes.

Fig. 8.14 Selected RP surface windspeed distributions for the Galveston-Houston, TX region for current and future climate. Region defined by 5 degree latitude-longitude-longitude box centered on Galveston, TX. Windspeeds are modeled one-minute sustained at 10 m AGL

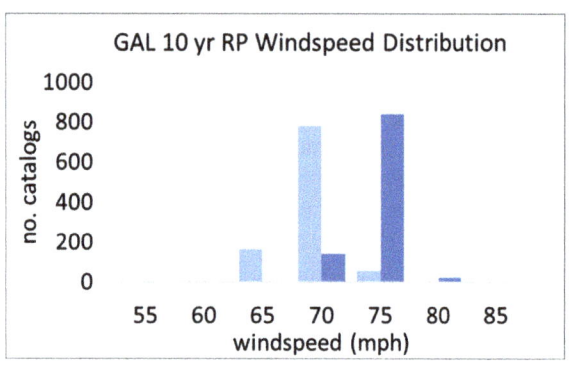

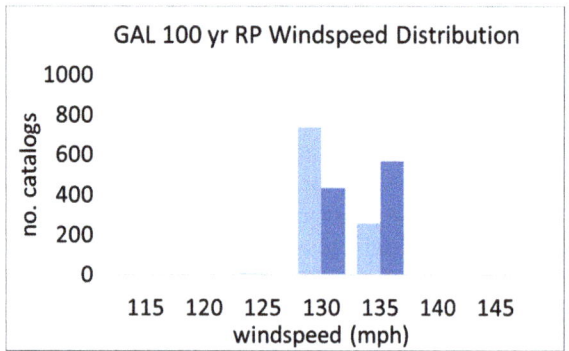

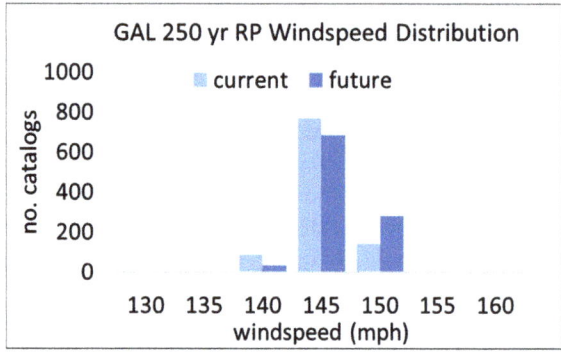

8.3.6 Miami Return Period Wind and Storm Surge Results

Although not quite as hurricane-prone as Galveston, the Miami region sees a high frequency of hurricane activity (14 landfalls within ~50 nmi of downtown since 1900). For Miami-Dade County, the contribution from storm surge to wind and surge losses, however, is slightly higher (32%) than for any coastal county in the Houston Region examined here, as shown in Fig. 8.15. The reasons for that include (a) Miami, unlike Houston, is located right at the coast, (b) it is an extensively built-

Fig. 8.15 Percent contribution (numbers) of storm surge to wind+surge AAL for Miami-Dade County, Florida for current climate

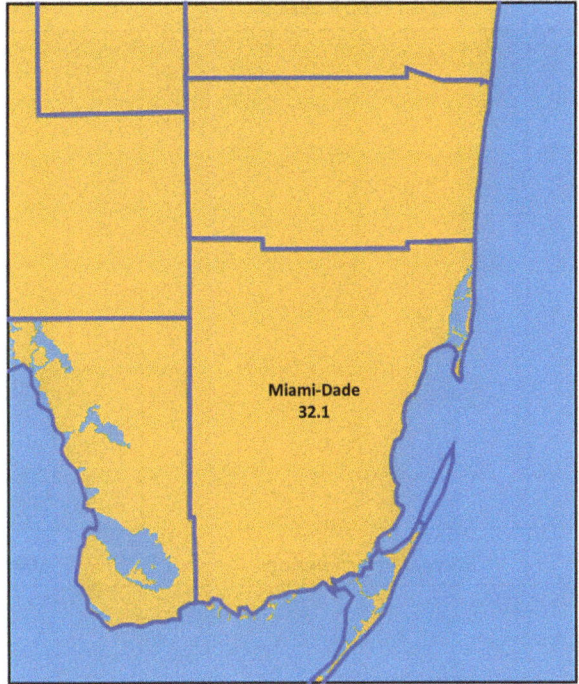

up region extending inland several tens of kilometers, and (c) the terrain is nearly flat with high water content nearby even under normal conditions because of the Everglades. Figure 8.16 shows selected return period inundation heights for the coastal portion of Miami-Dade County for current sea levels. It is useful to compare the results with the Galveston ones in Fig. 8.10. It is evident that despite the relatively flat terrain of the county as a whole, the stretch of coastal area from Miami northward is mostly spared, although that is not the case for the city of Miami Beach which essentially acts as a barrier island to Miami. South of Miami, even lower elevations (e.g., 2 m or less extending out to Interstate 1) and the presence of the Everglades south and west contribute significantly to the dramatic inland extent of surge. A comparison of Figs. 8.15 and 8.10 and Figs. 8.16 and 8.11 indicates that storm surge contributes more to loss despite the fact that the maximum storm surge inundation heights are lower at all return periods for Miami than for Galveston. While the lower RP storm surge heights may be related to the slightly lower frequency of activity for Miami and the low SLR projections (c.f. Table 8.1), it is more likely related to the steeper bathymetry and lesser degree of concavity of the southern Florida coastline. Not only is Galveston Bay a storm surge enhancing feature, but the entire coastline of Texas is conducive for that as well. The higher percent contribution for Miami could be related to the type of exposure in the county and requires a grid cell level analysis that is beyond the scope of this study.

The increases in storm surge loss from the two SLR scenarios, separately and combined with the increases in storm frequency, as well as the relative storm surge

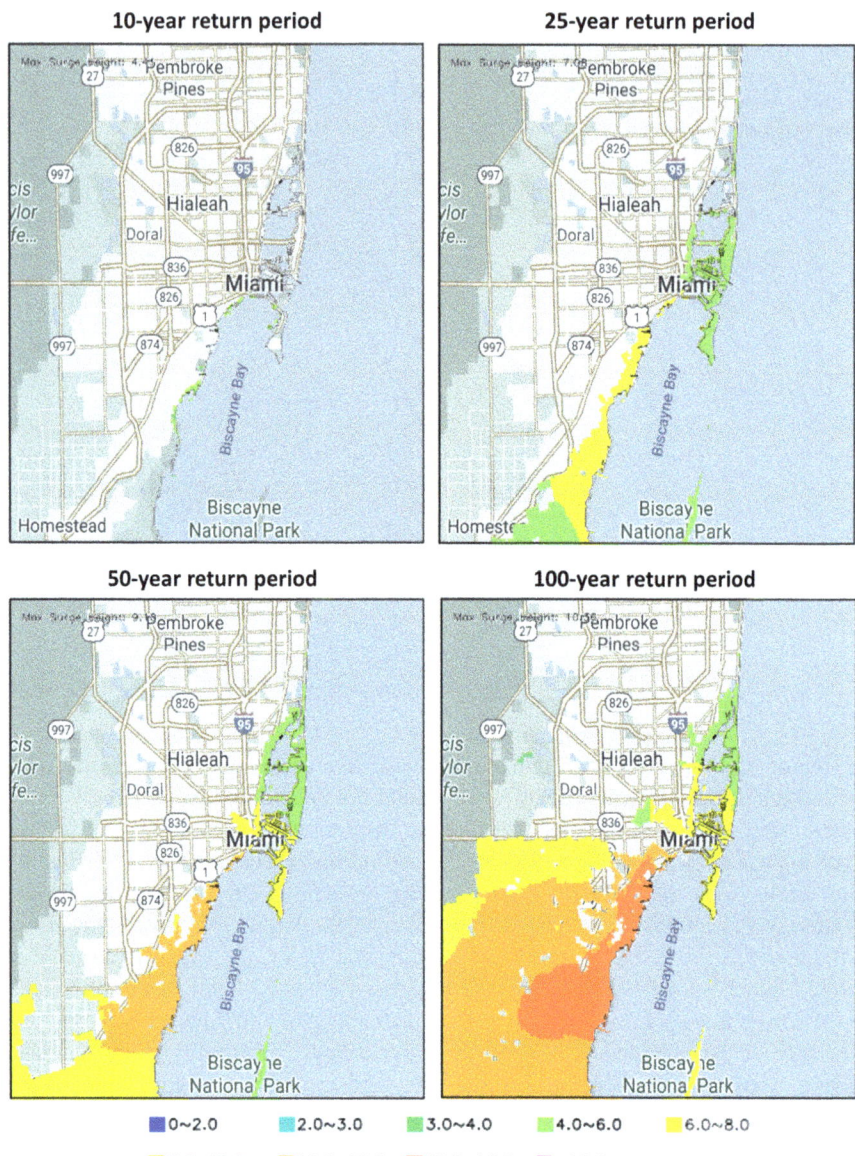

Fig. 8.16 Storm surge inundation height (feet) for Miami Region for return periods and current sea levels. Plotted domain spans 0.60° lat × 0.60° lon

loss compared to wind and surge loss are shown in Table 8.5. The increases in storm surge loss are comparable to those for Galveston, reaching 123% for the Intermediate-High SLR scenario combined with increases in storm activity. One difference is that for the Intermediate-Low SLR scenario, just the SLR by itself

8 Climate Change Impacts to Hurricane-Induced Wind and Storm Surge...

Table 8.5 Summary of impact of SLR and changes in storm frequency on storm surge loss for Miami Region

Scenario	Pct change in surge AAL	Pct contr surge to wind+surge AAL
Current climate	–	32.0
Incr freq	23	32.0
Int-lo SLR scenario	34	38.2
Int-lo SLR scenario + incr freq	64	36.7
Int-hi SLR scenario	83	46.0
Int-hi SLR scenario + incr freq	123	43.8

increases the storm surge loss by about 50% over that for storm activity; and with the 20% increase in storm activity the storm surge loss increase further doubles, resulting in a three-fold increase in storm surge loss from just an increase in storm activity. Recall from Table 8.1 that the projected SLR amounts for Miami are the lowest of the three regions examined in this study, but the increases in the relative contributions from storm surge loss to wind and surge loss are larger than what is shown in the results for Galveston and exceed 40% for the Intermediate-High SLR scenarios (both with and without increases in storm frequency). Finally, we note from Table 8.5 the impact or lack thereof of increased storm frequency on relative contribution of storm surge loss to wind and surge loss. Without SLR, the impact is imperceptible. The addition of SLR amplifies the effect so it is evident to some degree – more so than for the Galveston-Houston Region. The decreased contribution is the result of adding more category 3–5 storms that have smaller relative contributions than their weaker storm counterparts as shown in Fig. 8.9.

Figure 8.17 shows the RP storm surge inundation heights around Miami for select return periods for the two SLR scenarios with increased storm frequency, compared to those for current sea levels. Unlike the situation in Galveston, even the Intermediate-Low SLR scenario shows a dramatic increase in the size of the inundation footprint over that for current conditions, especially at the 50- and 100-year RPs, despite the fact that the sea level increase is the smallest for the three locations. Increase in depths (by virtue of the color changes) are also very evident. Differences in storm surge footprints for the Intermediate-Low scenario are shown in Fig. 8.18 for the 100-year RP, percent changes in footprint sizes are shown for other return periods, and other scenarios are shown in Table 8.6, and maximum storm surge heights for the scenarios at select return periods are shown in Table 8.7. The impacts of SLR and increased storm frequency are considerably larger than for the Galveston Houston Region despite the smaller increases in SLR, although the impacts of increased frequency appear to be less impactful. The increased footprint size from SLR can be attributed to the relative flatness of Miami. Despite the larger impacts on storm surge footprint size, it is difficult to comment on how storm surge losses increase because of changes in footprint size without a much more detailed analysis.

Figure 8.19 shows return period windspeed information for the Miami region in the same format as for Galveston in Fig. 8.14. Note that the return period windspeeds are higher than they are for Galveston (e.g., 10-year RP windspeed is 95 vs. 75 mph)

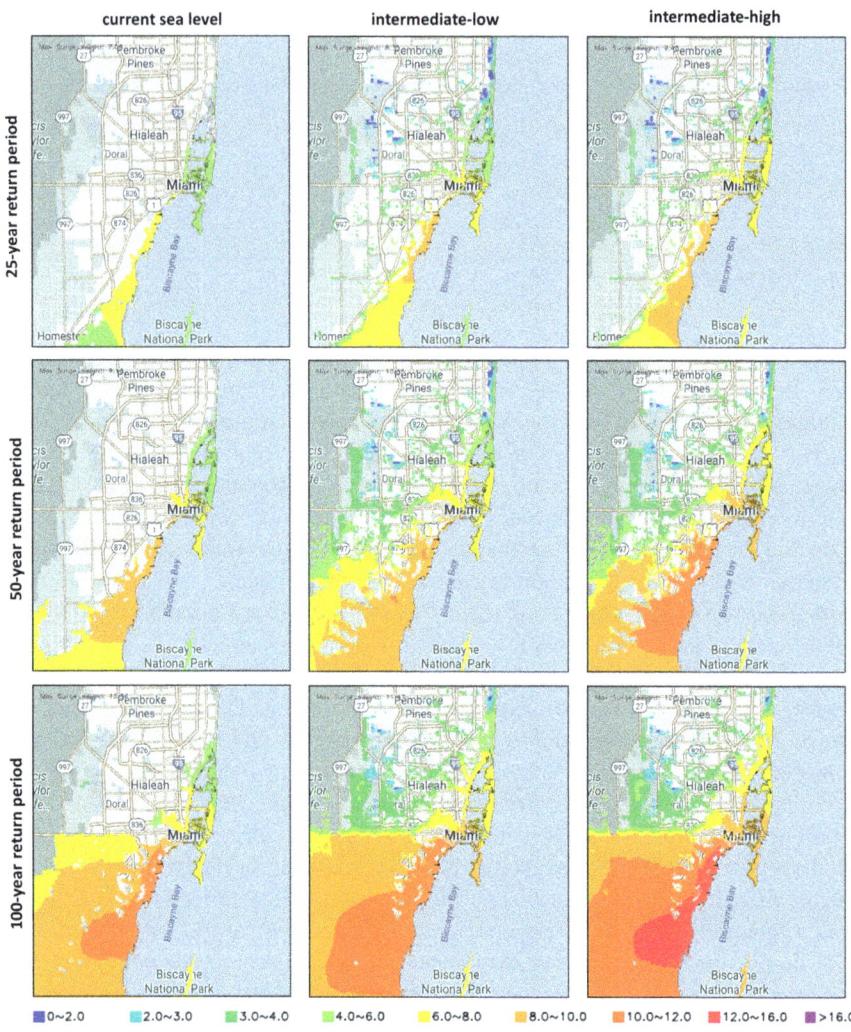

Fig. 8.17 Storm surge inundation heights (feet) for Miami Region for return periods and SLR scenrios as shown. Plotted domain spans 0.60° lat × 0.60° lon

despite the slightly lower storm frequency. Another difference is that there is less impact on the return period windspeeds from the increase in storm activity (note the most probable RP windspeed does not change at the 100- or 250-year RP). The lesser impact suggests that it is the exposure make up (e.g., buildings and contents) and, more specifically, the vulnerability of the buildings in Miami that cause a greater increase in wind loss than storm surge loss for an increase in storm activity.

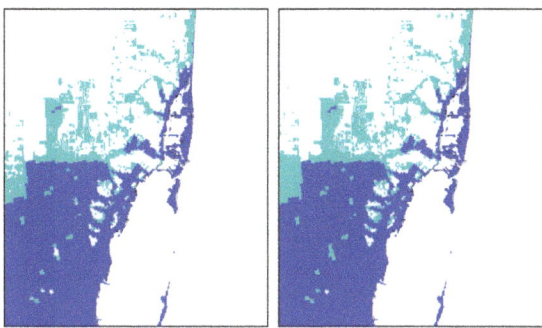

Fig. 8.18 Additional areas inundated (light blue) relative to current sea levels and storm frequency (dark blue) for Intermediate-Lo SLR scenario without frequency increases (left) and for Intermediate-Lo SLR scenario with frequency increases (right) for Miami region for 100 year RP. Base map removed for clarity. Plotted domain spans 0.60° lat × 0.60° lon

Table 8.6 Relative contribution to storm surge footprint area for Miami region for select return periods (top row in years) and for varous climate scenarios

Scenario	1000	500	250	100	50	25	10	5
Base	55,624	54,584	44,366	36,784	20,658	10,543	328	0
Int lo w/o freq incr	5.8	6.2	13.4	**16.2**	28.7	37.3	1062.5	–
Int lo w/ freq incr	6.3	7.0	20.2	**21.5**	54.2	59.3	1473.5	–
Int hi w/o freq incr	6.3	6.9	15.1	17.7	34.0	43.2	1223.5	–
Int hi w/ freq incr	6.9	7.5	21.6	22.8	59.2	65.2	1680.5	–

Numbers for base indicate the numbr of inundated cells for current conditions. Bold numbers correspond to the changes shown in Fig. 8.19

Table 8.7 Maximum storm surge inundation heights (feet) for Miami region for return periods (left column, years) and climate scenarios as shown

RP	Current SLR		Interm low		Interm high	
	w/o freq incr	w freq incr	w/o freq incr	w freq incr	w/o freq incr	w freq incr
5	0.00	0.00	0.63	0.63	1.67	1.67
10	4.41	4.83	5.03	5.45	6.09	6.51
25	7.08	7.72	7.70	8.35	8.76	9.40
50	9.10	9.42	9.73	10.05	10.77	11.10
100	10.36	10.84	10.98	11.47	12.03	12.52
250	11.89	12.34	12.52	12.86	13.56	13.91
500	12.83	13.32	13.45	13.94	14.51	14.99
1000	13.80	14.30	14.43	14.93	15.48	15.98

Current refers to current storm frequency, future refers to increased frequency for major category storms as descibed in the text

Fig. 8.19 Selected RP surface windspeed distributions for the Miami Region. Region defined by 5 degree latitude-longitude-longitude box centered on Miami, FL

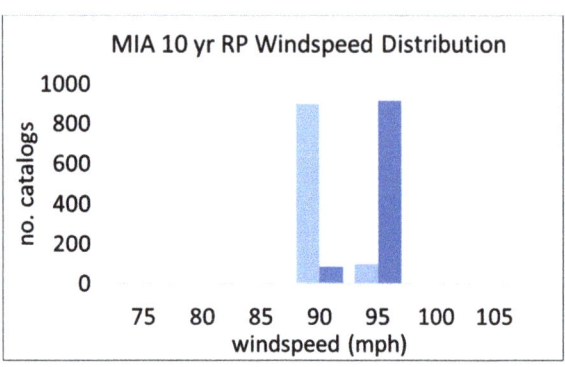

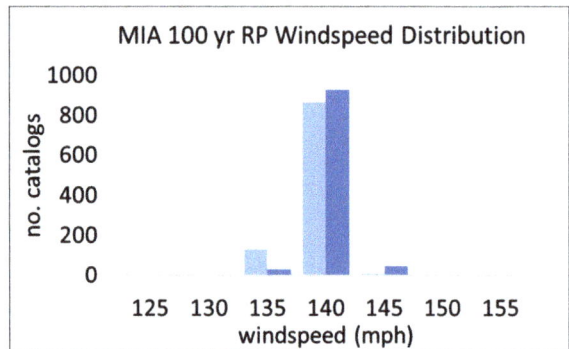

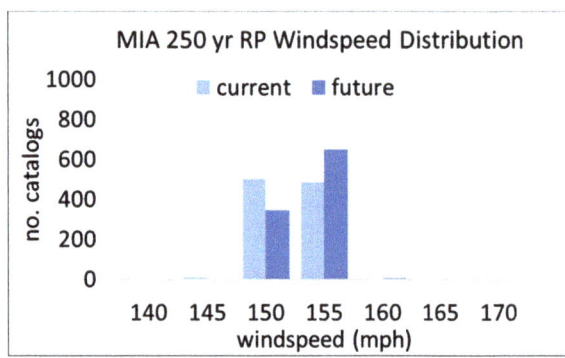

8.3.7 New York City Return Period Wind and Storm Surge Results

Only four hurricanes have made landfall in the New York City Region (50 nmi north or south of southern tip of New York County) since 1900, although many have passed by close enough to cause considerable damage (e.g., Hurricane Sandy in 2012). For this study, all stochastic storms that make landfall between southern Delaware and the northern tip of Long Island, as well as all loss-causing offshore

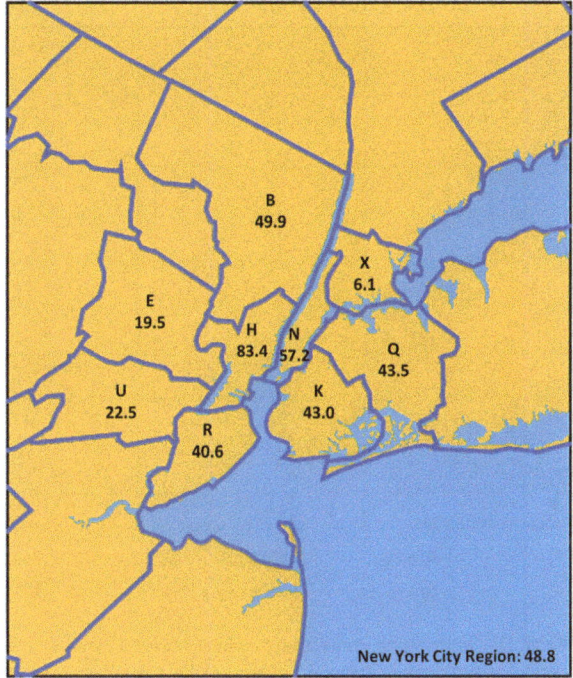

Fig. 8.20 Percent contribution (numbers) of storm surge to wind+surge AAL by county (indicated by first letter except for Bronx) for New York City Region including Bronx, New York, Queens, Kings, Richmond, Bergen, Hudson, Essex, and Union Counties for current climate. Nine county average for New York City Region is 48.8%

bypassing events, are included. Figure 8.20 shows the AIR Modeled result for the contribution of storm surge loss to wind and surge loss. Counties adjacent to the coast or to the Hudson River (where Bergen and Hudson Counties are to the west and New York and Bronx Counties are to the east) have half or more of the loss from storm surge (with the exception of Bronx). The region-average contribution to wind and surge AAL from storm surge is in fact nearly 50%. Like the case with Miami, this very high percentage is the result of exposure locations within the county, although that is where the similarity ends. The New York/New Jersey Coastline in the vicinity of New York City is highly intricate with concave bays, inlets, and narrow rivers nearby. The storm surge return period heights shown in Fig. 8.21 are not as high as what occurs around Miami or Galveston. In fact there is no storm surge footprint from hurricane activity at the 10- or 25- year return period, and storm surge heights are still below 8 feet at the Battery in Manhattan (southrn tip) at the 100-year return period. The high storm surge percentages (Fig. 8.20) come from the fact that wind speeds are also very low (as will be shown), and the buildings are built to withstand strong winds from other more frequent weather phenomena like severe thunderstorms and nor'easters (NYC Emergency Management 2014).

The percent changes in storm surge AAL are shown in Table 8.8 for the SLR scenarios with and without frequency increases. Sea level rise alone increases storm surge AAL by 34–103%.

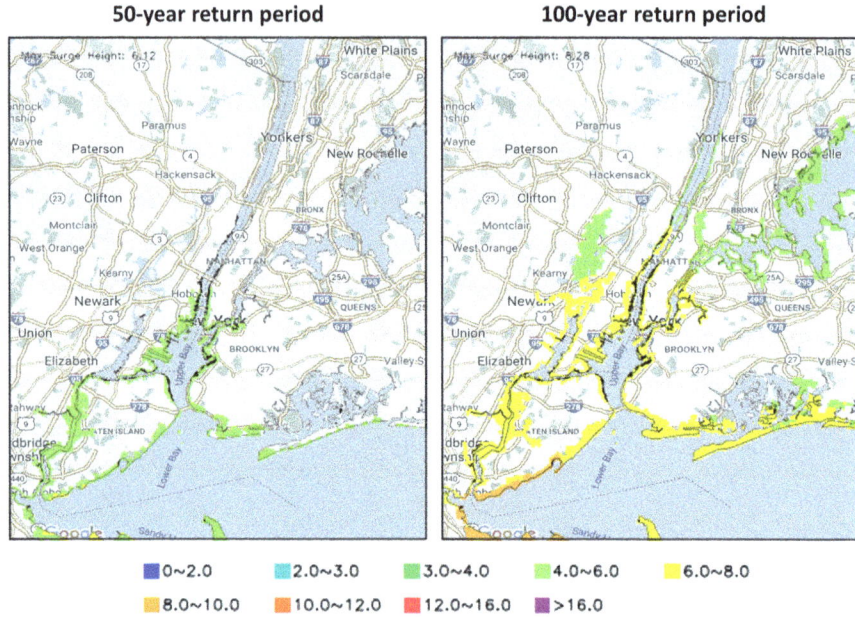

Fig. 8.21 Storm surge inundation height (feet) for Miami region for return periods and current sea levels. Plotted domain spans 0.72° lat × 0.72° lon

Table 8.8 Summary of impact of SLR and changes to storm frequency/intensity on storm surge loss for New York City Region

Scenario	Pct change in surge AAL	Pct contr surge to wind+surge AAL
Current climate	–	48.8
Incr freq	19	48.6
Int-lo SLR scenario	34	55.9
Int-lo SLR scenario + incr freq	64	51.4
Int-hi SLR scenario	103	65.8
Int-hi SLR scenario + incr freq	146	61.6

These are large numbers as well as a very large range, but the results are consistent with the high contributions from the current sea level result. The addition of increases in storm frequency boosts the range to 64–146%. The contribution from storm surge to the wind+surge AAL also increases with SLR: from 48.8% with no SLR to 56% for the Intermediate-Low SLR scenario to 66% for the Intermediate-High SLR scenario. The numbers decrease by several percent with the inclusion of increases in storm frequency. The decreases because of storm activity changes in fact are the largest of the three regions considered. Again, the decreases in relative storm surge contribution may be understood by recalling Fig. 8.9. Given that the northeast region curve is the most concave of all the regions shown, it is easier to see how an

8 Climate Change Impacts to Hurricane-Induced Wind and Storm Surge... 197

increased frequency of storms with relatively low storm surge contribution would lower the overall relative contribution of storm surge.

Additional understanding to changes in storm surge loss comes from the information in Figs. 8.22 and 8.23 and Tables 8.8 and 8.9. Figure 8.23 ilustrates the impacts of combined SLR and increased storm frequency for the Intermediate-Low and Intermediate-High SLR scenarios for the 50- and 100-year RP footprints. Note that at the 25-year RP, the region is still dry! At the 50-year RP, Manhattan remains dry even for the Intermediate-High SLR scenario. Figure 8.23 shows this more clearly – the increased flood in southern Manhattan is primarily the result of increased storm surge height. Although this is generally true for the region, increased flood area does exist in parts of coastal Long Island and northern New Jersey. Tables 8.9 and 8.10 indicate that significant increases in areal extent occur not only because of SLR but also because of increased storm frequency and have a similar impact on maximum storm surge inundation height.

Table 8.10, for example, shows that for the 100-year RP, the maximum surge height from just an increase in storm frequency is the same as that for just an increase in sea level from the Intermediate-Low scenario.

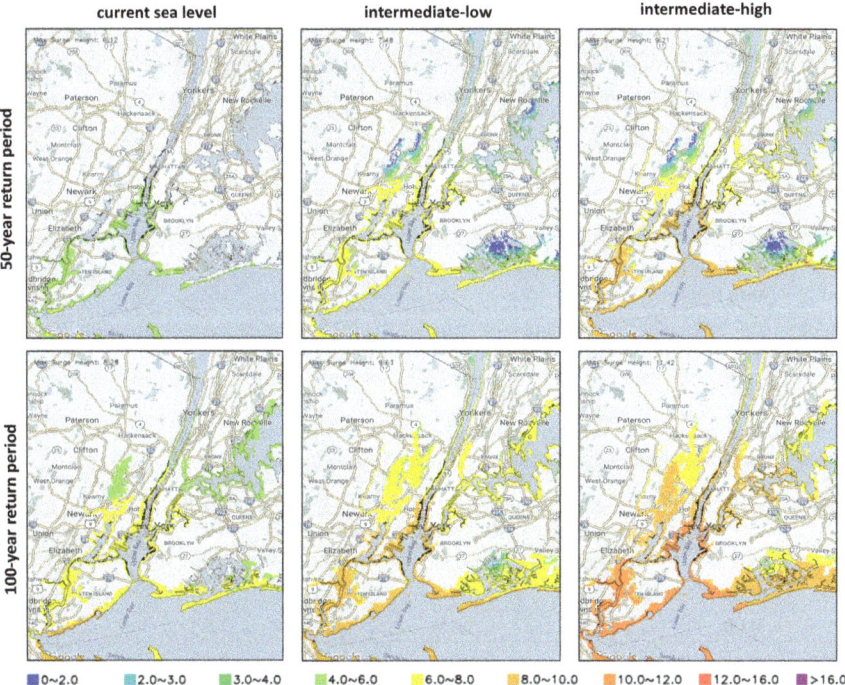

Fig. 8.22 Storm surge inundation heights (feet) for New York City Region for return periods and SLR scenrios as shown. Plotted domain spans 0.72° lat × 0.72° lon

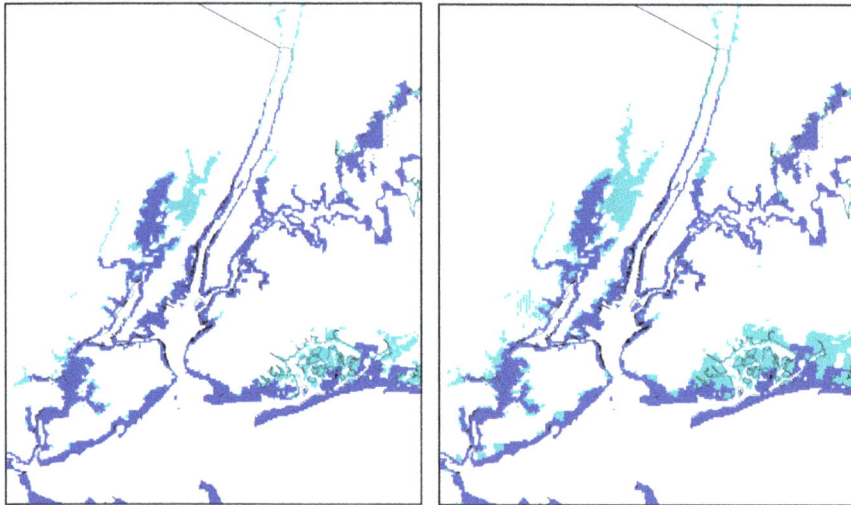

Fig. 8.23 Additional areas inundated (light blue) relative to current sea levels and storm frequency (dark blue) for Intermediate-Hi SLR scenario without frequency increases (left) and for Intermediate-Hi SLR scenario with frequency increases (right) for New York region for 100 year RP. Base map removed for clarity. Plotted domain spans 0.60° lat × 0.60° lon

Table 8.9 Relative contribution to storm surge footprint area for Miami region for select return periods (top row in years) and for varous climate scenarios

Scenario	1000	500	250	100	50	25	10	5
Base	19,138	17,994	16,105	10,277	4151	0	0	0
Int lo w/o freq incr	3.0	3.8	5.0	26.5	112.9	–	–	–
Int lo w/ freq incr	4.3	6.2	8.9	38.9	136.6	–	–	–
Int hi w/o freq incr	3.8	5.1	6.9	**29.7**	120.6	–	–	–
Int hi w/ freq incr	5.6	8.3	12.6	**49.2**	165.2	–	–	–

Numbers for base indicate the numbr of inundated cells for current conditions. Bold numbers correspond to the changes shown in Fig. 8.19

Table 8.10 Maximum storm surge inundation heights (feet) for Miami region for return periods (left column, years) and climate scenarios as shown

RP	Current SLR		Interm low		Interm high	
	w/o freq incr	w freq incr	w/o freq incr	w freq incr	w/o freq incr	w freq incr
5	0.00	0.00	0.00	0.00	0.00	0.00
10	0.00	0.00	0.69	0.69	2.03	2.03
25	0.00	0.00	0.69	0.69	2.03	2.03
50	6.12	6.78	6.81	7.48	8.15	9.21
100	8.28	8.94	8.97	9.63	10.31	11.42
250	10.86	11.74	11.55	12.43	12.89	13.93
500	12.67	13.48	13.36	14.17	14.70	15.81
1000	14.33	14.99	15.02	15.68	16.36	17.32

Current refers to current storm frequency, future refers to increased frequency for major category storms as descibed in the text

Fig. 8.24 Selected RP surface windspeed distributions for the New York City region. Region defined by 5 degree latitude-longitude-longitude box centered on Newy York City, NY

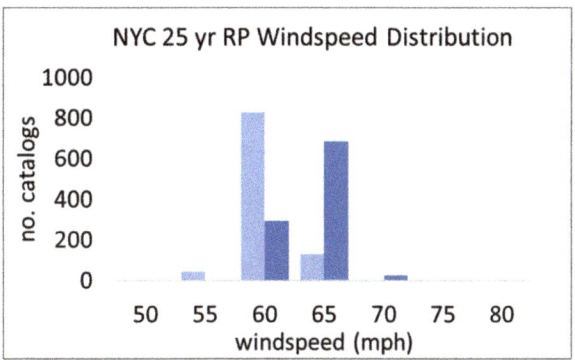

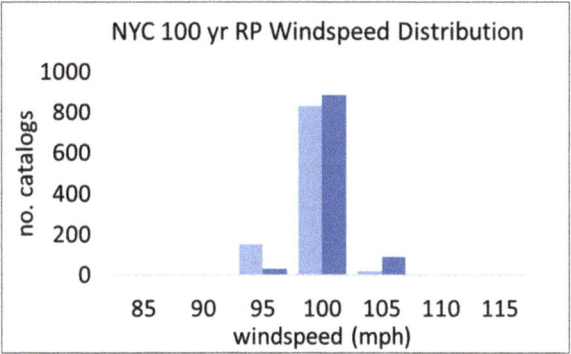

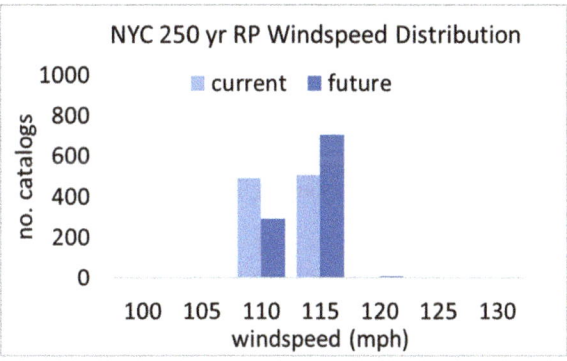

The New York City region experiences fewer and weaker storms than do the other two regions because of its northern location. Figure 8.24 shows return period windspeed information for the region obtained from this study. The 25-year RP windspeed result shows the most probable windspeed is below hurricane strength (74 mph) for the current climate and for the future climate scenario evaluated here will stay below hurricane strength. Higher RP windspeeds, beginning with the 50-year one (not shown), are above hurricane strength but will likely change less –

only by a couple miles per hour. It is notable that such small increases in windspeed are associated with a 20% increase in wind loss, but it is important to keep in mind that it is the increase in frequency that is really driving the increase – although, because of the way the sub-sampling is performed, the slight increases in intensity (from a return period perspective) are inherently tied to the frequency increase.

8.4 Discussion and Future Work

The loss results in the last section describe some of the potential impacts from climate change on hurricane activity; namely, that there could be a higher frequency of strong (i.e., major category strength) storms. Impacts of climate change on precipitation rate are not accounted for, although changes in precipitation-induced flooding do reflect increased storm frequency. Key Finding #1 from this study is that by mid-century, climate change could result in a 20% increase in U.S. hurricane loss from a 20% increase in Saffir Simpson category three or higher storms (where the categories are defined by landfalling central pressure.) While it seems somewhat coincidental that a twenty percent increase in the frequency of major category storms at landfall leads to an increase in expected loss twenty percent, the result can be explained by the fact that these strong storms contribute to the majority of the loss even though they only account for less than half of landfalling storm activity. It is also related to how the twenty percent increase in major storm frequency is achieved. Table 8.11 summarizes the contributions of various strength storms in terms of frequency and loss for the current climate in the AIR Hurricane Model, as well as for the future climate scenario used in this study.

The simplicity of the result for AAL is an artifact of the implied regional homogeneity for the climate change target used in this study. Moreover, it suggests that other changes to AAL from alternate changes to frequencies of categories can be estimated quickly without the need for generating entire climate change catalogs. For example, even though changes in weak category storms were not imposed as part of

Table 8.11 Summary of current and future climate landfall frequency as defined in this study and annual average loss contribution from all sub-perils for U.S. (percent contribution) by Saffir Simpson category defined in terms of landfalling central pressure

	Category 1	Category 2	Category 3	Category 4	Category 5	Total
Cp range (mb)	≥980	980–965	965–945	945–920	< 920	
Current climate						
Freq contr (%)	15.7	39.4	28.8	13.5	02.6	100.0
Loss contr (%)	01.2	14.9	31.9	38.9	13.1	100.0
Future climate						
Δ freq (%)	00.0	00.0	15.0	25.0	35.0	20.0
Loss contr (%)	01.2	14.9	36.7	48.6	17.7	119.1

For future climate landfall frequency changes are shown. Sea level rise is not included in this result

the climate change target, we can get a credible U.S.-wide estimate of the impacts using the information in Table 8.5 for an assumed reduction in weak storm frequency. A fifteen percent reduction in category 1 storms changes the 119.1% result to 118.9%. Importantly, the catalogs are necessary to evaluate changes in return period losses as well as other aspects of loss and hazard. The catalogs allow an evaluation of how specified changes in the frequencies of different intensities influence regional structure to the sub-peril losses. Finally, the process for creating the catalogs provides a framework by which to evaluate more complex future climate scenarios, e.g., from accounting for changes by region or accounting for changes in forward speed. While it is probably the case that the specific increase depends on how the twenty percent increase is achieved across the major strength storms, it is a straightforward exercise to obtain the range.

An equally important dimension of this study is that it provides information on changes in the relative contributions of storm surge loss from sea level rise. The result is not a simple one that can be accomplished in an excel-spread sheet, nor one that can even be addressed by sub-sampling. In fact, the analysis in this study required the creation of tens of thousands of new events with new storm surge footprints. Key Finding #2 is that sea level rise could likely contribute significantly to increased storm surge loss. Combined with increases from increased storm activity, storm surge loss could more than double, and the contribution of storm surge loss relative to loss from wind will also likely increase, but how much will depend on the amount of sea level rise relative to increases in storm frequency, as well as the geography of the region, and the resilience of the buildings and infrastructure to wind and water damage. Without additional coastal protection, adaptation, or retreat, rising sea levels will impact a larger proportion of land area, population, and global assets in the years ahead (Kirezci et al. 2020). This study has shown how storm surge footprints at different return periods may change because of SLR and because of increased storm frequency. That breakdown can allow a comparison of the relative impacts to loss from areas that already get flooded to those from new areas that get flooded because of SLR and increased storm frequency although it was not shown in this study. The information is of more than just academic interest because insurance companies want to know what new risks they can decide to take on or decline.

Significant utility stems from the second key finding and suggests how insurance underwriting practices may change, how optimal resilience strategies can be developed, and how and where money should be spent to mitigate future risk. Although not presented in this study, changes in the sizes and intensities of storm surge footprints for the different sea level rise scenarios can be evaluated quantitatively from output generated for this study to better inform decisions regarding increased sea-wall protection. Similarly, how to alter building codes to make buildings more resilient along the coast and inland and how much additional cost is involved (e.g., per building) to offset otherwise increased damage from increased storm activity is also information that can be obtained from further analyses.

In obtaining the results, several assumptions were made. The extent to which an extreme scenario like RCP 8.5 will be the emissions pathway that is followed at least through 2050 is perhaps at the top of the list (although the remainder of the list is not in any particular order of priority). Although it is sometimes referred to as a business-as-usual scenario (e.g., with no additional curbing of greenhouse emissions), it could unfortunately be achieved even with cutting greenhouse gasses if other catastrophic events occur such as collapsing Antarctic ice shelves still occur. That possibility, plus a conservative desire especially from the insurance industry to be prepared for the worst, is the reasoning behind our choice. How tropical cyclone activity responds to climate change is also uncertain. Despite the best science, it is still not known whether the total number of tropical cyclones per year will actually change and how. Although there has been general agreement that the total number of tropical cyclones will decrease, primarily because the number of weak storms is expected to decrease, some recent work has suggested that the total number of tropical cyclones may increase (Lee et al. 2020). Our choice of a climate change target is based on the general consensus that stronger storms will likely become more frequent (Hayhoe et al. 2018). While frequency changes for weaker storms cannot be ignored, we have shown that their contribution to total loss is relatively small, but that analysis assumes that the inland flood contribution from weaker storms does not change disproportionally – e.g., that weaker storms become wetter from slower forward speeds, overall size, etc. – in comparison to stronger storms. We have also made assumptions that relative landfall frequencies will not change, inland decay rates will not change (Li and Chakraborty 2020), storm tracks will not change, forward speeds, etc. will not change. Additionally, we did not account for potential changes in precipitation rate from climate change. While these may not be immaterial, there may be less confidence in how they will change relative to frequency or intensity. Lastly, we note potential limitation of the results from the subsampling approach used, that it is possible that future climate change may yield 10,000-year storms that are not contained within our 100,000-year catalog of current climate storms.

We plan to address potential changes in these other storm characteristics noted above and how they may affect loss in future studies. We also plan to complete the detailed sea level rise – storm surge analysis for the rest of the U.S. hurricane-affected coastline. The latter will include a more detailed evaluation of how footprint sizes change.

Acknowledgements We gratefully acknowledge the many comments and suggestions from two anonymous reviewers on an earlier draft of this manuscript.

References

AIR (2008) Climatological influences on hurricane activity: the AIR warm SST conditioned catalog. Available from: https://www.air-worldwide.com/SiteAssets/Publications/White-Papers/documents/Climatological-Influences-on-Hurricane-Activity%2D%2DThe-AIR-Warm-SST-Conditioned-Catalog

AIR (2021) The AIR hurricane model for the U.S. V1.0.0 as implemented in Touchstone® 2020: submitted in compliance with the 2019 standards of the Florida Commission on Hurricane Loss

Projection Methodology May 19, 2021. 452 pp. Available from The State Board of Administration, Tallahassee

Bacmeister JT, Reed KA, Hannay C et al (2018) Projected changes in tropical cyclone activity under future warming scenarios using a high-resolution climate model. Clim Change 146:547–560. https://doi.org/10.1007/s10584-016-1750-x

Bilskie MV, Hagen SC, Alizad K, Medeiros SC, Passeri DL, Needham HF, Cox A (2016) Dynamic simulation and numerical analysis of hurricane storm surge under sea level rise with geomorphologic changes along the northern Gulf of Mexico. Earth's Future 4:177–193. https://doi.org/10.1002/2015EF000347

Caesar L, McCarthy GD, Thornalley DJR, Cahill N, Rahmstorf S (2021) Current Atlantic Meridional Overturning Circulation weakest in last millennium. Nat Geosci 14:118–120

Camargo SJ (2013) Global and regional aspects of tropical cyclone activity in the CMIP5 models. J Clim 26:9880–9902. https://doi.org/10.1175/JCLI-D-12-00549.1

DeConto RM, Pollard D (2016) Contribution of Antarctica to past and future sea-level rise. Nature 531:591–597

Emanuel K (2021) Response of global tropical cyclone activity to increasing CO2: results from downscaling CMIP6 models. J Clim 34:57–70

Emanuel K, DesAutels C, Holloway C, Korty R (2004) Environmental control of tropical cyclone intensity. J Atmos Sci 61:843–858

Evans JL (1993) Sensitivity of Tropical Cyclone Intensity to Sea Surface Temperature. J Climate 6:1133–1140. https://doi.org/10.1175/1520-0442(1993)006<1133:SOTCIT>2.0.CO;2

Frederikse T, Landerer F, Caron L et al (2020) The causes of sea-level rise since 1900. Nature 584: 393–397. https://doi.org/10.1038/s41586-020-2591-3

Grenier R, Sousounis P, Schneyer J, Raizman D (2020) Quantifying the impact from climate change on U.S. hurricane risk. AIR Worldwide, 44 pp. Available from: How climate change could impact U.S. hurricane risk | Visualize | Verisk Analytics

Hall TM, Kossin JP (2019) Hurricane stalling along the North American coast and implications for rainfall. NPJ Clim Atmos Sci 2:17. https://doi.org/10.1038/s41612-019-0074-8

Harris DL (1957) The hurricane surge. Coast Eng Proc 1:5. https://doi.org/10.9753/icce.v6.5

Hayes C, Rowe J (2008) The AIR industry exposure databases. AIR Curr Oct 2008. https://www.air-worldwide.com/SiteAssets/Publications/AIR-Currents/attachments/AIR-Currents%2D%2DIED

Hayhoe K, Wuebbles DJ, Easterling DR, Fahey DW, Doherty S, Kossin JP, Sweet W, Vose R, Wehner M (2018) Our changing climate. In: Reidmiller DR, Avery CW, Easterling DR, Kunkel KE, Lewis KLM, Maycock TK, Stewart BC (eds) Impacts, risks, and adaptation in the United States: fourth national climate assessment, volume II. U.S. Global Change Research Program, Washington, DC, 72:144. https://doi.org/10.7930/NCA4.2018.CH

Hofer S, Lang C, Amory C et al (2020) Greater Greenland Ice Sheet contribution to global sea level rise in CMIP6. Nat Commun 11:6289. https://doi.org/10.1038/s41467-020-20011-8

IPCC (2013) Climate change 2013: the physical science basis. In: Stocker TF, Qin D, Plattner G-K, Tignor M, Allen SK, Boschung J, Nauels A, Xia Y, Bex V, Midgley PM (eds) Contribution of Working Group I to the fifth assessment report of the Intergovernmental Panel on Climate Change. Cambridge University Press, Cambridge and New York, p 1535. https://doi.org/10.1017/CBO9781107415324

Jelenianski CP, Chen J, Shaffer WA (1992) SLOSH: Sea, lake, and overland surges from hurricanes; NOAA technical report NWS 48; National Oceanic and Atmospheric Administration, U.S. Department of Commerce, Silver Spring, pp 1–71

Kirezci E, Young IR, Ranasinghe R, Muis S, Nicholls RJ, Lincke D, Hinkel J (2020) Projections of global-scale extreme sea levels and resulting episodic coastal flooding over the 21st century. Sci Rep 10:11629. https://doi.org/10.1038/s41598-020-67736-6

Knutson TR, Sirutis JJ, Vecchi GA, Garner S, Zhao M, Kim H-S, Bender M, Tuleya RE, Held IM, Villarini G (2013) Dynamical downscaling projections of twenty-first-century Atlantic hurricane activity: CMIP3 and CMIP5 model-based scenario. J Climate 26:6591–6617. https://doi.org/10.1175/JCLI-D-12-00539.1

Knutson TR, Sirutis JJ, Zhao M, Tuleya RE, Bender MA, Vecchi GA, Villarini G, Chavas D (2015) Global projections of intense tropical cyclone activity for the late twenty-first century from dynamical downscaling of CMIP5/RCP4.5 scenarios. J Climate 28. https://doi.org/10.1175/JCLI-D-15-0129.1

Knutson TR, Camargo SJ, Chan JCL, Emanuel K, Ho C, Kossin JP, Mohapatra M, Satoh M, Sugi M, Walsh K, Wu L (2019) Tropical cyclones and climate change assessment: part I. Detection and attribution. Bull Am Meteor Soc 100. https://doi.org/10.1175/BAMS-D-18-0189.1

Knutson TR, Camargo SJ, Chan JCL, Emanuel K, Ho C, Kossin JP, Mohapatra M, Satoh M, Sugi M, Walsh K, Wu L (2020) Tropical cyclones and climate change assessment: part II. Projected response to anthropogenic warming. Bull Am Meteor Soc 10. https://doi.org/10.1175/BAMS-D-18-0194.1

Kopp RE, Horton RM, Little CM, Mitrovica JX, Oppenheimer M, Rasmussen DJ, Strauss BH, Tebaldi C (2014) Probabilistic 21st and 22nd century sea-level projections at a global network of tide gauge sites. Earth's Future 2:287–306. https://doi.org/10.1002/2014EF000239

Kossin JP (2018) A global slowdown of tropical cyclone translation speed. Nature 558:104–107

Kossin JP, Emanuel KA, Vecchi GA (2014) The poleward migration of the location of tropical cyclone maximum intensity. Nature 509:349–352

Lee C-Y, Camargo SJ, Sobel AH, Tippett MK (2020) Statistical-dynamical downscaling projections of tropical cyclone activity in a warming climate: two diverging genesis scenarios. J Clim 33:4815–4834. https://doi.org/10.1175/JCLI-D-19-0452.1

Li L, Chakraborty P (2020) Slower decay of landfalling hurricanes in a warming world. Nature 587: 230–234. https://doi.org/10.1038/s41586-020-2867-7

Lin N, Emanuel K, Oppenheimer M, Vanmarcke E (2012) Physically based assessment of hurricane surge threat under climate change. Nat Clim Change 2462–2467. https://doi.org/10.1038/NCLIMATE1389

Little CM, Horton R, Kopp R et al (2015) Joint projections of US East Coast sea level and storm surge. Nat Clim Change 5:1114–1120

Liu M, Vecchi GA, Smith JA, Knutson TR (2019) Causes of large projected increases in hurricane precipitation rates with global warming. NPJ Clim Atmos Sci 2:38. https://doi.org/10.1038/s41612-019-0095-3

Lonfat M, Marks FD, Chen SS (2004) Precipitation distribution in tropical cyclones using the tropical rainfall measuring Mission (TRMM) microwave imager: a global perspective. Mon Wea Rev 132:1645–1660. https://doi.org/10.1175/1520-0493(2004)132<1645:PDITCU>2.0.CO;2

McInnes KL, Macadam I, Hubbert G, O'Grady J (2013) An assessment of current and future vulnerability to coastal inundation due to sea-level extremes in Victoria, southeast Australia. Int'l J Climatol 33:33–47

NAIC (2017) CIPR study: flood risk and insurance. Available online: https://content.naic.org/sites/default/files/inline-files/cipr_study_1704_flood_risk.pdf

NOAA Tides and Currents (2021). https://tidesandcurrents.noaa.gov/sltrends/sltrends.html

NYC Emergency Management (2014) New York City's risk landscape: a guide to hazard mitigation, 172 pp. Available online: nycs_risk_landscape_a_guide_to_hazard_mitigation_final.pdf

Pielke RA, Gratz J, Landsea CW, Collins D, Saunders MA, Musulin R (2008) Normalized hurricane damages in the United States: 1900–2005. Nat Hazards Rev 9:29–42

Roberts MJ, Camp J, Seddon J, Vidale PL, Hodges K, Vannière B, Mecking J, Haarsma R, Bellucci A, Scoccimarro E, Caron L-P, Chauvin F, Terray L, Valcke S, Moine M-P, Putrasahan D, Roberts CD, Senan R, Zarzycki C, Ullrich P, Yamada Y, Mizuta R, Kodama C, Fu D, Zhang Q, Danabasoglu G, Rosenbloom N, Wang H, Wu L (2020) Projected future changes in tropical cyclones using the CMIP6 HighResMIP multimodel ensemble. Geophys Res Lett 47:e2020GL088662. https://doi.org/10.1029/2020GL088662

Robinson E, Cipullo M, Sousounis P, Kafali C, Latchman S, Higgs S, Maisey P, Mitchell L (2017) UK windstorms and climate change. An Update to ABI Research Paper No. 19, 2009. pp 18. Available online: https://www.abi.org.uk/globalassets/files/publications/public/property/2017/abi_final_report.pdf

Schwalm CR, Glendon S, Duffy PB (2020) RCP8.5 tracks cumulative CO_2 emissions. Proc Natl Acad Sci 117:19656–19657

Sealya K, Strobl E (2017) A hurricane loss risk assessment of coastal properties in the Caribbean: Evidence from the Bahamas. Ocean Coast Manag 149:42–51

Sweet WV, Kopp RE, Weaver CP, Obeysekera J, Horton RM, Thieler ER, Zervas C (2017) Global and regional sea level rise scenarios for the United States. NOAA technical report NOS CO-OPS 083. NOAA/NOS Center for Operational Oceanographic Products and Services

Ting M, Kossin JP, Camargo SJ, et al. (2019) Past and future hurricane intensity change along the U.S. East Coast Sci Rep 9:7795. https://doi.org/10.1038/s41598-019-44252-w

USACE (2020) Economic guidance memorandum (EGM) 01-03, generic depth damage relationships. U.S. Army Corps of Engineers Memorandum, CECWPG4, Washington, DC

Villarini G, Vecchi GA (2013) Projected increases in North Atlantic tropical cyclone intensity from CMIP5 models. J Climate 26:3231–3240

Zhang K, Yi L, Liu H, Xu H, Shen J (2013) Comparison of three methods for estimating the sea level rise effect on storm surge flooding. Climatic Change 118:487–500. https://doi.org/10.1007/s10584-012-0645-8

Open Access This chapter is licensed under the terms of the Creative Commons Attribution 4.0 International License (http://creativecommons.org/licenses/by/4.0/), which permits use, sharing, adaptation, distribution and reproduction in any medium or format, as long as you give appropriate credit to the original author(s) and the source, provide a link to the Creative Commons license and indicate if changes were made.

The images or other third party material in this chapter are included in the chapter's Creative Commons license, unless indicated otherwise in a credit line to the material. If material is not included in the chapter's Creative Commons license and your intended use is not permitted by statutory regulation or exceeds the permitted use, you will need to obtain permission directly from the copyright holder.

Chapter 9
Downward Counterfactual Analysis in Insurance Tropical Cyclone Models: A Miami Case Study

Cameron J. Rye and Jessica A. Boyd

Abstract The insurance industry uses catastrophe models to assess and manage the risk from natural disasters such as tropical cyclones, floods, and wildfires. However, despite being designed to consider a credible range of future events, catastrophe models are ultimately calibrated on historical experience. This means that unexpected things can happen, either because risks that were overlooked or deemed immaterial turn out to be meaningful, or because black swans occur that scientists and insurers were not yet aware of. When faced with these types of extreme uncertainty, insurers can use downward counterfactual analysis to explore how historical events could have had more severe consequences (and help identify previously unknown or overlooked risks). In this chapter, we present a methodology for insurers to operationalise downward counterfactuals using tropical cyclone catastrophe models. The methodology is applied to three recent major hurricanes that were near misses for Miami—Matthew (2016), Irma (2017), and Dorian (2019). The results reveal downward counterfactuals that produce insured losses many times greater than what transpired, at up to 300x greater for Matthew, 25x for Irma, and 250x for Dorian. We argue that it is increasingly important for insurers to examine such near-miss events in a changing climate, particularly in disaster prone regions, like Miami, that might not have seen a large loss in recent years. By operationalising downward counterfactuals, insurers can increase risk awareness, stress-test risk management frameworks, and inform decision-making.

Keywords Insurance · Catastrophe · Modelling · Counterfactual · Hurricanes

---

C. J. Rye (✉) · J. A. Boyd
MS Amlin, The Leadenhall Building, London, UK
e-mail: Cameron.Rye@msamlin.com; Jess.Boyd@msamlin.com

© The Author(s) 2022
J. M. Collins, J. M. Done (eds.), *Hurricane Risk in a Changing Climate*,
Hurricane Risk 2, https://doi.org/10.1007/978-3-031-08568-0_9

9.1 Introduction

Risk management practices, such as those in the insurance industry, are often strongly shaped by historical events. For example, in the 1980s and 1990s, insurers were surprised when a series of large natural disasters struck the United States, Europe, and Japan in close succession (tropical cyclones Hugo in 1989, Mireille in 1991, and Andrew in 1992; European windstorms 87J in 1987 and Daria in 1990; and the Kobe earthquake in 1995). The outcome was several reinsurance firms filing for bankruptcy and an increased demand for detailed physically-based catastrophe models for managing the risk from natural disasters (Grossi and Kunreuther 2005; Jones et al. 2017). Today, catastrophe models form an integral component of insurance risk management frameworks in a number of countries and are frequently updated to reflect lessons learnt from new disasters.

However, past experience does not always fully prepare us for the future. Despite being designed to consider a credible range of future events, catastrophe models are ultimately calibrated on historical experience (Lin et al. 2020). This means that unexpected things can happen, either because black swans (unpredictable or unforeseen events) occur that scientists and insurers were not yet aware of (Taleb 2007), or because risks that were overlooked or deemed immaterial turn out to be meaningful. For example, Hurricane Katrina made landfall in New Orleans in 2005, resulting in US$65 billion (2005 dollars) in insurance claims, making it the most expensive natural catastrophe for the global insurance industry to date (Swiss Re 2020). The severity of the disaster was in a large part due to a storm surge of up to 20 ft, which led to the failure of levees and flooding of 80% of the city (Knabb et al. 2005). This event was not a black swan—the historical record contains several instances of levee failures (e.g. Dunbar et al. 1999). But catastrophe models used at the time overlooked the risk as they did not consider the prospect of levee failure. The possibility of significant flood damage in New Orleans (and the subsequent displacement of the city's inhabitants) had therefore not been considered by most insurers.

When faced with these types of extreme uncertainty, insurers can use counter-factual analysis to explore alternative histories (e.g. Woo et al. 2017; Woo 2018, 2019; Lin et al. 2020). In particular, *downward* counterfactual thinking provides a framework for considering how historical events could have had more severe consequences, with a view to identifying disasters (such as Katrina) before they occur. For example, an insurance firm may investigate how near-miss weather events—which are only footnotes in the historical record—could have led to large economic losses had they turned out slightly differently. A multitude of disasters is theoretically possible because history represents just a single realisation of the underlying climatic variability; alternative realisations could have led to different outcomes and different ex-post decisions being made. In this way, lateral thinking using downward counterfactuals can help with the identification of previously unknown or overlooked risks, which are not fully visible in the historical record, and may not be adequately represented in existing catastrophe models.

One application of downward counterfactual thinking that has yet to be fully explored by the insurance industry is climate change. The current generation of insurance catastrophe models are built and calibrated with historical hazard and loss data, so they reflect the recent past rather than the present or future (Golnaraghi et al. 2018). Given this limitation, insurers have turned to scenario analysis—often using probabilistic climate model projections—to explore how future changes in the frequency and/or severity of extreme weather events could impact financial losses (e.g. PRA 2019; CISL 2020; Rye et al. 2021). However, uncertainties in predicting future weather extremes at the regional scale mean that such scenarios often hinder rather than support decision-making (e.g. Fiedler et al. 2021). Downward counterfactual thinking can provide insurers with an alternative approach that focuses on individual events without being burdened by the uncertainties that come with weather and climate prediction. Thinking in terms of events is beneficial because it is more in-line with how humans are known to perceive and respond to risk (Shepherd et al. 2018). The practicality of an event-oriented approach for climate change decision-making has been demonstrated through event attribution studies, which aim to assess the effect of climate change on individual historical catastrophes (e.g. Schwab et al. 2017). But unlike downward counterfactuals, attribution investigations tend to focus on high-impact historical events such as Hurricane Harvey in 2017 (Van Oldenborgh et al. 2017), while low-impact or near-miss events are largely ignored.

We argue that in a changing climate it is increasingly important for insurers to examine near-miss events and contemplate what could have been. Focus should be placed on areas that are particularly prone to disasters, such as Miami, but might not have seen a large loss in recent years so may now have a different risk profile due to factors such as urban growth and sea level rise. As a result, people may not be fully aware of the potential risk, since we know from behavioural science that humans have cognitive biases that mean they tend to emphasise the importance of historical experience (or the lack thereof) in estimating future events (Kahneman 2011). Although catastrophe models simulate a wide range of natural disasters, the emphasis is mostly placed on loss probabilities rather than on specific event outcomes, which means cognitive biases can still exist despite the use of these models. It should be noted that deterministic catastrophe scenarios are often used in the insurance industry for regulatory stress-testing (e.g. Lloyd's Realistic Disaster Scenarios, see Sect. 7.3), but these focus on a limited number of hypothetical events and are not directly related to historical disasters in the same way that downward counterfactuals are.

In this chapter we consider Miami, Florida, as a case study because the region has not seen a major hurricane landfall since Andrew in 1992. The Miami metropolitan area has experienced substantial urban development over the last 30 years, and with much of the land near sea level, there are concerns for Miami's resilience under a changing climate (e.g. Tompkins and Deconcini 2014). Insurance claims for Andrew in 1992 totalled US$15.5 billion and a reoccurrence of the storm today would result in an insured loss in the region of US$50-60 billion (Swiss Re 2020). We present a methodology for insurers to operationalise downward counterfactual analysis using tropical cyclone catastrophe models. This is demonstrated for three recent major hurricanes—Matthew (2016), Irma (2017), and Dorian (2019)—which were all, at one point in time, forecast to strike Miami and produce significant economic damages. Fortunately, the actual storm tracks were more favourable to Miami, which escaped the worst outcomes. We do not attempt to quantify the role of climate change in these events—that is best left to event attribution scientists (Allen 2003). Instead, our aim is to demonstrate how insurers can use downward counterfactual analysis as a tool for managing risk in a changing climate, especially in situations where cognitive biases may exist.

For insurers to operationalise counterfactual analysis, a pragmatic solution is required that facilitates decision-making. For this reason, we adopt a 'storyline' approach (Shepherd et al. 2018) which focuses on understanding event outcomes, not event likelihoods. A storyline can be defined as "a physically self-consistent unfolding of past events, or of plausible future events" (Shepherd et al. 2018). Storylines can be viewed as conditional scenarios that aim to understand the consequences of an event or situation, assuming it has occurred. For example, after identifying a downward counterfactual for Hurricane Matthew, a storyline could be developed to consider the business implications of the event (e.g. solvency) and identify risk management actions that could improve future resilience. The overall outcome is a set of deterministic scenarios (storylines) that can be used by insurers to increase risk awareness, stress-test risk management frameworks, and inform decision-making.

9.2 Catastrophe Modelling

A "natural catastrophe" can be broadly defined as an extreme event resulting from a natural process—such as a tropical cyclone or earthquake—that exceeds the capability of those affected to manage the consequences. Catastrophe models are tools designed for the insurance industry (but also increasingly used in other domains such as the public sector) to quantify the financial risks arising from such events (Jones et al. 2017). They simulate the frequency, severity, and location of natural disasters over a specified time period—usually 100,000 years—with each modelled year representing a possible realisation of "next year". This is achieved by considering the interactions between four core components (Fig. 9.1):

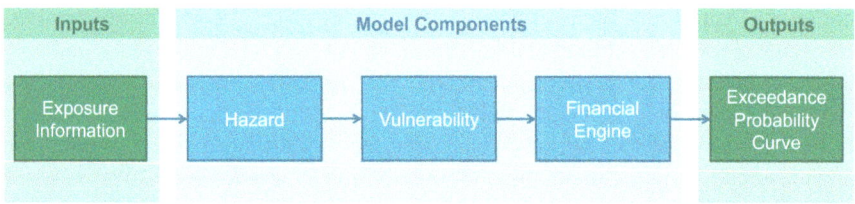

Fig. 9.1 The main components, inputs and outputs of a catastrophe model

- **Exposure.** The primary input to a catastrophe model is information on the assets (exposure) being insured. The data typically comprise detail on the location, type, and physical characteristics—such as construction and year built—of each asset, along with information about the insurance terms and conditions such as deductibles.
- **Hazard.** The hazard module comprises a stochastic "event set" (e.g. Hall and Jewson 2007), which represents a wide range of plausible events, from small events which have minimal impacts to major disasters that cause widespread damage over entire regions. For each event, a hazard footprint is created, which provides information on the intensity (e.g. flood depth or wind speed) at each point within the affected area.
- **Vulnerability.** Vulnerability models (known as vulnerability curves) are used to convert between hazard intensity and physical damage (e.g. Khanduri and Morrow 2003). These curves are often built using historical claims data, engineering principles, and expert judgement. In most catastrophe models, damageability varies depending on exposure characteristics such as construction type, occupancy, and year of construction.
- **Financial Loss.** A financial engine is used to translate physical damage into a monetary loss. This accounts for the value of insured assets as well as any insurance terms and conditions. The primary output of a catastrophe model is an exceedance probability (EP) curve, which provides the insurers with the annual probability of exceeding certain levels of loss (Fig. 9.2).

Catastrophe models simulate a wide range of physically plausible events that have not been observed in history, which enables insurers to undertake a comprehensive analysis of the risks they face from natural disasters. However, an over-reliance on models can lead to gaps in the assessment of risk. This is because surprise events can occur that are not adequately represented in catastrophe models (e.g. Hurricane Katrina), or represented in the models but dismissed as unlikely due to cognitive biases that place more weight on historical experience (Kahneman 2011; Shepherd et al. 2018).

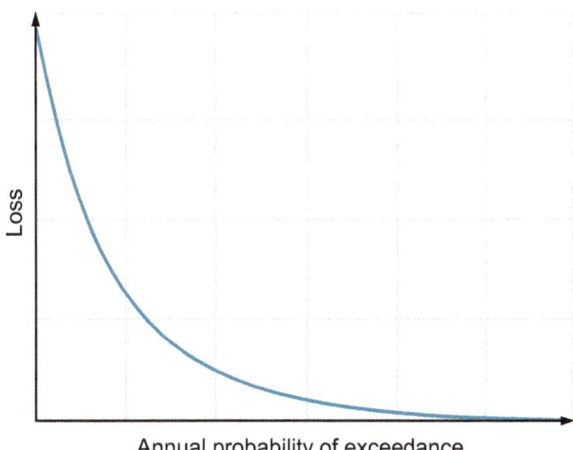

Fig. 9.2 A schematic of an exceedance probability curve, which is the primary output of a catastrophe model and provides an annual probability of exceeding certain levels of loss

9.3 Counterfactual Disaster Risk Analysis

A counterfactual is a "what if" exercise designed to explore hypothetical alternatives to historical events by modifying them in some way (Woo et al. 2017). For example, "what if national governments had acted sooner to stop the spread of the COVID-19 global pandemic?" (Born et al. 2021). This is an example of *upward* counterfactual thinking, which considers how things could have turned out for the better with the benefit of hindsight. But according to experts in psychology, it is much less common to consider the antithetical scenario that involves *downward* counterfactual thinking to explore how an outcome could have had more severe consequences (Roese 1997)—"in what ways could the pandemic have been made worse?" This is because mitigating actions are often only taken in direct response to disasters that have actually occurred, rather than in response to what might have been (Shepherd et al. 2018).

Insurers often adjust risk management practices after large natural disasters. For instance, in 2011, extensive flooding in Thailand shut down manufacturing production, impacting global supply chains and resulting in US\$12 billion in insured losses at the time (Lloyd's 2012). This event led to many insurance firms improving their management of flood exposures outside of the United States and Europe, which were generally not modelled (and often poorly monitored) at the time. The advantage of downward counterfactual thinking is that it can improve resilience by providing foresight on risks that fall outside of realm of current expectations. Similar meteorological conditions that led to the 2011 Thai floods had occurred before in 1995, and therefore, a downward counterfactual analysis could have foreseen the risk (Woo et al. 2017).

Downward counterfactual analysis is ultimately a lateral thinking exercise (De Bono 1977) that involves exploring the phase space of a disaster—the 'space' in which all possible outcomes are represented, with each outcome corresponding to a unique point in the phase space. Searching the disaster phase space for counterfactuals can be considered analogous to traditional numerical methods used to find the minimum or maximum of an objective function (e.g. Nelder and Mead 1965). This involves producing a trajectory of system perturbations along a downward path of increasing impact relative to the original historical event (Woo 2019, 2021). The search is terminated when further iterations no longer lead to new events with worse outcomes, or the computational requirements are prohibitive. Figure 9.3 shows a schematic of the phase space of a disaster. In this simple example, the historic disaster is shown by an asterisk, and the characteristics of the disaster that can be varied are the landfall location (distance along the coast, x-axis) and hurricane intensity (y-axis). If the historic event were to make landfall at the same intensity but closer to an area of high population density (moved rightwards towards the dashed line in the centre of this figure), the resulting loss severity could be higher. Similarly, if the hurricane intensity were to increase but the landfall location remained the same (moved upwards in this figure), the loss could also increase.

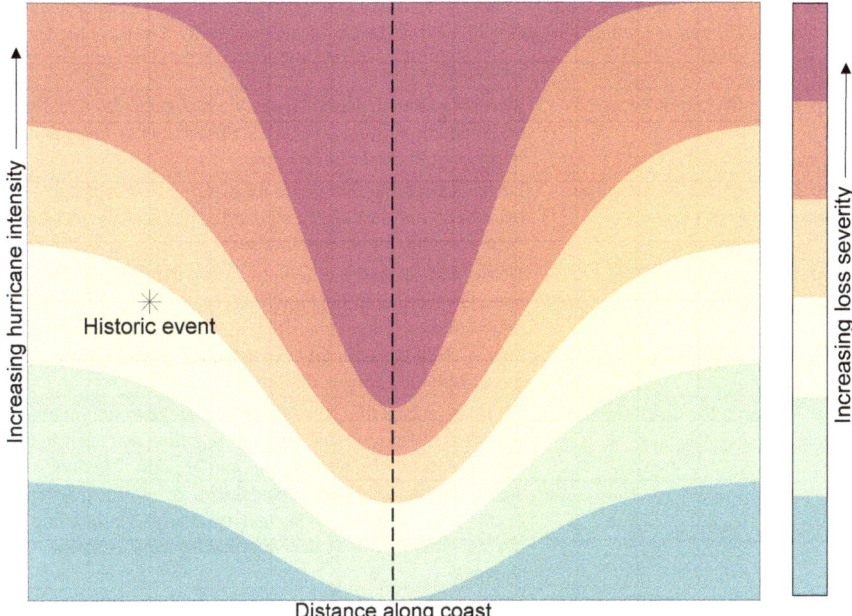

Fig. 9.3 A schematic of the phase space of a historical disaster, where perturbations of the characteristics can lead to more or less severe consequences. The dashed line represents a landfall on a major population centre

Exploring the phase space of a natural disaster and quantifying the financial impacts of each outcome can be a time-consuming exercise. For this reason, Lin et al. (2020) have proposed a guiding framework that sets out the conditions of the search (Table 9.1). The framework involves identifying a past event which may or may not have caused catastrophic damage, and then perturbing some of the event parameters to explore small changes that could result in worse consequences. Perturbations can be applied to a wide range of different parameters such as geography, hazard, exposure, compound risks, and socio-economic conditions. The more parameters that are perturbed, the larger the search space, and the greater the resources required for the downward thinking exercise. The search continues until one or more pre-defined end-of-search criteria are reached. Both the parameter perturbations and end-of-search criteria should be defined such that the final set of scenarios are physically plausible.

9.4 Matthew, Irma, and Dorian

We explore unrealised downward counterfactuals for three recent hurricanes—Matthew (2016), Irma (2017), and Dorian (2019). These storms provide an interesting case study because historical wind swathes show that Miami was spared

Table 9.1 The six step framework for identifying downward counterfactual events as defined by Lin et al. (2020) that is used as a basis for the counterfactual analysis presented in this chapter

Step	Description
Step 1: Identify a past event	Identify a factual, historical event which provides a realistic and relatable starting point. Describe, model or estimate the impacts from the original event.
Step 2: Define the disaster phase space	Define acceptable changes to the historical event parameters in order to ensure that the counterfactual analysis remains both plausible and computationally feasible.
Step 3: Define an end-of-search criteria	Define end-of-search criteria to ensure that the search does not continue indefinitely and that the resulting counterfactual scenarios remain plausible. In some cases, the modeller may wish to consider all counterfactual possibilities within the acceptable changes defined in Step 2.
Step 4: Search the disaster phase space	Apply an acceptable counterfactual change to the input historical event to reveal an event that does not exist in the historic record. The changes that can be applied include, but are not limited to: a geographical shift in hazard, cascading events (e.g. triggering of secondary hazards), coinciding events, human error or decision-making, and exposure changes.
Step 5: Compare to the historic consequence	Compare the counterfactual consequence to the historic outcome to assess whether the potential outcome is worse or better than the actual outcome.
Step 6: Criteria to continue or end counterfactual search	If the end-of-search criteria is met, then the counterfactual search ends. If it has not been met, Steps 4-6 are repeated until the search is complete.

hurricane strength winds in all three events (Fig. 9.4). Table 9.2 shows the insured losses incurred from each of the three events; note that while these events (particularly Irma) produced significant insured losses in the United States, for the purposes of our study they are physical near-misses for Miami. It is not hard to conceive of counterfactual realisations in which all three storms had worse outcomes from small changes to the hurricanes' paths. Given the lack of recent large loss experience in Miami, as well as the expected impacts of climate change on sea levels (e.g. Wdowinski et al. 2016) and hurricane activity (e.g. Knutson et al. 2020), a downward counterfactual analysis is warranted to raise awareness of potential future hurricane losses and stress test risk management frameworks.

9.4.1 Matthew

Hurricane Matthew originated from a tropical wave off the west coast of Africa that developed into a tropical storm east of the Lesser Antilles on 28th September 2016. The system underwent rapid intensification and reached Category 5 strength by 1st October at the lowest latitude ever recorded in the Atlantic Basin. Matthew made

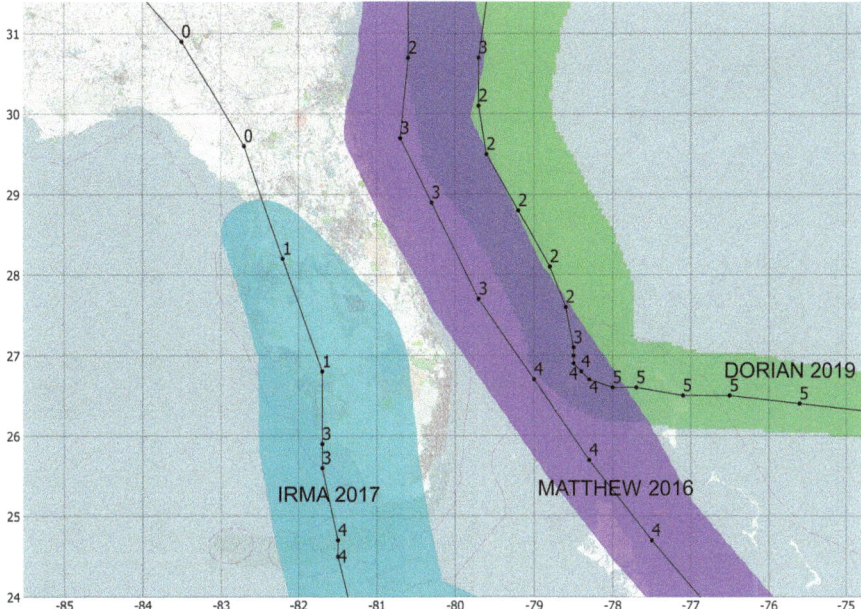

Fig. 9.4 The hurricane tracks of Matthew, Irma and Dorian and their Saffir Simpson categories at six-hourly timesteps along the track. Also shown are the estimated hurricane-force wind swathes (1-minute sustained windspeeds of over 74 mph). Data from the National Hurricane Center (NOAA NHC 2021). Basemap © OpenStreetMap contributors

Table 9.2 United States nominal gross industry insured losses from Matthew, Irma and Dorian as documented by Aon Benfield. Note that insured loss estimates are also available from other sources (e.g. Property Claims Services) but these are not available in the public domain and so are not included in this study

Hurricane	Year	Contiguous United States Gross Industry Insured Loss (US$ billion)	Source
Matthew	2016	4	Aon Benfield (2017)
Irma	2017	25	Aon Benfield (Personal communication, 2021)
Dorian	2019	1	Aon Benfield (2020)

landfall in Haiti, Cuba, and the northern Bahamas as a Category 4 hurricane (see Fig. 9.5). Although some forecasts predicted that Matthew would make landfall in Miami, the storm instead remained offshore and moved northwards, parallel to the eastern coast of Florida, before making a final landfall in South Carolina at Category 1 strength (NOAA 2016). Whilst the strong winds associated with the bypassing track of Matthew caused some damage on the east coast of Florida and led to a mainland United states insured loss of around US$4 billion (Table 9.2), the effects were minimal compared to the potential impact of a landfall in Miami if the hurricane eye had crossed the coastline.

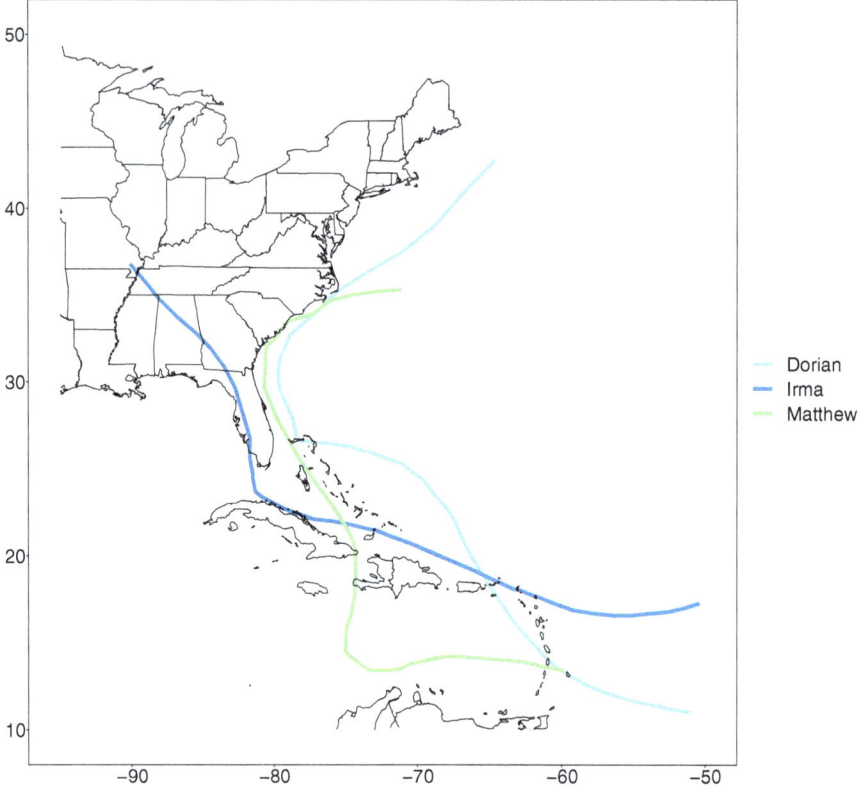

Fig. 9.5 The actual tracks of Matthew, Irma and Dorian. Data from the National Hurricane Center. (NOAA NHC 2021)

9.4.2 Irma

The following year, Irma formed from a tropical wave near Cape Verde on 30th August 2017. Irma rapidly reached a peak intensity of Category 5 strength only two days after genesis; this rate of intensification is rare and only achieved by about 1 in 30 Atlantic tropical cyclones and was not well captured in the early forecasts (NOAA 2017). Irma underwent a second period of rapid intensification as it moved towards Barbuda and made landfall there on 6th September (see Fig. 9.5). Irma made further landfalls between 6th and 9th September whilst traversing westward through the Caribbean islands. The forward speed of Irma then decreased and the hurricane turned northwards, which was captured in many forecasts. However, Irma moved further west of many of the predicted tracks which delayed the northward turn toward Florida and avoided a landfall or close bypass in Miami. Irma made further landfalls in the Florida Keys at Category 4 strength and in southwestern Florida at Category 3 strength, both of which are relatively sparsely populated compared to Miami (NOAA 2017). The overall insured loss to the mainland United States from Hurricane Irma was around US$25 billion (Table 9.2). As Irma was the only hurricane of the three to make landfall in Florida, the insured losses for this storm are significantly higher than those for Matthew and Dorian.

9.4.3 Dorian

Two years later, Dorian developed into a tropical storm on 24th August 2019 and reached hurricane strength on 27th August while moving over the U.S. Virgin Islands. Between 28th and 30th August, many forecasts predicted that the hurricane eye would pass directly over Miami; a state of emergency was declared for the whole of Florida on 29th August. Dorian underwent a period of rapid intensification on around 31st August and reached Category 5 strength before continuing on a westward trajectory and making a first landfall in the Bahamas on 1st September as the strongest hurricane in modern records to make landfall here. Although Dorian was still moving towards Miami-Dade County after landfall in the Bahamas, it changed direction sharply, and the eye of the storm remained around 100 miles from the Florida coastline as it traversed northwards (see Fig. 9.5). As a result, Florida experienced tropical storm force winds but remained relatively unscathed compared to the potential impacts of a direct landfall. Dorian later made landfall in North Carolina, which experienced Category 1 strength winds over land (NOAA 2019) and total United States insured losses reached around US$1 billion (Table 9.2).

9.5 Methodology

We use the six-step framework outlined in Lin et al. (2020) (Table 9.1) to demonstrate how insurers can operationalise downward counterfactual analysis by using tropical cyclone catastrophe models. This is achieved by first utilising operational ensemble weather forecasts to define a disaster phase space from which counterfactuals can be drawn (Steps 1–2). A Dynamic Time Warping (DTW) similarity algorithm (Berndt and Clifford 1994) is then employed to select a subset of stochastic storm tracks from a catastrophe model which have similar properties to those within the phase space (Steps 3–4). Finally, the catastrophe model is used to quantify the insured loss impact of each stochastic (counterfactual) event that has been identified (Step 5). To ensure computational efficiency, the search stops once a pre-defined number of downward counterfactuals have been identified (Step 6). The advantage of this approach is that it can be easily incorporated into existing insurance risk-management frameworks—which often involve the use of catastrophe models—to provide a set of downward counterfactuals in near real-time. Note that most catastrophe models are proprietary and cannot be edited by end-users, hence the need to select similar stochastic tracks from the catastrophe model, rather than using the operational ensemble forecast tracks directly.

Version 17 of the RMS North Atlantic Hurricane model (RMS 2021) is used to illustrate the methodology, although any stochastic tropical cyclone catastrophe model could be used. The RMS model is one of several models that have been approved by the Florida Commission on Hurricane Loss Projection Methodology (FCHLPM), which aims to protect homeowners and insurers by setting standards to rigorously evaluate model methodologies. This includes specifications on the historical "Base Hurricane Storm Set" that hurricane catastrophe models must be calibrated and validated against in order to be approved by the FCHLPM. The RMS model comprises tens of thousands of physically plausible stochastic storm tracks that make landfall in or bypass the coastlines of the Gulf of Mexico, Florida, and United States Eastern Seaboard. For each event, the model simulates the financial impacts of wind and storm surge damage (using the RMS recommended default model settings), as well as post-event loss amplification (PLA), which includes factors such as demand surge and claims inflation. Damage resulting from precipitation-induced flooding is not included.

As detailed earlier, we use a 'storyline' approach to evaluate the downward counterfactuals that are identified using the RMS model (Shepherd et al. 2018). Each counterfactual is considered a physically plausible and self-consistent future event. We do not assign a priori probabilities to the scenarios; instead, emphasis is placed on the event outcomes and the implications for the insurance industry (or individual insurer). The benefit of a storyline approach is that it presents risk in an event-oriented way, which is how most people perceive and respond to risk (Shepherd et al. 2018). This improves risk awareness and facilitates decision-making without being burdened by the uncertainties that come with weather and climate prediction.

9.5.1 Step 1: Identify Past Events

The first step is to identity one or more historical events of interest (in our case Matthew, Irma, and Dorian). The observed parameters of each event—such as the storm intensity or landfall location—provide the starting point for the downward counterfactual search. The search also requires an observed outcome against which the unrealised counterfactual outcomes can be compared. For this, we use the reported gross industry insured loss of each historical event (Table 9.2).

9.5.2 Step 2: Define Disaster phase space parameters

As detailed by Lin et al. (2020), acceptable event perturbations should be defined upfront to ensure that: (1) the resultant downward counterfactuals are physically plausible; and (2) the search is computationally feasible. For Matthew, Irma, and Dorian, the disaster phase space is constrained using two criteria:

Table 9.3 Forecast data sources used in the counterfactual analysis

Hurricane	Forecast initialisation time	Number of ensemble members	
		ECMWF	GEFS
Matthew	04/10/2016 00:00	51	21
Irma	08/09/2017 00:00	51	21
Dorian	31/08/2019 00:00	51	21

- **Storm track.** One of the best ways of defining the disaster phase space for a windstorm is to use ensemble weather forecasts. This is because the ensemble members represent a set of physically realistic perturbations to the historical event given the underlying atmospheric conditions. For Matthew, Irma, and Dorian we use operational ensemble weather forecasts from the European Centre for Medium-Range Weather Forecasts (ECMWF) and the Global Ensemble Forecast System (GEFS). The data were downloaded from NCAR/UCAR Research Data Archive (THORPEX 2021). For each historical event there are 51 ensemble members available from ECMWF and 21 from GEFS, and the location of the centre of the hurricane is provided at six-hourly timesteps (Table 9.3). Forecasts initialised between three and four days prior to a potential landfall are used to define the disaster phase space to ensure a sufficient range of realistic storm tracks.
- **Landfall intensity.** Since the focus of a downward counterfactual search is to identify more severe events, the phase space is restricted to only include RMS stochastic tracks that make landfall somewhere in the United States at Category 3 or above (111mph+ sustained winds). This is physically realistic because all three events were major hurricanes on their approach to the United States.

Finally, prior to commencing the search, it is important to apply expert judgement to remove any outlier forecast ensemble members that may produce erroneous matches with the catastrophe model storm tracks. Out of the three historical events considered in this study, only one ensemble member for Hurricane Matthew is removed, which is shown in Fig. 9.6. This track was excluded from the phase space to prevent unrealistic matches with catastrophe model tracks that pass into the Gulf of Mexico.

9.5.3 Step 3: Define End-of-Search Criteria

The search for counterfactual events is completed in this example once 70 stochastic tracks per hurricane are selected from the catastrophe model. This number is chosen as it produces a wide range of outcomes that reflect the variety in forecast tracks (72 ensemble members), while ensuring that the search is computationally feasible. Selecting significantly more than 70 tracks would provide a wider range of outcomes, but this would also lead to counterfactual events that are not as good matches to the forecast tracks.

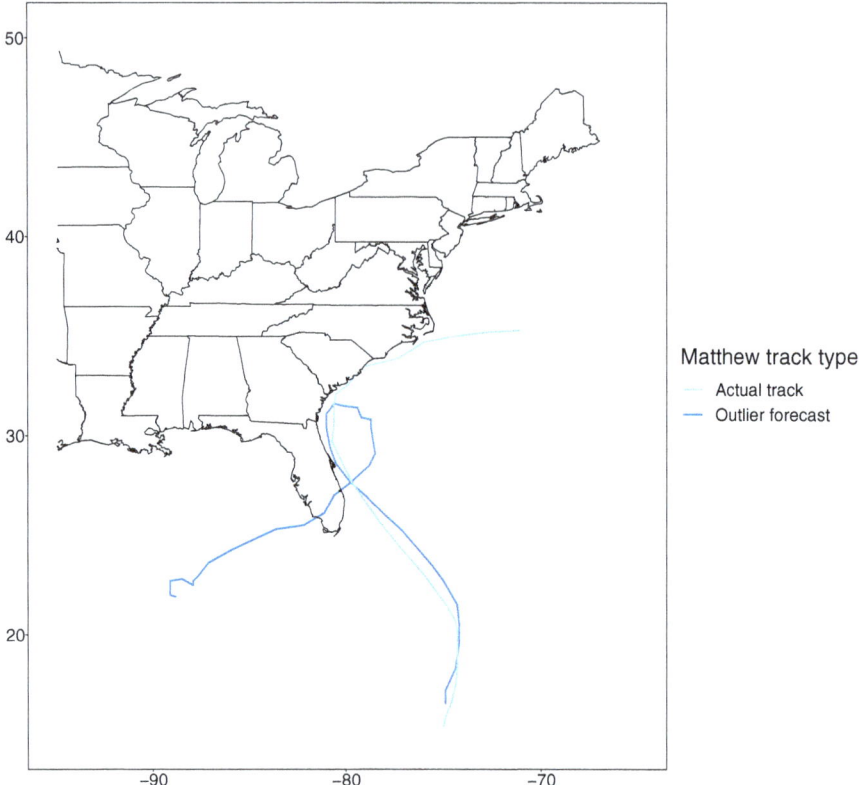

Fig. 9.6 The outlying Hurricane Matthew ECMWF forecast track that is removed from the analysis (dark blue) because it causes unrealistic counterfactual events to be selected. Also shown is the actual Hurricane Matthew track (light blue)

9.5.4 Step 4: Search the Disaster Phase Space

An iterative search algorithm is used to identify physically plausible counterfactuals within the RMS stochastic model. For each forecast ensemble member, the search algorithm loops through all stochastic events in the RMS model and calculates the dynamic time warping distance between each pair of tracks. DTW is a commonly used algorithm for quantifying the similarity between two temporal sequences which may vary in speed (Berndt and Clifford 1994). The sequences are "warped" non-linearly in the time dimension, and the Euclidean distances between pairs of data points are calculated. The optimal match is the one that has the lowest total Euclidean distance.

For computational efficiency and to ensure good matches in the area of interest, the stochastic tracks are first clipped to a bounding box two degrees wider than the extent of all the forecast tracks in the east, west, and south directions. To the north, the bounding box is set to 45 degrees north to avoid the matching algorithm placing

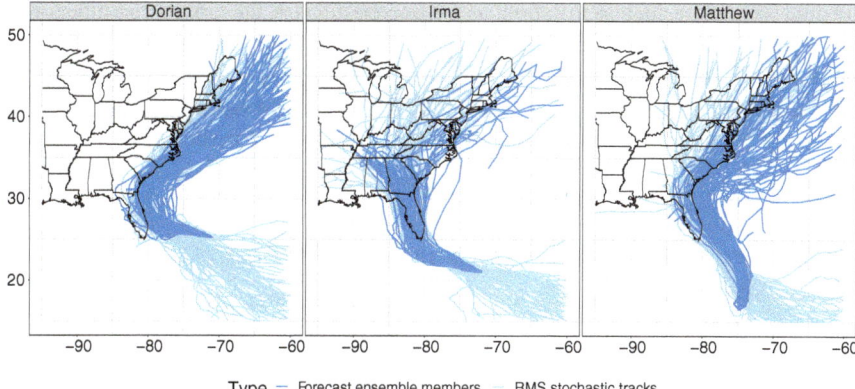

Fig. 9.7 The ECMWF and GEFS ensemble members (dark blue) and selected RMS stochastic tracks (light blue) for Dorian, Irma and Matthew

undue emphasis on the portion of the track that cannot cause damage in the mainland United States. Stochastic tracks that enter and exit the bounding box more than once are omitted, as these tracks often produce erroneously good DTW scores, matching well within the box, but deviating significantly from the forecast ensemble outside the box (e.g. looping into the Gulf of Mexico).

The 70 stochastic tracks with the best match to the set of forecast tracks (i.e., lowest DTW distances) are selected per hurricane. It is possible for a stochastic track to appear more than once in the top 70 if it produces the lowest DTW match for two or more forecasts. Therefore, if duplicate stochastic storms appear in the top 70, these are removed, and additional storms are included until 70 unique tracks have been identified. Figure 9.7 shows the resultant selected stochastic tracks for the three historical events.

It should be noted that in Lin et al. (2020)'s framework, the phase space search involves identifying a set of counterfactual events that form a 'chain', each one a perturbation of the previous event. However, in our study we apply the framework independently to each forecast ensemble member. This results in the identification of a set of counterfactuals that are related (each ensemble member is a perturbation of the forecast initial conditions), but the events themselves are not explicit perturbations of one another.

9.5.5 Step 5: Compare to the Historic Consequence

For each of the selected stochastic tracks shown in Fig. 9.7, the modelled gross industry losses are calculated using the RMS model. The actual insured losses for each event (Table 9.2) are used for comparison to the modelled counterfactual event losses. As all events have occurred recently, reported losses are not on-levelled to account for factors such as inflation, as the uncertainties in the reported numbers will be larger than the differences due to real-term monetary value adjustment over such a short time frame. In contrast to Lin et al. (2020)'s framework, the algorithm presented here is not prevented from returning upward counterfactuals, in which the counterfactual outcome is more favourable than the historical outcome.

9.5.6 Step 6: Criteria to Continue or End Counterfactual Search

The downward counterfactual search is terminated when the end-of-search criteria defined in Step 3 have been met. As noted above, this is when the best 70 DTW matches have been identified. In the unlikely event that 70 matches cannot be found, the search will end once the algorithm has iterated over all forecast ensemble members and each catastrophe model stochastic event track.

9.6 Results

9.6.1 Individual Scenarios

The results of the downward counterfactual searches for Matthew (a), Irma (b), and Dorian (c) are presented in a series of scatter plots in Fig. 9.8. Each light blue point represents a separate counterfactual scenario, which are plotted in ascending order of loss (y-axis). The x-axis represents the counterfactual loss normalised by the reported event loss (Table 9.2), which is also shown on each plot as a dark blue point. In accordance with a storyline approach, each counterfactual is considered a physically plausible outcome without assigning likelihoods. Note that the normalised gross industry losses produced by the RMS model reflect average industry practices (including insurance take-up rates and terms and conditions, such as deductibles). As detailed in Sect. 9.5, modelled losses also include post-event loss amplification which accounts for factors such as economic demand surge and claims inflation.

In total, 68, 61, and 70 downward counterfactuals are identified for Matthew, Irma, and Dorian, respectively. As noted above, we do not explicitly prevent the algorithm from returning upward counterfactuals. For example, in the case of Irma, nine upward counterfactuals are identified, which produce lower losses than the original event. The large number of downward counterfactual scenarios that were identified highlights that all three historical events could have produced significantly worse outcomes had they turned out slightly differently. In the worst-case counterfactuals, insured losses are nearly 300 times the reported loss for Hurricane Matthew, 25 times higher for Hurricane Irma, and over 250 times higher for Hurricane Dorian. Note that the reported insured loss for Irma (US$25 billion, Table 9.2) is an order of

9 Downward Counterfactual Analysis in Insurance Tropical Cyclone Models:... 223

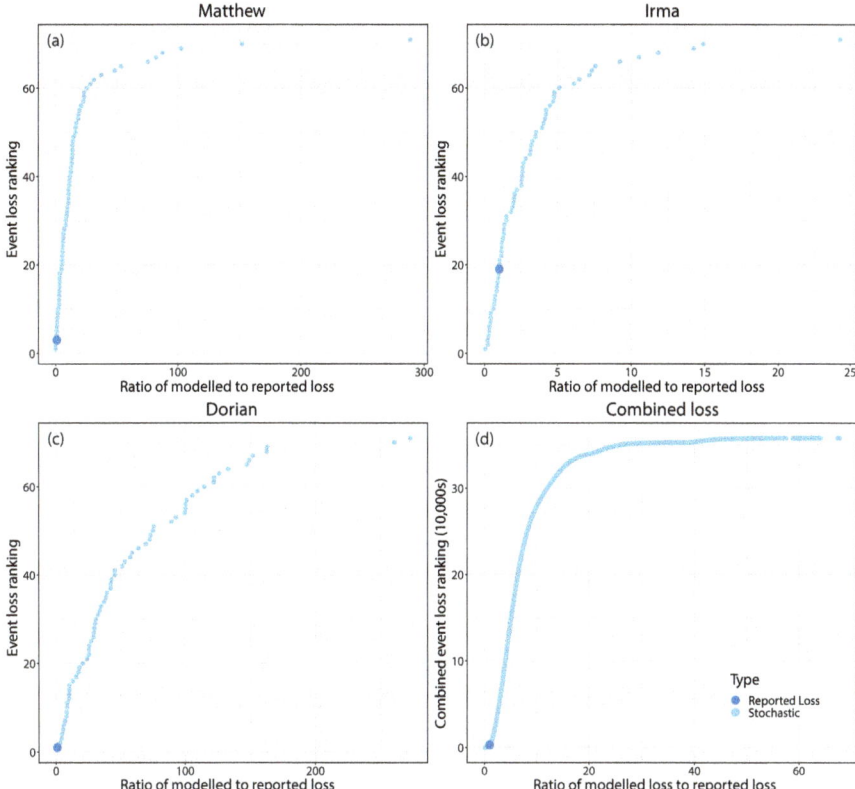

Fig. 9.8 Counterfactual scenarios for (**a**) Matthew, (**b**) Irma and (**c**) Dorian. Each light blue point represents the loss associated with an RMS stochastic track; the top 70 tracks with the lowest DTW distance are shown. The losses are normalised by the historical reported loss (x-axis) and ordered according to increasing event severity (y-axis). Subplot (**d**) shows the range of losses from all possible combinations of the three hurricane losses. The reported loss is shown in dark blue

magnitude larger than the reported losses for Matthew and Dorian. This explains why the normalised worst-case scenario of Irma is an order of magnitude lower than the other two events.

Figure 9.9 shows the tracks of the five worst downward counterfactuals for each hurricane, with the single worst outcome highlighted in dark blue. All of the worst-case scenarios involve significant wind and storm surge damage to the Miami metropolitan region, which has a population of over six million inhabitants (United States Census Bureau 2019). The worst-case counterfactuals for Irma and Matthew also transit much of the eastern coast of Florida, causing damage in the heavily populated cities of Orlando and Jacksonville. Also note that for both Dorian and Irma, some of the worst five outcomes impact the city of Tampa on the western coast of Florida, and many of the events make landfall a second time in states north of Florida—all of which contributes to the cumulative damage and loss for these events.

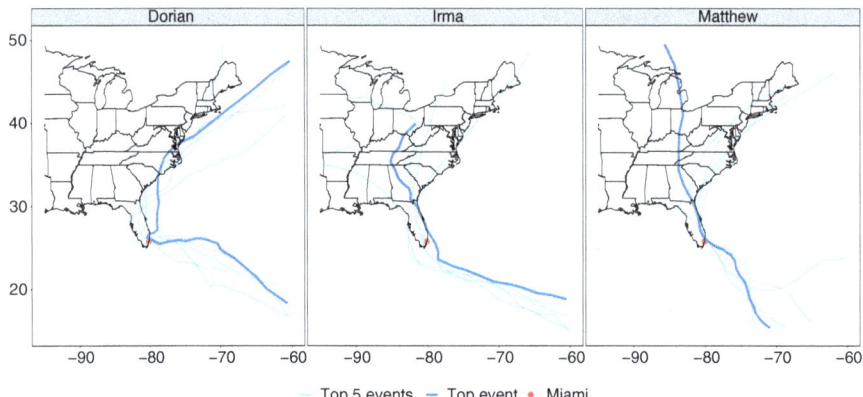

Fig. 9.9 The top five downward counterfactuals by gross industry loss for each historical hurricane (light blue) and the top overall loss event (dark blue). The location of the City of Miami is marked in red

9.6.2 Combined Scenarios

Figure 9.8d shows all possible counterfactual combinations for the three historical events. In total, 361,934 possible loss combinations are identified, the largest of which is the sum of the three worst-case outcomes for each individual storm. This shows that the combined loss for all three hurricanes could have been up to 70 times larger than was observed. Overall, 22% of the combinations had losses that were 10 times greater than the combined historical loss, and 2% had losses that were 25 times greater (as noted earlier, these outcomes represent storylines, not likelihoods). Many of the worst-case outcomes involve three direct hits on Miami which, while unlikely, was possible given the atmospheric conditions at the time of each event. It should be caveated that this analysis assumes that each event is independent, meaning that full economic and societal recovery occurs between each landfall. In reality, if three hurricanes were to impact Florida over a short period of time, each storm could create antecedent conditions that may affect subsequent events. For example, the loss from a given storm could be amplified if the local economy was under strain from a previous event, leading to inflated labour, material, and alternative accommodation costs.

9.6.3 Climate Change

Under climate change, the intensity of some hurricanes is likely to increase (e.g. Knutson et al. 2020). Therefore, it is possible that hurricanes similar to Matthew, Irma, and Dorian may make landfall in Florida with a higher intensity

Table 9.4 Modelled industry mean gross loss normalised by the reported loss for each historic storm per hurricane category across the 70 selected stochastic events. The numbers represent how many times greater the modelled mean loss is than the reported loss

Hurricane	Mean loss normalised by reported loss		
	Category 3	Category 4	Category 5
Matthew	25	56	126
Irma	2	3	6
Dorian	6	17	110

than occurred historically. To explore the impact of storm severity on insured losses, Table 9.4 shows the mean counterfactual loss normalised by the reported loss across each major hurricane category (3, 4 and 5). As can be seen, the loss increases significantly with increasing hurricane category. For Matthew, Category 3 counterfactuals cause average losses of seven times the reported loss, whilst Category 5 counterfactuals cause average losses of 110 times the actual loss. For Irma, the factors are less extreme, with counterfactuals on average causing up to 6 times more loss for Category 5 storms. For Dorian, the mean counterfactual loss is 25 times larger than the actual loss for Category 3 counterfactuals, and 126 times the actual loss for Category 5 counterfactuals. This is not unexpected since stronger storms are known to cause greater damage; however, it does raise awareness of how increasing storm severity due to climate change could affect insured losses. Note that this is only one example; climate change could affect tropical cyclones in several ways other than wind severity, such as precipitation intensity, forward speed, or event frequency. These are all factors that could be considered in further research (see Sect. 9.7.4).

9.7 Discussion

Our analysis has shown that Matthew, Irma, and Dorian could have, in many scenarios, had much worse outcomes for the United States. This in itself is not unexpected—catastrophe models have been used by the insurance industry for more than three decades to manage the risk from physically plausible disasters that have not yet occurred. However, behavioural science tells us that most humans have problems perceiving risk that falls outside historical experience (availability bias) or occurred a long time ago (recency bias), even when quantitative information—such as from catastrophe models—is available (Shepherd et al. 2018). Downward counterfactual thinking can overcome this problem by highlighting situations in which risks have become distorted by near misses or good fortune, such as in Miami (Woo et al. 2017; Woo 2019). We have presented a methodology building on the work of Lin et al. (2020) that allows insurers to explore such downward counterfactuals using tropical cyclone catastrophe models. Our approach has three key benefits for insurers, each of which will now be discussed in the context of our Miami case study: (1) increased risk awareness; (2) operationalised counterfactuals within risk management frameworks; and (3) improved decision-making in the face of extreme uncertainty, including climate change.

9.7.1 Risk Awareness

What if Matthew, Irma, and Dorian had all hit Miami? This is a scenario that was possible given the underlying atmospheric conditions at the time of each event. Yet for most insurers this is not something to which they would have given much contemplation; a similar set of events has not occurred in living memory and is therefore considered unlikely. However, as argued by Woo et al. (2017), in order to avoid future surprises, it is important for insurers to consider the potential implications of unlikely, but possible, disasters. This is particularly true in a changing world, where urbanisation, economic growth, and climate change are constantly altering the risk profile. Using downward counterfactual analysis to focus on event (storyline) outcomes, rather than probabilities, helps raise awareness by providing tangible information to decision-makers (Shepherd et al. 2018). For example, consider a storyline in which the combined loss from Matthew, Irma, and Dorian was 33 times larger (which was possible according to our analysis in Fig. 9.8d). This would have resulted in insurance claims of around US$1 trillion. Average annual insured losses from global tropical cyclones for the period 2000–2018 have been around US$20 billion (Aon Benfield 2018). A loss in excess of US$1 trillion would have therefore put significant strain on the insurance industry, and likely the global economy (Mahalingham et al. 2018). This illustrates the importance of raising awareness in situations where there is a known risk, but the last major disaster (Hurricane Andrew) was a long time ago and therefore cognitive biases may exist.

In addition to direct financial impacts, downward counterfactuals can also be used to contemplate the wider implications for the insurance industry. For example, in 2004 and 2005, unusually warm sea surface temperatures in the North Atlantic produced a record number of hurricanes, which led to several large, insured losses from Hurricanes Katrina, Wilma, and Ivan (Virmani and Weisberg 2006). As a direct reaction to this, catastrophe model vendors introduced alternative "near-term" views of risk in order to quantify expected losses during more active seasons (e.g. Jewson et al. 2009). The 2004/5 hurricane seasons also led to wide-ranging changes to insurance policies, including an increase in premiums and more stringent underwriting practices, such as higher deductibles and sub-limits (Guy Carpenter 2014). It is likely that far worse disruption would occur following a cumulative US$1 trillion loss over the space of just a few years. This could include the Floridian government stepping in to legislate to protect home and business owners, as occurred in California following the 2017/18 wildfires when the State Senate ruled that insurers must grant up to 36 months of additional living expenses and offer to renew policies for up to two years (Senate Bill 894 2018).

Similarly, counterfactuals can be used to consider the impact on the perception of climate change risk, both within the insurance industry and among the general public. For example, following Hurricane Irma in 2017 there was significant media attention on the extent to which climate change had contributed to the severity of the event (e.g. Carbon Brief 2017). Subsequent scientific investigations have found that while there is evidence that the precipitation from Irma contained a climate change signal, the wind hazard did not (Patricola and Wehner 2018). If Matthew, Irma, and Dorian had all hit Miami in close succession, the media reaction would almost certainly have been far greater (regardless of the role of climate change in the losses). Contemplating the implications of changing public perceptions around climate change risk following such a large set of disasters might be considered by some to be a leap into the unknown. However, it is not too difficult to envision a situation in which climate change is pushed to the forefront of insurance and government agendas, leading to greater regulation, changing insurance products and risk pricing, more investment in research, and an increased focus on mitigation and adaption.

9.7.2 Operationalisation

Following a natural disaster, most insurance companies will have an "event response" process to produce an early loss estimate, which is shared with key stakeholders both internally (e.g. business planning) and externally (e.g. regulators, rating agencies, markets). The magnitude of the event will usually determine the level of event response, with large disasters (e.g. a major hurricane landfall in the United States) receiving the most attention. The lessons learnt following a large disaster—for example from claims data—will often feed into risk management activities, such as the validation and adjustment of catastrophe models (Jones et al. 2017). However, this ex-post process is very much an *upward* counterfactual thought exercise that focuses on the worst events. Near-misses are often ignored, both during the event response and the post-event analysis.

The operationalisation of counterfactual analysis could therefore provide significant value to insurers by brining downward scenarios into risk management and decision-making frameworks. While the methodology presented in this chapter uses historical weather forecasts, it could easily be extended to include real-time data as an event unfolds. This would enable downward counterfactuals to be included in event response processes. For example, in addition to asking the question "what is our best guess of the loss for this event?", insurers can also ask "could the loss be worse?", thereby enabling real-time stress-testing of portfolios. This also facilitates post-event analysis, where counterfactuals could be used to validate and adjust catastrophe models for lessons learnt. It is worth noting that operationalisation is only possible because catastrophes contain a wide range of physically plausible events and are already integrated into existing insurance risk management frameworks. Without catastrophe models, the process would be far more arduous to undertake in real-time.

9.7.3 Decision-Making

Downward counterfactuals that focus on event outcomes can improve insurance decision-making by providing conditional statements that lead to tangible business impacts (Shepherd et al. 2018). This is opposed to probabilistic statements that cover a wide range of scenarios and are therefore less discernible. Focusing on specific events can help insurers identify vulnerabilities and develop risk-mitigating business strategies (Woo et al. 2017). For example, consider the question: "If Matthew, Irma, and Dorian had all hit Miami as major hurricanes, how would this have affected capitalisation?". This provides a specific scenario in which capital can be stress-tested and mitigating actions taken if weaknesses are found. This is the opposite of an upward counterfactual approach, which only considers reacting after a disaster has occurred (by which point it might be too late).

Deterministic scenarios are already widely used in the insurance industry for decision-making and regulatory purposes. For example, following the large natural catastrophes of the 1980s and 1990s, Lloyd's of London introduced Realistic Disaster Scenarios (RDSs), which are designed to stress-test insurance portfolios to plausible high-loss events of low probability (e.g. Lloyd's 2021). Downward counterfactuals could therefore easily be integrated into existing insurance decision frameworks. A distinction should of course be made between existing deterministic scenarios (which are often hypothetical) and counterfactuals (which are grounded in history).

An event-orientated approach to decision-making is particularly useful in situations of extreme uncertainty, where an event may not have occurred in living memory and is therefore hard to imagine, even for subject matter experts. Climate change is a good example of this—even those that are familiar with the science often struggle to make decisions because it is difficult to conceptualise what the future will look like (Weber 2006). Given the short observational record and changing socio-economic/demographic factors over time, it is often difficult to attribute loss trends to climate change (e.g. Hoeppe 2016). Therefore, using near-misses to "fill-in" history can add significant value in a changing climate, particularly for cities like Miami that have not experience a large disaster in several decades.

9.7.4 Caveats and Further Research

The results presented in this study are conditional both on the reported historical losses (Table 9.2) and the RMS catastrophe model. The reported losses may include sources of loss that are not simulated by the catastrophe model. For example, loss adjustment expenses (associated with investigating and settling insurance claims) are included in the reported loss numbers, but not in the RMS model. This does not undermine our findings for Matthew, Irma, and Dorian, since any uncertainty in the reported losses is lower than the magnitude of range in downward counterfactuals, but it should be borne in mind when interpreting the results. It should also be noted that whilst RMS simulate a wide range of physically plausible hurricanes, it is possible that events (e.g. black swans) or sources of loss exist that are not represented in the catastrophe model.

The work presented here could be extended by widening the disaster phase space to consider events that fall outside the realm of the catastrophe model. This could be achieved by considering extreme outcomes under the present-day climate (e.g. compound events, Woo 2021) or by incorporating aspects of future climate change not currently represented in the model (e.g. sea level rise). The methodology could also be applied to different catastrophe model vendors, as well as different perils (e.g. flooding) and geographic regions. Another area of further research would be to investigate the sensitivity of the results to additional parameters that have an impact on potential damage such as the translational speed, intensity along the track and tidal state at landfall. Obtaining these details would require close collaboration with catastrophe model vendors, as this information is not provided to users as standard in the models.

9.8 Conclusions

We have presented a methodology for insurers to operationalise downward counterfactual analysis using tropical cyclone catastrophe models. Downward counterfactuals provide insurers with a way of exploring how historical events could have turned out for the worse. We combine this with a 'storyline' approach, which focuses on describing and understanding specific event outcomes, rather than prescribing likelihoods. The methodology was applied to three recent major hurricanes that were near misses for Miami—Matthew (2016), Irma (2017), and Dorian (2019). The results revealed downward counterfactuals that produced insured losses many times greater than what transpired—Matthew (300x), Irma (25x), Dorian (250x), and up to 70x for all three combined. Downward counterfactuals are an important tool that should be used by insurers to complement catastrophe models. They provide a set of deterministic scenarios that can be used to increase risk awareness, stress-test risk management frameworks and inform decision-making. This is particularly true in situations where cognitive biases may exist due to a lack of recent loss experience, such as in Miami, and therefore people may not be fully aware of the potential risk due to factors such as urban growth and climate change. This work will also have applications outside of the insurance industry and will therefore be of interest all readers concerned with tropical cyclone risk in a changing climate.

9.9 Competing Interests

C.J.R. and J.A.B. are employees of MS Amlin — a global speciality insurer and reinsurer that is part of the MS&AD Insurance Group. C.J.R. and J.A.B. have contributed to this chapter in their own capacity; any views expressed in this chapter are their own and not those of MS Amlin or MS&AD. The THORPEX dataset was used for non-commercial research purposes only.

References

Allen M (2003) Liability for climate change. Nature 421:891–892
Aon Benfield (2017) 2016 Annual global climate and catastrophe report. Available at: http://thoughtleadership.aonbenfield.com/Documents/20170117-ab-if-annual-climate-catastrophe-report.pdf. Accessed 27 May 2021
Aon Benfield (2018) 2017 Annual global climate and catastrophe report. Available at: http://thoughtleadership.aon.com/Documents/20180124-ab-if-annual-report-weather-climate-2017.pdf. Accessed 27 May 2021
Aon Benfield (2020) 2019 Annual global climate and catastrophe report. Available at: http://thoughtleadership.aon.com/Documents/20200122-if-natcat2020.pdf?utm_source=ceros&utm_medium=storypage&utm_campaign=natcat20. Accessed 27 May 2021
Berndt DJ, Clifford J (1994) Using dynamic time warping to find patterns in time series, vol. 10, pp 359–370. Available at: https://www.aaai.org/Papers/Workshops/1994/WS-94-03/WS94-03-031.pdf. Accessed 27 May 2021
Born B, Dietrich AM, Müller GJ (2021) The lockdown effect: a counterfactual for Sweden. Plos One 16
Carbon Brief (2017) Media reaction: Hurricane Irma and climate change. Available at: https://www.carbonbrief.org/media-reaction-hurricane-irma-climate-change/. Accessed 27 May 2021
CISL (2020) Climatewise physical risk framework. Available at: https://eprints.soas.ac.uk/33510/1/case_studies_of_environmental_risk_analysis_methodologies.pdf. Accessed 27 May 2021
De Bono E (1977) Lateral thinking. Penguin Press, London
Dunbar JB, Torrey III VH, Wakeley LD (1999) A case history of embankment failure: geological and geotechnical aspects of the Celotex levee failure, New Orleans, Louisiana. Available at: https://apps.dtic.mil/sti/pdfs/ADA375282.pdf. Accessed 27 May 2021
Fiedler T et al (2021) Business risk and the emergence of climate analytics. Nat Clim Change 11: 87–94
Golnaraghi M et al (2018) Managing physical climate risk: leveraging innovations in catastrophe risk modelling. Geneva Association-International Association for the Study of Insurance
Grossi P, Kunreuther H (2005) Catastrophe modeling: a new approach to managing risk. Springer
Guy Carpenter (2014) Hurricane seasons that changed the industry: landmark 2005 Hurricane season. Available at: https://www.gccapitalideas.com/2014/10/25/hurricane-seasons-that-changed-the-industry-landmark-2005-hurricane-season/. Accessed 27 May 2021
Hall TM, Jewson S (2007) Statistical modelling of North Atlantic tropical cyclone tracks. Tellus A Dyn Meteorol Oceanogr 59:486–498
Hoeppe P (2016) Trends in weather related disasters–consequences for insurers and society. Weather Clim Extrem 11:70–79
Jewson S et al (2009) Five year prediction of the number of hurricanes that make United States landfall. In: Elsner JB, Jagger TH (eds) Hurricanes and climate change. Springer, Boston, pp 73–99

Jones M, Mitchell-Wallace K, Foote M, Hillier J (2017) Fundamentals. In: Natural catastrophe risk management and modelling: a practitioner's guide

Kahneman D (2011) Thinking, fast and slow. Macmillan

Khanduri AC, Morrow GC (2003) Vulnerability of buildings to windstorms and insurance loss estimation. J Wind Eng Ind Aerodyn 91:455–467

Knabb RD, Rhome JR, Brown RD (2005) Tropical cyclone report: Hurricane Katrina. Available at: https://www.nhc.noaa.gov/data/tcr/AL122005_Katrina.pdf. Accessed 27 May 2021

Knutson T et al (2020) Tropical cyclones and climate change assessment: part II: projected response to anthropogenic warming. Bull Am Meteorol Soc 101

Lin YC et al (2020) Modeling downward counterfactual events: unrealized disasters and why they matter. Earth Sci 8:1–16

Lloyd's (2012) Lloyd's global underinsurance report. Available at: https://assets.lloyds.com/assets/pdf-global-underinsurance-report-global-underinsurance-report/1/pdf-global-underinsurance-report-global-underinsurance-report.pdf. Accessed 27 May 2021

Lloyd's (2021) RDS Scenario specification 2021. Available at: https://www.lloyds.com/conducting-business/underwriting/realistic-disaster-scenarios. Accessed: 27 May 2021

Mahalingham A et al (2018) Impacts of severe natural catastrophes on financial markets. Cambridge Centre for Risk Studies

Nelder JA, Mead R (1965) A simplex method for function minimization. Comput J 7:308–313

NOAA (2016) Tropical cyclone report: Hurricane Matthew. Available at: https://www.nhc.noaa.gov/data/tcr/AL142016_Matthew.pdf. Accessed 27 May 2021

NOAA (2017) Tropical cyclone report: Hurricane Irma. Available at: https://www.nhc.noaa.gov/data/tcr/AL112017_Irma.pdf. Accessed 27 May 2021.

NOAA (2019) Tropical cyclone report: Hurricane Dorian. Available at: https://www.nhc.noaa.gov/data/tcr/AL052019_Dorian.pdf. Accessed 27 May 2021

NOAA NHC (2021) NHC data in GIS formats. Available at: https://www.nhc.noaa.gov/gis. Accessed 27 May 2021

Patricola CM, Wehner MF (2018) Anthropogenic influences on major tropical cyclone events. Nature 563:339–346

PRA (2019) General insurance stress test 2019. Available at: https://www.bankofengland.co.uk/-/media/boe/files/prudential-regulation/letter/2019/general-insurance-stress-test-2019-scenario-specification-guidelines-and-instructions.pdf. Accessed 27 May 2021

RMS (2021) North Atlantic Hurricane. Available at: https://www.rms.com/models/cyclone-hurricane-typhoon/north-atlantic-hurricane. Accessed 27 May 2021

Roese NJ (1997) Counterfactual thinking. Psychol Bull 121:133

Rye CJ, Boyd JA, Mitchell A (2021) Normative approach to risk management for insurers. Nat Clim Change:1–4

Schwab M, Meinke I, Vanderlinden J-P, von Storch H (2017) Regional decision-makers as potential users of extreme weather event attribution-case studies from the German Baltic Sea coast and the Greater Paris area. Weather Clim Extrem 18:1–7

Senate Bill 894 (2018) Senate Bill No. 894 Chapter 618. Available at: https://leginfo.legislature.ca.gov/faces/billNavClient.xhtml?bill_id=201720180SB894. Accessed 27 May 2021

Shepherd TG et al (2018) Storylines: an alternative approach to representing uncertainty in physical aspects of climate change. Clim Change 151:555–571

Swiss Re (2020) 15 years after Katrina: Would we be prepared today?. Available at: https://www.swissre.com/dam/jcr:a835acae-c433-4bdb-96d1-a154dd6b88ea/hurrican-katrina-brochure-usletter-web.pdf. Accessed 27 May 2021

Taleb NN (2007) The black swan: the impact of the highly improbable, vol 2. Random House

THORPEX (2021) THORPEX Interactive Grand Global Ensemble (TIGGE) Model Tropical Cyclone track data. https://doi.org/10.5065/D6GH9GSZ. Accessed 27 May2021

Tompkins F, Deconcini C (2014) Sea-level rise and its impact on Miami-Dade county. Available at: https://www.wri.org/research/sea-level-rise-and-its-impact-miami-dade-county. Accessed 27 May 2021

United States Census Bureau (2019) QuickFacts: Palm Beach County, Florida; Broward County, Florida; Miami-Dade County, Florida. Available at: https://www.census.gov/quickfacts/fact/table/palmbeachcountyflorida,browardcountyflorida,miamidadecountyflorida/POP0 60210. Accessed 27 May 2021

Van Oldenborgh GJ et al (2017) Attribution of extreme rainfall from Hurricane Harvey, August 2017. Environ Res Lett 12

Virmani JI, Weisberg RH (2006) The 2005 hurricane season: AN echo of the past or a harbinger of the future? Geophys Res Lett 33

Wdowinski S, Bray R, Kirtman BP, Wu Z (2016) Increasing flooding hazard in coastal communities due to rising sea level: case study of Miami Beach, Florida. Ocean Coastal Manag 126:1–8

Weber EU (2006) Experience-based and description-based perceptions of long-term risk: why global warming does not scare us (yet). Clim Change 77:103–120

Woo G (2018) Counterfactual disaster risk analysis. Variance J 2:279–291

Woo G (2019) Downward counterfactual search for extreme events. Front Earth Sci 7:340

Woo G (2021) A counterfactual perspective on compound weather risk. Weather Clim Extrem 32: 1–6

Woo G, Maynard T, Seria J (2017) Reimagining history: counterfactual risk analysis. Available at: https://assets.lloyds.com/assets/reimagining-history-report/1/Reimagining-history-report.pdf. Accessed 27 May 2021

Open Access This chapter is licensed under the terms of the Creative Commons Attribution 4.0 International License (http://creativecommons.org/licenses/by/4.0/), which permits use, sharing, adaptation, distribution and reproduction in any medium or format, as long as you give appropriate credit to the original author(s) and the source, provide a link to the Creative Commons license and indicate if changes were made.

The images or other third party material in this chapter are included in the chapter's Creative Commons license, unless indicated otherwise in a credit line to the material. If material is not included in the chapter's Creative Commons license and your intended use is not permitted by statutory regulation or exceeds the permitted use, you will need to obtain permission directly from the copyright holder.

Chapter 10
Identifying Limitations when Deriving Probabilistic Views of North Atlantic Hurricane Hazard from Counterfactual Ensemble NWP Re-forecasts

Tom J. Philp, Adrian J. Champion, Kevin I. Hodges, Catherine Pigott, Andrew MacFarlane, George Wragg, and Steve Zhao

Abstract Downward counterfactual analysis – or quantitatively estimating how our observed history could have been worse – is increasingly being used by the re/insurance industry to identify, quantify, and mitigate against as-yet-unrealised "grey-swan" catastrophic events. While useful for informing site-specific adaptation strategies, the extraction of probabilistic information remains intangible from such downside-only focused analytics. We hypothesise that combined upward and downward counterfactual analysis (i.e., how history could have been either better or worse) may allow us to obtain probabilistic information from counterfactual research if it can be applied objectively and without bias.

Here we test this concept of objective counterfactual analysis by investigating how initial-condition-driven track variability of events in our North Atlantic Hurricane (NAHU) record may affect present-day probabilistic views of US landfall risk. To do this, we create 10,000 counterfactual NAHU histories from NCEP GEFS v2 initial-condition ensemble reforecast data for the period 1985-2016 and compare the

T. J. Philp (✉)
Maximum Information, London, UK
e-mail: tom@maximuminformation.com

A. J. Champion
University of Exeter, Exeter, UK
e-mail: adrian.champion@aon.com

K. I. Hodges
National Centre for Atmospheric Science, Department of Meteorology, University of Reading, Reading, UK
e-mail: k.i.hodges@reading.ac.uk

C. Pigott · G. Wragg · S. Zhao
AXA XL, London, UK
e-mail: catherine.pigott@axaxl.com; george.wragg@axaxl.com; steve.zhao@axaxl.com

A. MacFarlane
AXA XL, Hamilton, Bermuda
e-mail: Andrew.macfarlane@axaxl.com

© The Author(s) 2022
J. M. Collins, J. M. Done (eds.), *Hurricane Risk in a Changing Climate*, Hurricane Risk 2, https://doi.org/10.1007/978-3-031-08568-0_10

statistics of these counterfactual histories to a model-based version of our single observational history.

While the methodology presented herein attempts to produce the histories as objectively as possible, there is clear – and, ultimately, intuitively understandable – systematic underprediction of US NAHU landfall frequency in the counterfactual histories. This limits the ability to use the data in real-world applications at present. However, even with this systematic under-prediction, it is interesting to note both the magnitude of volatility and spatial variability in hurricane landfalls in single cities and wider regions along the US coastline, which speaks to the potential value of objective counterfactual analysis once methods have evolved.

Keywords Counterfactual · Ensemble · Landfall · Probabilistic · Risk

10.1 Introduction

For centuries, the global re/insurance industry has estimated future risk by applying statistical methods to historical loss data (Halley 1693), with the early methods growing into an expansive research discipline now known as actuarial science.

While the advent of catastrophe modelling in the late 1980s and early 1990s saw an evolution of traditional actuarial loss-based methods toward the incorporation of explicit scientific information from a wide variety of non-loss focused sources, probabilistic views of catastrophe risk have largely remained driven by historical observations.

In very recent years, methods to derive views of probabilistic atmospheric risk directly from climate models have begun to be developed (Carozza and Boudreault 2021; Jones et al. 2020). However, it is well known that ingesting climate model information coherently into the historical, observation-based, and usually site-specific, views of risk that are widely prevalent in the industry is likely to bring its own suite of problems, both scientific and philosophical (Frigg et al. 2015).

Although extremely non-trivial, there is substantial demand to overcome this challenge given increasing societal concern surrounding the need to quantify the impact of a shifting climate on the frequency and intensity of catastrophic atmospheric perils, from both acute and chronic onset perspectives. This demand is evidenced, for example, by recent climate disclosure guidance issued by the Bank of England to financial institutions (Bank of England 2019).

It is important to note that although the challenge of building the connection between forward-looking climate and catastrophe models is yet to be fully overcome, the re/insurance industry already looks to alternative methods to help improve its ability to use information about extreme weather events from climate projections as coherently and appropriately as possible. For example, for many years the industry has employed the concept of "downward counterfactual" thinking to sensitivity test observationally derived views of risk; the same type of thinking exists in other financial markets, usually under the auspice of more generically-termed "financial stress tests". As this downward counterfactual term is not yet widely used in catastrophe risk management, we explicitly define it here as "a thought about the past where the outcome was worse than what actually happened", with the definition taken directly from the introductory discussion provided in Woo (2019), itself following the formative definition provided by Roese (1997).

While this language is relatively new to the risk management industry, methods to incorporate the underlying ideas in risk modelling are not. The industry understands well that – for rare and high impact events in particular – historical observations may only paint a partial picture of future risk. Thus, deterministic "what-if?" scenario analyses have been employed, which ultimately attempt to foresee those "grey swans" that do not appear in our observed loss record, yet can easily be imagined, rationally estimated, and thus mitigated against.

At present, this "downward counterfactual" or "what-if" thinking is applied through such practical actions as formal reporting on Realistic Disaster Scenarios (RDSs), which help to steer re/insurers on capital requirements, as well as quantifying estimates such as maximum possible/probable loss in the case of an extreme event happening. As an example, in the context of North Atlantic Hurricanes (NAHUs), one of the Lloyd's RDSs for 2021 poses the scenario: "A North-East US hurricane, immediately followed by a South Carolina hurricane" (Lloyd's 2021).

While already embedded in risk management processes, in a joint report on the topic, Lloyd's and RMS concluded that including more "downward counterfactuals" in analyses of risk – that is, specifically re-imagining "how historical near misses might have become major disasters" – is likely to "bring benefits to insurers" (Lloyd's 2017). For example, what if Hurricanes Matthew (2016) or Dorian (2019) had rolled onshore in downtown Miami, as opposed to only grazing the South Florida coastline? Would the building codes and practices put in place since the highly destructive Hurricane Andrew (1992) work to limit outsized losses in this region? And how have population dynamics in the past twenty years altered the shape of risk in that area? While the Lloyd's RDSs begin to touch on questions like this, they are often less focused on re-imagining actual historical events. Other industry-led research is beginning to unpick the potential value of this type of targeted question (Chap. 9).

Thus, while more focus on downward counterfactual analysis is certainly likely to be useful, it is not necessarily novel. Arguably, it relegates the potential value of the work to deterministic applications only and doesn't help to address the problem of extracting potentially valuable and untapped probabilistic information from such types of analyses.

At present, probabilistic information for catastrophe risk quantification in the re/insurance industry is driven by stochastic modelling, itself underpinned almost solely by historical event data. The stochastic modelling process acts to fill in holes such that our spatial picture of risk is smoother (and more realistic) than if drawn from raw observations. However, the statistics of the historical dataset are preserved during the stochastic modelling process, at least to some extent. While this preservation of underlying statistics may be highly desirable for well-observed perils and long historical records, it is likely less desirable in places with sparse observational datasets (as is the case with many extreme weather perils). The preservation of statistics in these sparsely observed areas may lead to an erroneous view of risk purely by virtue of random chance or luck in the observed historical period. Thus, identifying further information sources that can help to remove random luck in risk quantification and to ultimately facilitate more optimized risk selection may prove highly valuable. It is hypothesized here that counterfactual analysis may provide an opportunity to build additional reliability on baseline stochastic modelling.

For probabilistic counterfactual applications to be achieved, we must move away from the single-sided "downward" philosophy of counterfactual analysis, toward one where both "upward" and "downward" counterfactuals are considered simultaneously. If it were possible that this upward and downward analysis could be done objectively, we may be able to extract information such that our stochastic, observation-based methods for risk quantification become more robust and reliable. At present, questions about validity of stochastic modelling output can only be asked from a subjective "model completeness" perspective – i.e., do we believe that the stochastic process is filling in the gaps in history appropriately?

To practically combine the two worlds of future focused climate-catastrophe modelling with objective upward and downward counterfactual analysis, here we propose to use NWP ensemble re-forecasts to create multiple counterfactual NAHU histories, and to compare them to the observed historical record of NAHUs. At present, the authors are only aware of one other study that attempts to extract probabilistic information from re-forecast data in the case of Tropical Cyclones (Ng and Leckebusch 2021). This study, while similar in fundamental philosophy, is somewhat tangential to the applications hypothesized here. Ng and Leckebusch (2021) utilize a multi-model archive to create basin-wide counterfactual climatologies of key catastrophe risk related variables, such as return periods of windspeeds. Here we intend to accomplish something slightly different, focusing only on identifying historical NAHUs in re-forecast data and attempting to use these data to evaluate – however loosely – the probabilities of intersection with the US coastline of these historical storms. If the methods were to prove successful, it would allow targeted adjustment of the stochastic track sets that typically drive contemporary NAHU catastrophe models.

Thus, beyond the counterfactual aspect of the analysis, we also hope this work will help to lay the foundation to connect traditional historical observation derived views of catastrophe risk to weather and climate model derived views of risk, and highlight where limitations currently exist that may limit the ability to do this optimally.

10.2 Potential Real-World Applications

While beyond the scope of this experimental study, it is envisaged that a truly objective implementation of counterfactual thinking would allow us to quantifiably deconstruct our observational history, and begin to ask questions such as:

1. Do counterfactual histories suggest that a specific country or city has been particularly lucky or unlucky given our single historical record of landfalling hurricanes, in a way that the hole-filling of traditional stochastic catastrophe modelling may not reveal?
2. Was any single highly impactful historical event extremely unusual, even after the stochastic process attempted to fill in gaps, or do we see these events regularly in our NWP modelling?
3. Were highly active years such as 2005 and 2020 extreme outliers, as a traditional observation-derived stochastic view would suggest, or are they more common than they appear in our historical record?
4. Were particularly active/inactive periods in our history (e.g., Hall and Hereid, 2015) random or structural in a way that dynamical climate and NWP models may reveal? Relatedly, are there correlations in active/inactive seasons following one another, and is the assumption that individual hurricane seasons are independent, as is often derived during stochastic modelling of NAHUs, a good one?
5. Considering the US coastline in aggregate, where does our observed history sit in the distribution of possible outcomes, and what does the relative uncertainty between stochastic modelling and counterfactual analysis look like?

Some of the above questions can be seen to fall under the more general area of climate attribution research; it is important to note that it is necessary to use counterfactual analysis in the climate attribution process to allow questions on non-events to be asked, and thus for probabilistic views on events to be gleaned (van Oldenborgh et al. 2021). For example, without explicit counterfactual data, it is possible to ask "was the 2012 Superstorm Sandy event made more likely by climate change?", but it is impossible to ask the philosophical reverse of that question - i.e., "was the non-occurrence of event X made more likely by climate change?" – because there is no concrete event to begin to quantify changes from.

Further than facilitating the derivation of probabilistic information from climate attribution research, the first four examples in the list above speak to investigating the potential to "hedge" risk – with (1) and (2) focused on spatial hedging, and (3) and (4) focused on hedging of temporal aspects, or certain parts of an Exceedance-Probability (EP) curve.

The spatial distribution of risk is an important business consideration when a (re)insurer constructs its portfolio geographically. Understanding a portfolio's areas of high or low levels of accumulation are key to determining whether to take on or hedge additional risk in a specific geographic area, be that a city, county, or state. The assessment of portfolio shape is typically determined using stochastic catastrophe models and calculating key statistics such as the average annual loss (mean expected loss) and a range of return periods for the aggregate loss distribution. Deterministic scenarios, such as the RDSs mentioned previously, also provide context against (re)insurers' risk appetite statements regarding the maximum downside they wish to be exposed to under given conditions, but are limited in the sense that they fail to give statistics relating to the loss distribution. Identification of bias or error within the stochastic tools used to make this assessment is important in determining the underwriting strategy and capital management and influences decisions regarding geographic portfolio mix, technical pricing assessment and any mechanisms used to mitigate risk.

(Re)insurers will typically mitigate their gross risk and manage their capital with the use of ceded reinsurance protections. Various structures and vehicles can be used to take the risk carried by the original insurer and pass it further up the risk chain to another party in exchange for a premium. Some of these transactions focus on the severity of potential losses, looking to cap the downside from a single event. In this case, understanding the probability of large events occurring is crucial. Whether a particular geography, or indeed an area in aggregate, has been "lucky" or "unlucky" historically feeds directly into the real-world decision regarding how best to manage capital through risk reduction; any objective information that can help better inform this concept of historical luck would be greatly welcomed.

In addition to spatial conditions, the frequency distribution is another important parameter for managing capital in the form of ceded reinsurance. An example of this in the context of a core business decision is deciding when to buy cover that will pay out dependent on the number of landfalling storms in a hurricane season. There is a cost and benefit to the number selected, and determining the preferred choice relies on assessing the probability of these protections being required. Hence, the application of counterfactual analysis in this area could have very tangible use in the (re)insurance market.

Finally, this type of analysis may seek to present opportunities to (re)insurers to take on risk that – when looked at with traditional stochastic methods – appears not to fit within their target risk profile, due to price or aggregation of risk in a particular area. In turn, the analysis could help to benefit consumers of (re)insurance in areas that have been "unlucky" in the past, by allowing risk to be reassessed and viewed with a counterfactual perspective.

10.3 Methods & Data

Ten thousand counterfactual NAHU histories are generated from ensemble reforecast NWP data for the period 1985–2016, along with one "best-estimate" reforecast history that represents our historical baseline, from which to draw fair comparisons. This section details the data selection and processing methods.

To circumvent any issues regarding tropical cyclogenesis biases in NWP output (e.g., Halperin et al. 2013), the selection of the counterfactual ensemble data is restricted to finding alternative tracks only of the NAHUs that are recorded in our observational history. This immediately imposes a limitation on the output – and is a key divergent point from the setup of Ng and Leckebusch (2021) – as we will not capture certain types of counterfactual realities of NAHUs. For example, we are unable to capture NAHUs that didn't form in our observed reality but may have done in a counterfactual one. Therefore, each one of the counterfactual histories will always contain the same number of tracks as has been observed in our history.

Ultimately, this means that the study is restricted to looking at track uncertainty of historical storms only, as opposed to a more complete track uncertainty coupled with genesis uncertainty. This means that, for the time being, the results would be unable to unpick some of the broader questions presented earlier. The authors acknowledge that this is likely to severely limit the real-world application of the data at present, and stress that the results should be seen as experimental at this point. However, given that the study is exploratory, and primarily aims to uncover potential limitations with this type of application, we feel that adding in the cyclogenesis aspect at this point has the potential to add a needless layer of complexity to experimental methods and results.

10.3.1 Data Selection

The counterfactual histories are generated from NWP reforecast data, as opposed to historical operational NWP forecast ensemble data, because a reforecast allows for a consistent dataset whilst also using a contemporary NWP model across the entire historical period of the study.

The NOAA PSL Global Ensemble Forecast System (GEFS) reforecast v2 (Hamill et al. 2013) is utilized as the reference reforecast dataset. This initializes daily (00z timestep) an 11-member (1 control + 10 initial condition ensemble members) forecast, with a running period of 16 days, approximately 40–54-km spatial horizontal resolution, and 6-hourly temporal resolution. The data used in the study runs from 1985 to 2016.

Given the need to match to historically observed NAHUs, two more datasets are employed, namely:

i. the observation-based International Best Track Archive for Climate Stewardship (IBTrACS), which enables the identification of historical NAHUs and their real-world intensities.
ii. the National Center for Environmental Protection's Climate Forecast System Reanalysis (NCEP-CFSR) (Saha et al. 2010), for allowing track matching between the model world of the NWP reforecasts and the observation world of the IBTrACS data.

Importantly, both datasets are available for the entire period of the GEFS reforecast data.

10.3.2 Tracking and Storm Matching Part 1: Reanalysis to Observations

Tropical Cyclones are identified and tracked in the NCEP-CFSR reanalysis using the TRACK algorithm (Hodges 1994, 1995; Hoskins and Hodges 2002). Initially all cyclonic systems are tracked in the NCEP-CFSR and the tracks are then matched to, and filtered by, the IBTrACS tracks using mean separation matching (Hodges et al. 2017). Any amount of temporal overlap and a mean separation distance of 5 degrees (geodesic) for the overlap periods causes a track match.

10.3.3 Tracking and Storm Matching Part 2: Reforecast to Matched Reanalysis

The TRACK algorithm is then applied to the GEFS reforecast. These tracks are subsequently matched to the previously matched and filtered reanalysis tracks using mean separation matching: the reforecast tracks are matched to the reanalysis tracks using the first day of the forecast track that overlaps with the analysis track to within a 4-degree (geodesic) radius, and the reforecast tracks have their first point within the first 3 days of the forecast (Hodges and Klingaman 2019; Froude et al. 2007).

This allows for the TCs to be found in the NWP reforecast data before they are identified in the observations. The resultant combined data files contain up to 12 tracks per historical track per day (due to the daily GEFS initialization) – one reanalysis track of a historical storm, one control track of that storm from the GEFS reforecast, and up to 10 GEFS initial condition ensemble members (the number of GEFS ensemble members being dependent on whether the perturbed ensemble members continue to develop the storm or not).

10.3.4 Tracking and Storm Matching Part 3: Reforecast Tracks to Observational Tracks

A final matching and filtering step of the reanalysis and reforecast tracks to the IBTrACS data is undertaken to confirm the tracking and matching process has been successful, and to ensure that a historical hurricane "name" is attached to the reforecast and reanalysis track data. First, a match occurs if there is at least one timestamp that the NCEP-CFSR and the IBTrACS track are within 1 degree (geodesic) of each other. The variable used from the NCEP CFSR tracks to define the center of the storm is the latitude & longitude of the maximum 850hPa vorticity center. Using this method across the entire study period, there are ten unmatched IBTrACS storms with the criteria at 1 degree. The spatial-matching region is therefore relaxed successively: five more NAHUs are matched within 2 degrees, and two more are matched within 5 degrees. Three storms remain unmatched, namely Matthew (2004), Zeta (2005), and Barbara (2013). These storms are thus, in effect, filtered from the historical dataset, both for our "observational" model history, and also for our alternative history creation.

At this step in the process the combined track files contain up to 12 tracks per historical storm and, because of the daily initialization of the 11-member GEFS reforecast, there is a new combined track file created each day that an observational track exists. For example, in our observed record, Hurricane Andrew (1992) formed on August 16th, 1992, and dissipated on August 29th, 1992. Thus, with 14 days of existence, and with a maximum of 12 storm tracks per day, we have a maximum of 168 tracks across the 14 combined files for Hurricane Andrew. However, the reanalysis track for a single storm will be identical in each of the daily files for a single historical storm, as it is merely a reference track from a single model run that will have been truncated to start on the date of the GEFS initialization. Thus, the number of different GEFS tracks that are theoretically available for selection into the counterfactual histories is 11 multiplied by the number of days a single storm is active. In the case of Hurricane Andrew, this would be a maximum of 154 distinct tracks that could be selected from for addition into the counterfactual histories.

10.3.5 Creation of Extended Landmasses for Track Selection into Counterfactual Histories

There are many potential ways to construct the counterfactual histories from the track files, and it is at minimum difficult, but arguably impossible, to completely remove all levels of subjectivity from this process. While it would be possible to collate all of the daily GEFS tracks for a single storm and simply randomly sample from them, we realize that this may introduce structural issues. For example, it is likely that sampling for Hurricane Andrew (1992) from GEFS data initialized at a point shortly before US landfall, versus Hurricane Katrina (2005) initialized in the Atlantic basin's Main Development Region, would introduce structural biases. Conversely, if generating counterfactual histories by selecting an ensemble member at only the time and date that the storm first appears in the IBTrACS observed data, we limit ourselves to very little data (i.e., we remove the potential to use all of the ensemble reforecasts created after the start date of the storm in the record, and thus only have the ability to select from a maximum of 11 GEFS versions of Andrew), while at the same time we may also introduce steering biases that are present in the more cyclogenesis prone regions (e.g., Main Development Region) of the Atlantic. And further, if we introduced a single "spatial barrier" that the storm would have to cross for it to be included in selection (for example, if we stated that the storm would have to cross the 55th Meridian West for it to be included in the sample selection), we would still be constrained both by limiting data and by introducing a potentially difficult to untangle structural issue.

To attempt to combat any region-specific model bias issues, utilize as much data as possible, and have the ability to piece together any unforeseen structural issues, we introduce a novel methodology in which multiple theoretical "extended" landmasses are generated at various distances from the US coastline. Once a historical storm crosses the line of the theoretical extended landmass, the GEFS tracks for that storm will be available for an ensemble member selection to create the alternative histories. For example, on the date that Hurricane Andrew crosses the 300-km extended landmass in the observational data, the GEFS reforecast data initialized on that day becomes available for selection into the counterfactual histories. While this means that we will only retain the storms that have occurred in our history, it's important to note that this method still allows for large divergence of the tracks – for example, because the GEFS reforecast data is free-running and not constrained by observations, some or all of the Hurricane Andrew tracks in the forecast may not make actual landfall in the US. It is this aspect of the analysis that we hope will begin to allow us to probabilistically re-evaluate our history, and even probabilistically re-evaluate specific historical events.

Ideally the number of landmasses should be so numerous as to use as much of the ensemble data as possible, without being so exhaustive as to cause overly cumbersome re-selection of data that has already been used. Four different "extended landmasses" are created using QGIS and converted to a grid with a resolution of 0.1 degrees. These landmasses are generated at 300-km intervals at distances between 400-km and 1300-km from the North American coastline. While the choice of extended landmass distances will always carry some level of arbitrariness, they are here chosen given knowledge about NAHU translation speeds and daily initialization limitation of the GEFS data. With a mean NAHU translation speed between 18 and 25-km/h (Kim et al. 2020), it is likely that the average NAHU will travel approximately 432–600 km distance in 24 h, which thus represents the maximum distance that would be reasonable to employ between extended landmasses. Given the further reality that NAHUs neither travel perpendicularly to extended landmass contours, nor in straight lines, the distance between extended landmasses is reduced to 300 km.

10.3.6 Counterfactual History Creation

2,500 histories per extended landmass are created, producing a total of 10,000 counterfactual histories. With the period of analysis, this creates 320,000 years (or NAHU seasons) of data. It is important to note that, because of the finite number of GEFS ensemble storms per observed historical event, generating this many histories is likely to cause multiple selections of the same ensemble storm on occasion. Thus, the 2500 histories at each extended landmass cannot be considered entirely independent of one another. However, the method remains desirable for risk management because:

i. this re-selection of the same ensemble member multiple times potentially better allows the "worst-case scenario" of what the continuous chain of the most deleterious events in a single season could have been.
ii. The method produces 10,000 versions of each individual historical year. This is a somewhat standard number for the minimum number of years desirable for a

catastrophe model stochastic set (Jewson et al. 2019), and allows us to delve deeply into key loss years, such as 2005, while retaining probabilistic rigor.

To create the histories, the reanalysis tracks that make landfall with respect to the theoretical extended landmass (i.e., by crossing the imaginary line of the extended landmass) and have been successfully matched by name to a historical IBTrACS Tropical Cyclone are identified. Tracks that form whilst already over the extended landmass are also included so as not to impose a filtering of the storm number by virtue of their point of genesis. Thus, the date on which they cross the line of, or first form on, the extended landmass is used for ensemble selection. For example, for the 1300 km landmass, Hurricane Andrew's (1992) GEFS ensemble member will be selected for the initialization date that it crosses the line of said extended landmass. Hurricane Wilma (2005), however, formed in the Caribbean Sea, and thus technically never crosses the 1300 km landmass line because it already forms *on* the extended landmass. Therefore, Wilma (2005) is kept in the 1300 km landmass selection on the date that it forms.

Thus, an ensemble track is randomly selected from the GEFS reforecast data for each storm name on the date that it forms (if that formation point exists on the extended landmass), or on the date that it first crosses the line of the extended landmass.

A further filtering of the data occurs at this point: only an initial condition ensemble member (i.e., not the GEFS control member) can be selected for inclusion in the counterfactual histories. This decision was taken to remove the concept that some member selections may have been "better-estimate" (in the case of picking the control vs an initial condition member), and thus have introduced a probabilistic bias for some alternative history tracks.

10.3.7 GEFS Based Observational History

For comparison of the counterfactual histories to "reality", it is obvious that a direct evaluation between the counterfactual alternative histories and the IBTrACS data would be unfair; the limitations imposed by resolution and, relatedly, incomplete physics will make the GEFS model NAHUs, both in terms of track and intensity, look different from the observational IBTrACS history. Thus, the differences between the reforecast and observational data are minimized by creating a GEFS reforecast model view of our observed reality from which to make these comparisons.

However, this again is not a trivial task. The reforecast data are not constrained by observations while the forecasts are running, and thus the tracks of the GEFS data are likely to vary from the orientation of both the tracks seen in the observational history and in the reanalysis. We therefore use a "0km" landmass – in effect, just the US coastline – to generate the GEFS model-based history at as close to a timestep as possible from the GEFS initialization, and we only take the control run (i.e., the unperturbed model run, which in this instance could be considered to be the model best estimate) from the reforecast for this landmass. This is because, with the GEFS model initialized so close to land (i.e., at the timestep before it makes landfall), the GEFS model does not have a chance to materially impact the track or intensity of the storm. Thus, we would likely end up with a cluster of events at landfall in most situations. Further, this 0-km data is only intended to act as the benchmark data from which to understand relativities in the counterfactual histories, so we only need a single best estimate from which to do this.

While the creation of the 0-km benchmark GEFS data means that IBTrACS-derived observational US landfall locations should match fairly well with the 0-km reforecast data, the storm could still be some way away from landfall because of the timestep limitation enforced by the reforecast data. A cubic-spline interpolation is thus applied to the GEFS and IBTrACS data to up-sample the tracks to 15-minute temporal resolution, which allows for close temporal matching of the two datasets at the precise point of landfall. This same interpolation is later applied to landfalling storms in the alternative histories to allow for fair comparisons. The choice of cubic-spline interpolation here follows similar temporal resampling studies (e.g., Baudouin et al. 2019).

10.3.8 Intensity Downscaling

As mentioned previously, resolution and incomplete model physics data mean that windspeeds are likely to be systematically different between the model data and the IBTrACS observational data. While the potential issue is analytically negated by creating a model-based observational history from which to draw direct comparisons, reporting coherently on impact-based narratives for risk-focused communities is difficult to accomplish without attachment to easily understood intensity metrics, such as the Saffir-Simpson scale.

Thus, a simple statistical downscaling is applied to the histories to bias-correct them toward the usually higher observational intensities. The mean windspeed of all category 1+ hurricanes at landfall in the IBTrACS data is calculated. Using the names as matched in Sect. 10.3.4, the same hurricanes are then extracted from the control run of GEFS from the 0-km landmass. The hurricanes in GEFS are then interpolated to match the timing of the landfall in IBTrACS. The mean windspeed of these GEFS landfalls is then calculated. The percentage difference between the two is calculated, and this single factor scaling uplift is then applied to all GEFS data. The authors acknowledge that this is an overly simplistic method for generating accurate intensities across all Saffir-Simpson categories, but we purposefully keep the method simplistic here so that we can better focus on questions relating to the counterfactual analysis methods.

In Fig. 10.1, the red bars show the counts of US landfalls per Saffir-Simpson category from the raw GEFS 0-km history, while the blue bars show the counts per

Fig. 10.1 Counts of landfalling NAHUs in the GEFS 0-km reforecast history, split by Saffir-Simpson Category, pre-(red) and post- (blue) statistical downscaling to bias correct toward IBTrACS intensities. As can be seen, in the pre-downscaled data there are virtually no Major Hurricane (cat 3+) landfalls in the entire 1985–2016 study period; the presence of Major Hurricanes can thus be said to be better represented in the post-downscaling data. Further analysis can be found in Fig. 10.2

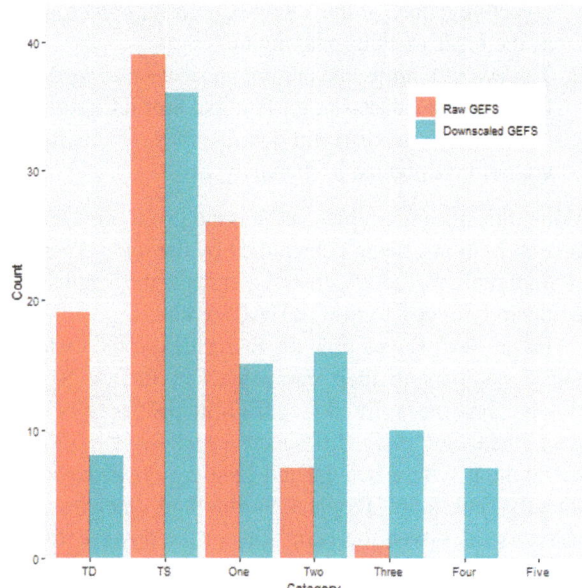

category after the downscaling has been applied. While crude, the overall results and narratives are unlikely to be negatively impacted given that we are primarily looking at the relative analytics of model-derived data for both our observational history and our alternative history.

10.4 Results

10.4.1 GEFS vs IBTrACS Observational History Differences

Before analyzing the differences between the statistics of the counterfactual histories and the GEFS-derived "observational" history, it is also important to note that the matching of these two landfalling data are not perfect. For example, 10 Tropical Cyclones that appeared as being US landfalling in the GEFS data did not have a corresponding landfall in the IBTrACS data. Upon unpicking, there were two key reasons for the mismatches:

1. IBTrACS has a human element to the track distance recorded. For example, it is up to a human forecaster to decide when a Tropical Cyclone has come into being, and when it has dissipated, with the human forecaster usually having to decide during an ongoing live event (though this can be corrected/adjusted later – but with no less subjectivity). Conversely, the GEFS tracks are constrained by the

objective nature of the TRACK algorithm, which defines the center of the storm as the local vorticity maximum.
2. The eye of a hurricane coming very close to land and only crossing the coastline in one of the datasets (e.g., Tropical Storm Cristobal, 2008). The relative diameter of the eye in the different datasets may add complicating consequences here that are not investigated in the analysis.

While this does not represent an issue given our GEFS model created history based on observations, being constrained by the same criteria as the alternative histories, it is important for us to stress again that the results presented here should not be compared directly to raw IBTrACS data.

Figure 10.2 shows how the intensities of the matched and downscaled storms in GEFS compare to their equivalents in IBTrACS. It can be seen that while the downscaling has uplifted the GEFS intensities to being comparable to the IBTrACS data, there are clear differences, such as the IBTrACS data having a skewed distribution with a heavier tail than the downscaled GEFS data. IBTrACS consequently sees more Tropical Storms and category 1 hurricanes, as well as more category 5s. Given the simplicity of the downscaling, the effects of which can be seen in these results, it was decided that only hurricane intensity storms would be analyzed. No further sub-division by intensity (e.g., between minor and major hurricane) is made in any of the analyses.

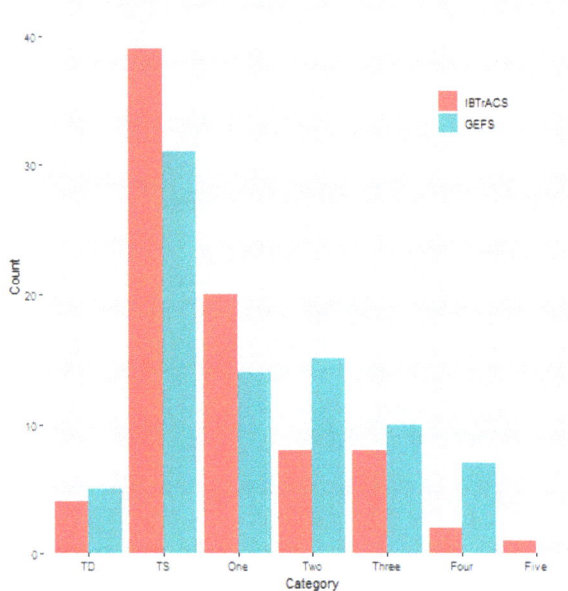

Fig. 10.2 Comparison of Counts, split by Saffir-Simpson Category, of Matched NAHU US Landfalls (0-km landmass) in the IBTrACS data and the Downscaled GEFS data. It can be seen that while overall counts per category are somewhat comparable, the statistical downscaling is not perfect, and leads to a less peaked + quicker decaying distribution than has been observed. For the relativity comparison of this study however, it is envisaged that this will not have significant impacts

10.4.2 Extended Landmass Histories: US NAHU Landfalls, All Categories

Table 10.1 shows the counts of NAHU landfalls in the US from the 0-km landmass data, and the mean number of landfalls per history per extended landmass distance. Figure 10.3 shows extended landmass histograms for the frequency of alternative histories split by the number of US landfalling storms.

What is immediately obvious from the table is that the mean landfalls for each of the extended landmasses are between approximately 41–52% of the observed (0-km landmass distance) landfall number. Secondly, the table and histogram plots show a consistent drop in US landfalling rates as the extended landmass is pushed further and further away from the US coastline. At 400 km, the mean number of US landfalling hurricanes in the histories is 28.5, and this drops away gradually until the mean of 22.8 at 1300 km. The range of the data is quite stable throughout, usually approximately +/− 8 hurricanes. It can therefore be said that this is a systematic effect that occurs consistently across extended landmass distances.

While finding such a systematic effect was an unexpected result, it is an intuitive one. On average, as one moves further from a coastline, the probability of a hurricane making landfall decreases (Brettschneider 2008). Thus, this is purely a probabilistic artefact that is imposed by the ensemble member selection from the extended landmass generation method. This has wide-ranging impacts on reliable information and potential conclusions that can be drawn from the results. Further discussion of this is therefore picked up in Sect. 10.5.

10.4.3 City Specific Investigation

The box and whisker plots in Fig. 10.4 show the number of hurricanes making landfall in the metropolitan areas of the cities Houston, Miami, New Orleans, New York City and Tampa over the historical period used in the study for all extended landmass distances aggregated together. While this figure was intended to facilitate the unpicking of the relative risk of the different cities – both from a mean activity and an extreme activity perspective – it is clear from the histograms in Fig. 10.4 that probabilistic impacts of the extended landmass selection method will be distorting these results. For example, Miami and New York City, both being

Table 10.1 No. of US NAHU landfalls (Cat 1–5) per landmass distance from the GEFS histories. It should be noted that the 0-km landmass is the actual US coastline and is our GEFS based version of the observational history. It is therefore a single history and not a mean count. The 400, 700, 1000 and 1300 km landmasses are all mean numbers of US NAHU landfalls generated from the counterfactual GEFS histories at each of these distances

Landmass distance	0-km	400-km	700-km	1000-km	1300-km
Hurricane count	48	28.5	25.2	23.4	22.8

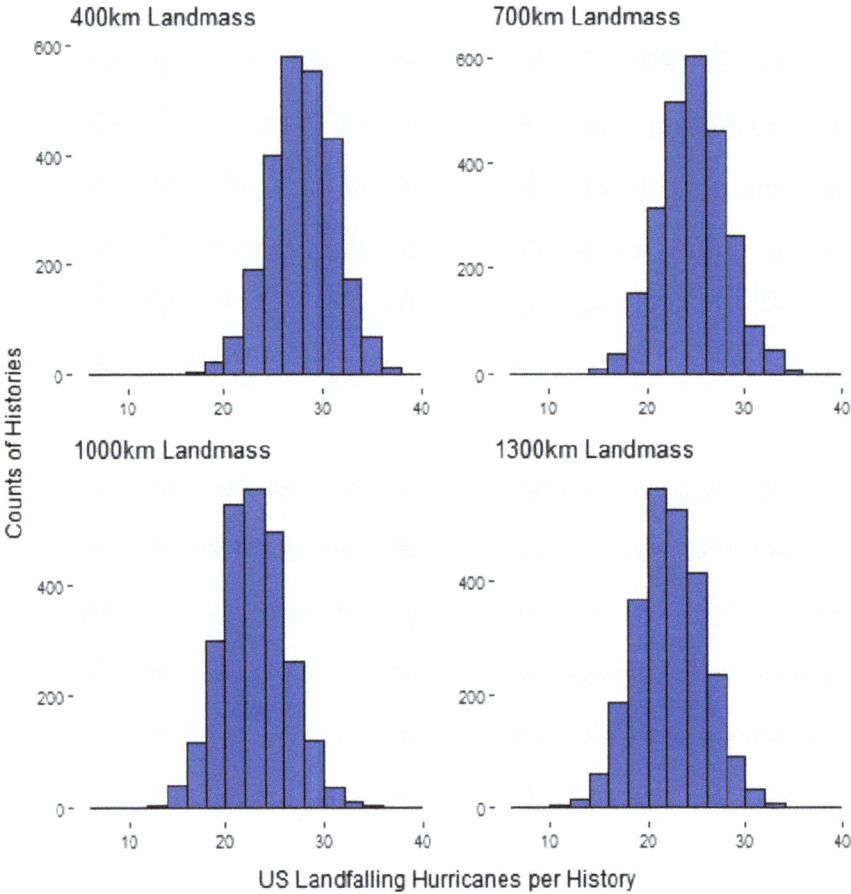

Fig. 10.3 Histograms for the counts of US hurricane landfalls in each of the alternative histories for the period 1985–2016, split by distance of the extended landmasses (top left: 400-km extended landmass, top right: 700-km extended landmass, bottom left: 1000-km extended landmass, bottom right: 1300-km extended landmass). The 0-km history is not shown because it is a single number, as opposed to a distribution

situated on the East Coast of the US, are much closer to the extended landmass ensemble selection lines than the cities of Houston and New Orleans in the Gulf of Mexico. Thus, there are likely to be probabilistic impacts that weight the east coast cities to seeing more landfalls in the alternative histories than the Gulf coast cities. Further, it is likely that there will be impacts even between the two east coast cities.

There are also likely region-specific issues with the simplistic downscaling when it comes to the North-East region. The sub/post/extra-tropical structure of the cyclones that make landfall here tend to have a broader wind field than their tropical counterparts, which weather forecast models more adequately capture the upper bounds of intensities of (Hodges & Emerton, 2015). Thus, the number of hurricanes in these regions may be over-inflated compared to the cities in the more tropical regions.

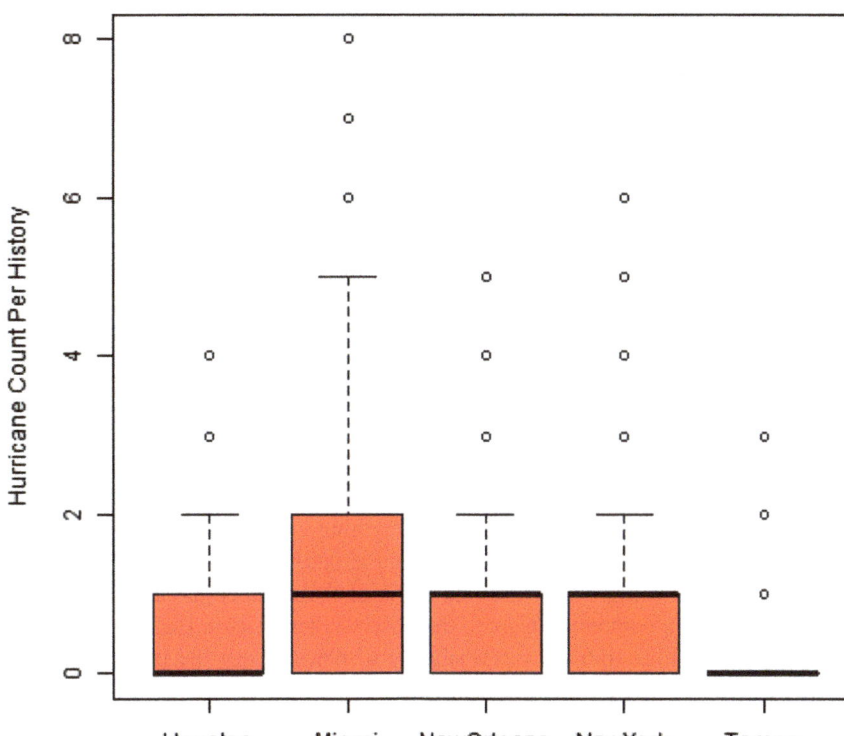

Fig. 10.4 Box and whisker plots for counts of hurricanes landfalling in the cities of Houston, Miami, New Orleans, New York City, and Tampa from the counterfactual GEFS histories. Data has been aggregated across all extended landmasses

We therefore encourage the cities in this part of the analysis to be viewed from a single-city variability perspective, as it is hypothesized that this aspect of the results still has the potential to be informative, if only from a fairly subjective basis. For example, it is very intriguing to notice that there seems to have been the potential for Miami to have been struck by up to 8 hurricanes in the historical period 1985–2016 – and this is for histories constructed from data that could be considered probabilistically weighted low, given what we can see for the numbers of landfalls from the counterfactual extended landmass histories relative to the observational history. The historical number of hurricanes observed over this period landfalling in Miami was just one.

10.4.4 Gate-Rate Maps

A key reason for stochastic modelling of Tropical Cyclone tracks is to appropriately fill gaps and extend the variability of our observational history. This type of variability is highlighted in the landfalling data that can be seen in Fig. 10.5, with various hotspots of landfalling activity in stochastic TC event sets likely driven by events that have occurred in a short historical period. While the stochastic track modelling process does help to overcome this hotspot issue, historical statistics are still preserved – at least to some extent – by the modelling process, and so it is always unknown if the issue is fully overcome. It is hypothesized that the counterfactual modelling process introduced here could help to unpick this issue because it offers the potential to include an independent data source in identification of potential hotspots. However, before this could be achieved, the probabilistic artefact from the extended landmass method would need to be overcome. At present, we know that the extended landmass selection method introduces a low bias toward landfalling US NAHUs.

To attempt to investigate whether the alternative histories can help to glean a better picture on local variability, Fig. 10.6 shows hurricane frequency "gate-rate" maps for the US that are generated by the alternative histories aggregated together. This type of gate-rate map is a relatively standard output in evaluations of catastrophe model output. While difficult to glean anything concrete in this instance, it is interesting to note that parts of south Florida and parts of the Gulf of Mexico look comparatively much higher hazard than some of their immediate adjoining gates. Additionally, when comparing back to the observed landfalls in Fig. 10.5, the peak landfall gates in Fig. 10.6 are often slightly displaced from the peak regions of observed landfalls. For example, the highest landfalling gate in the Gulf of Mexico would likely be centered around New Orleans if derived from observations, but from

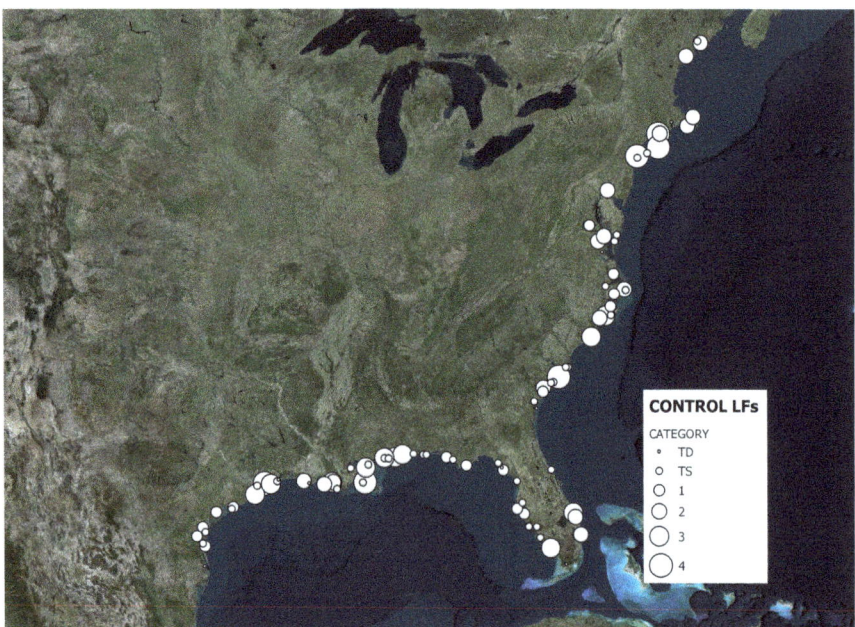

Fig. 10.5 Observed hurricane landfalls from the 0-km history (i.e., the best-estimate GEFS model-based observed history) for the period 1985–2016

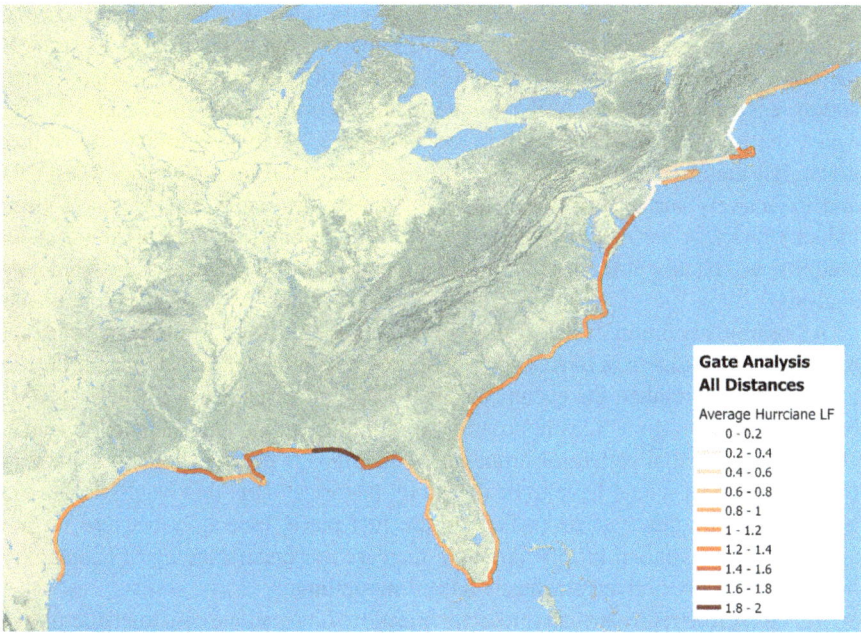

Fig. 10.6 Average number of NAHU landfalls per gate per counterfactual history for all extended landmasses

the alternative histories seems to be shifted further east toward the Florida panhandle. Similarly, from observations, the highest landfalling region in Florida looks like it would be on the mid-east coast of the state, while the alternative histories would suggest the highest hazard gate is the southern-most gate.

While it is difficult to know whether these conclusions hold, they raise important questions about how much trust should be put in probabilities derived from our single observed history, and thus in the stochastically derived risk estimates. This was the premise from which we hypothesized that these analyses could have real-world value. However, we have to stress at this point that the results should not be used practically given the biases apparent from the sampling issues driven by the extended landmass track selection process. Having said this, we believe that, from the foundational work presented herein, if the biases can be overcome there is large potential value in the use of these data from which to build probabilistic views of risk that are independent from traditional stochastic modelling techniques.

10.5 Discussion

Most importantly, there is a simple probabilistic US landfall artefact that falls out of the alternative history building from the method employed herein. As we move further and further from land, there is, on average, a lower and lower chance of a storm making landfall. For example, if we were to select a random storm near the Caribbean, it might have a 60% chance of US landfall, but if we selected a storm off the coast of Africa, it might have a 30% chance of making landfall. Thus, when adding ensemble members to the counterfactual histories from further and further away, the probability of US landfall, on average, is weighted further and further down. It is therefore currently impossible to use the data to compare to our 0-km data and objectively make statements such as "across the entire US, we have been unlucky/lucky in our observational history". The probabilistic artefact imposes an inability to reliably infer spatial information across the US coastline from this analysis.

A possible evolution of the method to overcome this issue could be to construct extended landmasses that have a consistent objective probability of US landfall, and to attempt to normalize the event rates of the ensemble history at a single "probability" distance to the US, as opposed to an arbitrary geographical/spatial distance. However, the spatial pattern of probabilistic landfalls is neither simple nor static in time, and it itself would have to be driven by historical data; this would negate the benefit of doing this type of analysis in the first place, especially because we are aiming to move beyond historical data to increase our understanding of uncertainty in observationally derived stochastic hazard modelling.

This therefore represents an important question: how can we construct alternative histories from model data in such a way to allow us to derive novel probabilistic information? While this question may seem somewhat narrow in scope with this single application (i.e., generating counterfactual NAHU histories) in mind, the impacts of this are much more wide-ranging, and likely extend to the modelling of any other atmospheric peril, as well as more longer-term climate change risk-oriented questions.

Thus, while these limitations currently exist in the results, it seems likely that overcoming them has the potential to unlock many opportunities for enhancing views of atmospheric hazard for many risk-focused practitioners.

10.6 Suggestions for Future Research

As the discussion section presented, if the valuable applied aspects of this work are to be achieved, it is the probabilistic challenge presented in previous sections that needs significant attention. Further than this, there are two other immediate avenues of research that would be valuable to address given the limitations already imposed.

The first is in addressing the tropical cyclogenesis aspect to include non-observed hurricanes in alternative histories. While it could easily be included in the methodology above, it will introduce another complex issue that will need to be overcome before the data are fully coherent. The second would be to add alternative NWP models into the study to see how much model difference drives variability in the results.

References

Bank of England (2019) The 2021 biennial exploratory scenario on the financial risks from climate change. Accessed Online May 19th, 2021, from https://www.bankofengland.co.uk/-/media/boe/files/paper/2019/the-2021-biennial-exploratory-scenario-on-the-financial-risks-from-climate-change.pdf

Baudouin JP, Caron L-P, Boudreault M (2019) Impact of reanalysis boundary conditions on downscaled Atlantic hurricane activity. Clim Dyn 52:3709–3727

Brettschneider B (2008) Climatological hurricane landfall probability for the United States. J Appl Meteorol Climatol 47:704–716

Carozza D, Boudreault M (2021) A global flood risk modeling framework built with climate models and machine learning. J Adv Model Earth Syst 13:e2020MS002221. https://doi.org/10.1029/2020MS002221

Frigg R, Thompson E, Werndl C (2015) Philosophy of climate science part II: modelling climate change. Philos Compass 10(12):965–977. https://doi.org/10.1111/phc3.12297

Froude LSR, Bengtsson L, Hodges KI (2007) The predictability of extratropical storm tracks and the sensitivity of their prediction to the observing system. Mon Weather Rev 135:315–333

Hall T, Hereid K (2015) The frequency and duration of U.S. hurricane droughts. Geophys Res Lett 42:3482–3485. https://doi.org/10.1002/2015GL063652

Halley E (1693) An estimate of the degrees of mortality of mankind, drawn from the curious tables of the births and funerals at the City of Breslaw, with an attempt to ascertain the price of annuities upon lives. Philos Trans 17:596–610

Halperin DJ, Fuelberg HE, Hart RE, Cossuth JH, Sura P, Pasch RJ (2013) An evaluation of tropical cyclone genesis forecasts from global numerical models. Weather Forecast 28(6):1423–1445

Hamill TM, Bates GT, Whitaker JS, Murray DR, Fiorino M, Galarneau TJ Jr, Zhu Y, Lapenta W (2013) NOAA's second-generation global medium-range ensemble forecast dataset. Bull Am Meteorol Soc 94:1553–1565. https://doi.org/10.1175/BAMS-D-12-00014.1

Hodges KI (1994) A general method for tracking analysis and its application to meteorological data. Mon Weather Rev 122:2573–2586

Hodges KI (1995) Feature tracking on the unit sphere. Mon Weather Rev 123:3458–3465

Hodges KI, Emerton R (2015) The Prediction of Northern Hemisphere tropical cyclone extended life cycles by the ECMWF ensemble and deterministic prediction systems. Part I: tropical cyclone stage. Mon Weather Rev 143:5091–5114

Hodges KI, Klingaman NP (2019) Prediction errors of tropical cyclones in the western north pacific in the met office global forecast model. Weather Forecast 34:1189–1209

Hodges KI, Cobb A, Vidale PL (2017) How well are tropical cyclones represented in reanalysis datasets? J Clim 30:5243–5264

Hoskins BJ, Hodges KI (2002) New perspectives on the Northern Hemisphere winter storm tracks. J Atmos Sci 59:1041–1061

Jewson S, Barnes C, Cusack S, Bellone E (2019) Adjusting catastrophe model ensembles using importance sampling, with application to damage estimation for varying levels of hurricane activity. Meteorol Appl. https://doi.org/10.1002/met.1839

Jones S, Raven E, Toothill J (2020) Assessing the global risk of climate change to re/insurers using catastrophe models and hazard maps, EGU General Assembly 2020, Online, 4–8 May 2020, EGU2020-5323. https://doi.org/10.5194/egusphere-egu2020-5323

Kim S-H, Moon I-J, Chu P-S (2020) An increase in global trends of tropical cyclone translation speed since 1982 and its physical causes. Environ Res Lett 15:094084

Lloyd's (2017) Reimagining history: counterfactual risk analysis. Accessed Online May 19th, 2021, from https://assets.lloyds.com/assets/reimagining-history-report/1/Reimagining-historyreport.pdf

Lloyd's (2021) Realistic disaster scenarios, scenario specification 2021. Accessed Online May 19th, 2021, from https://assets.lloyds.com/media/e73cc2f7-a535-4eaf-8196-fedcf5e1432c/2%20RDS%20Scenario%20Specification%20%20January%202021.pdf

Ng KS, Leckebusch GC (2021) A new view on the risk of typhoon occurrence in the western North Pacific. Nat Hazards Earth Syst Sci 21:663–682

Roese NJ (1997) Counterfactual thinking. Psychol Bull 121:131–148

Saha S, Moorthi S, Pan H-L et al (2010) The NCEP climate forecast system reanalysis. Bull Am Meteorol Soc 91:1015–1057

van Oldenborgh GJ, van der Wiel K, Kew S et al (2021) Pathways and pitfalls in extreme event attribution. Clim Change 166. https://doi.org/10.1007/s10584-021-03071-7

Woo G (2019) Downward counterfactual search for extreme events. Front Earth Sci 7. https://doi.org/10.3389/feart.2019.00340

Open Access This chapter is licensed under the terms of the Creative Commons Attribution 4.0 International License (http://creativecommons.org/licenses/by/4.0/), which permits use, sharing, adaptation, distribution and reproduction in any medium or format, as long as you give appropriate credit to the original author(s) and the source, provide a link to the Creative Commons license and indicate if changes were made.

The images or other third party material in this chapter are included in the chapter's Creative Commons license, unless indicated otherwise in a credit line to the material. If material is not included in the chapter's Creative Commons license and your intended use is not permitted by statutory regulation or exceeds the permitted use, you will need to obtain permission directly from the copyright holder.

Chapter 11
Estimating Tropical Cyclone Vulnerability: A Review of Different Open-Source Approaches

Katy M. Wilson, Jane W. Baldwin, and Rachel M. Young

Abstract Tropical cyclone (TC) risk assessments are critical for disaster preparedness and response. Alongside hazard and exposure, accurate TC risk assessment requires understanding the vulnerability of populations and assets. In this chapter, we examine multiple methods that have been used to assess and quantify TC vulnerability with a focus on open-source methods. We separately discuss structural, economic, and social (or demographic) vulnerability approaches. Structural vulnerability assesses the susceptibility of buildings to be affected by their exposure to hazards; in this section, we provide a detailed overview of how FEMA's Hazus model quantifies damages by utilizing engineering principles. Economic vulnerability employs regression analysis to relate wind speeds to damages; this discussion explores typical functional forms used to represent vulnerability in such analysis and efforts to constrain parameters in these functions. Finally, social approaches use demographic data to characterize the varying susceptibility of populations to TC risk; we provide some representative examples of this methodology. We conclude with a comparative discussion of these three classes of methods, suggest directions for future work, and ask whether the different approaches can be combined to yield a more holistic view of both the human and structural aspects of TC vulnerability.

Keywords Tropical cyclone risk assessment · Hurricane vulnerability · Structural vulnerability · Economic vulnerability · Climate vulnerability · Climate risk assessment

K. M. Wilson
Graduate School of Arts and Sciences, Columbia University, New York, NY, USA
e-mail: katy.wilson@columbia.edu

J. W. Baldwin (✉)
University of California Irvine, Irvine, CA, USA

Lamont-Doherty Earth Observatory, Columbia University, Palisades, NY, USA
e-mail: jane.baldwin@uci.edu

R. M. Young
School of Public and International Affairs, Princeton University, Princeton, NJ, USA
e-mail: rmyoung@princeton.edu

© The Author(s) 2022
J. M. Collins, J. M. Done (eds.), *Hurricane Risk in a Changing Climate*,
Hurricane Risk 2, https://doi.org/10.1007/978-3-031-08568-0_11

11.1 Introduction

Tropical cyclones are generally associated with sustained high winds, storm surges, and heavy rainfall, and cause widespread damage and fatalities upon landfall. TCs pose the greatest threats to coastal environments and populations, and have cost more human lives than any other natural disaster (Shultz et al. 2005; Li and Li 2013). Compounding on existing hazards, the frequency of more intense TCs and height of TC-induced storm surge is predicted to increase with climate change (Holland and Bruyére 2014; Little et al. 2015). While it is not possible to prevent TCs, their impacts can be minimized by implementing risk management techniques.

To effectively reduce TC risk, quantitative risk assessment of expected TC damages is essential. The effects of TCs are mediated by a combination of physical attributes of the storms and their interactions with human behavior (Shultz et al. 2005). This risk is often categorized into three components: hazard, exposure, and vulnerability. Hazard is the physical characteristics of the disaster, such as surge, wind, and rain of TCs, which may worsen with climate change (Holland and Bruyére 2014; IPCC 2021). Exposure is the human, structural, or agricultural assets in the hazard-prone area; exposure to TCs is generally increasing as populations continue to move to and develop near coasts (UN 2017; Crowell et al. 2010). Finally, and most relevant to this review, vulnerability is the susceptibility to be harmed by a disaster, including both the degree to which physical assets will be damaged for a given hazard and the capacity of systems to deal with and adapt to a disaster (IPCC 2014). Attributes influencing vulnerability can include building design and construction, forecasting, preparedness, emergency response systems (Shultz et al. 2005), and the strength of social support systems and economic resources (Hallegatte et al. 2017). Change in these factors can alter vulnerability in space and time. Accurate quantification of TC vulnerability is critical to assess risk and ultimately better direct aid and investment to reduce vulnerability and enhance communities' resilience to these natural hazards.

This chapter reviews different methods for assessing vulnerability to TCs, with a particular focus on open-source methods of particular use to the scientific community. Our goal is to provide a reference for scientists perhaps more familiar with the hazard-side of risk modeling, while also identifying opportunities for interdisciplinary synergy and advancement in TC vulnerability assessment.

There are some existing frameworks to characterize vulnerability methods. O'Brien et al. (2007) categorize climate vulnerability into two approaches: outcome and contextual. Outcome vulnerability is the residual of hazard and exposure in determining an impact and is commonly described quantitatively. In contrast, contextual vulnerability is based on a multidimensional view of how climate interacts with society, considering political, institutional, economic, and social structures, and is typically described qualitatively. The authors argue that these two interpretations of vulnerability are fundamentally different. Outcome vulnerability adopts a scientific framing where climate change is a problem of human impacts on the global climate system. This approach is quantifiable through climate models and generally draws a clear distinction between nature and society. Contextual vulnerability assumes a human-security framing that views climate change as a process that affects humans and focuses on the consequences of climate on societies. O'Brien et al. (2007) state the two categorizations prioritize different knowledge types and, consequently, have different interpretations and applications that mean the approaches are complementary but, in many ways, cannot be combined.

This chapter will primarily focus on what O'Brien et al. (2007) define as outcome vulnerability, with a brief discussion of contextual approaches. As multiple outcome approaches are discussed, this chapter will use alternative disciplinary-based categorizations to indicate the distinctions. These are: structural, economic, and social. For both structural and economic methods the target of vulnerability modelling is to calculate the damage ratio, i.e., what fraction of the relationship between total losses to total exposed value is destroyed for a given hazard. However, these methods are differentiated in their level of spatial detail and how explicitly damage processes are modeled. Structural vulnerability approaches are based on how the characteristics of local buildings perform under TC hazards and largely exist in the civil engineering literature. This method utilizes specific information about the local building stock (Pita et al. 2013, 2015). However, as detailed building information is difficult to acquire in many parts of the world, it is not always an appropriate method. Economic-based work to quantify vulnerability exists at the intersection of macro-economics and natural hazards research, and represents vulnerability by relating different severities of a hazard (e.g., wind speed, storm surge, flooding) to proportion of exposed value damaged. Often, this regression-based method is more appropriate when data is limited, as the damage is aggregated over a region rather than at the property scale. In contrast, social vulnerability approaches use social information to define the susceptibility of local populations to climate impacts and related risks (Soares et al. 2012) and are mostly found in the geography literature. As discussed in this review, structural and economic approaches are forms of outcome vulnerability, whereas social approaches are more closely aligned with contextual vulnerability and are less targeted at calculating damage ratios.

This review focuses on open-source techniques for assessing TC vulnerability, both because we seek to aid academic research on this topic, but also because, by definition, such techniques can be understood in detail based on the peer-reviewed literature. However, there also exists substantial proprietary work to assess TC vulnerability. As windstorms are included in various insurance policies, the insurance industry has been at the forefront of examining the impacts of wind on property. The insurance industry frequently uses models to assess the risk to their insurance portfolios during TCs (Khanduri and Morrow 2003) and generate damage functions (or "loss functions") to determine monetary loss for a given TC or other disaster event (Watson and Johnson 2004). These damage functions are typically based on building loss data from insurance companies' post-disaster and help determine insurance premiums for customers in at-risk zones. Models used by the insurance industry are largely proprietary and details are rarely published in the public literature (Watson and Johnson 2004; Pita et al. 2015). The exception occurs when insurance companies are mandated to publish model results and sensitivity analysis by state governments. For instance, hurricane models in Florida are evaluated by a team of five experts who have open access to the rationale, inputs, outputs, and assumptions of these industry models (Pita et al. 2015), but cannot share any content except the sensitivity analysis (see Iman et al. 2002a, b; 2005a; b). Therefore, open-source models are important alternatives as they allow open and transparent dialogue about the methodologies, in addition to providing risk quantification for institutions and municipalities unable to purchase data from insurance companies and other catastrophe modeling firms.

There has been some prior work to review different approaches to estimating TC vulnerability. Notably, Ward et al. (2020) more broadly surveyed global natural hazard risk assessment, and in doing so, explored some economic-based TC vulnerability assessments. Conversely, Watson and Johnson's (2004) review primarily explored loss estimation methods in the insurance industry. Pita et al. (2013, 2015) focused specifically on TC vulnerability, but mainly explored structural-based methods. This review is differentiated by seeking to synthesize a broader range of approaches to characterizing TC vulnerability. We believe such a wide-ranging approach is useful to highlight the differences between various methods and identify opportunities for interdisciplinary collaboration. We will discuss multiple disciplinary approaches that have been utilized to characterize vulnerability, exploring each methodology and surrounding debates in turn. The rest of the review is structured as follows: Sect. 11.2 explores structural vulnerability methods, with a detailed discussion of the Federal Emergency Management Agency's (FEMA) Hazus model; Sect. 11.3 discusses economic approaches to quantifying vulnerability, examining first functional forms for damage based on wind and other TC-related hazards, and then approaches to constrain these functional forms' parameters based on historical damages; Sect. 11.4 presents attempts to incorporate several physical hazards into vulnerability modeling; Sect. 11.5 describes social approaches that assess vulnerability as a result of population-level socioeconomic characteristics; finally, in Sects. 11.6 and 11.7, we conclude with a summary of our findings and suggest potential ways forward to improve TC vulnerability estimates and in turn TC risk assessment.

11.2 Structural Vulnerability Approaches

Structural methods use local data, including specifics on the composition of buildings, to model vulnerability under specific TC conditions. As a result of the building specific modelling, structural approaches are well suited to provide a detailed representation of a place's vulnerability when data availability permits. Estimating structural vulnerability requires determining building performance during TC-related hazards (Pita et al. 2013). Such methods use engineering principles of load and resistance to determine vulnerability. As a representative example of this class of approaches, in this section we describe how a leading model, FEMA's Hazus, quantifies the performance of buildings under TC hazards (FEMA 2012). We emphasize that this model emerges from much prior engineering work, and continues to be employed in new ways and built upon today. The publicly available model has numerous benefits as a research tool, but also helps the insurance industry in the ratemaking process, as the detailed analysis enables insurance to be tailored to building types (Pita et al. 2013).

The development of Hazus was initiated in the early 1990s with the aim of creating a natural hazard loss estimation software tool for agencies tasked with mitigation and decision making (Schneider and Schauer 2006). The software, released in 1997, was first developed for earthquake loss estimation, and other natural hazards including hurricanes and floods have since been included. The software includes the most up-to-date USA national datasets, building classification systems, and uses geographical information system (GIS) technology. In keeping with the scope of this chapter, this discussion of Hazus will focus on TC damage associated with winds. However, as Hazus is a multi-hazard model, it does have the ability to incorporate multiple hazards associated with TCs, including flooding and storm surges (see Sect. 11.4 for more detail).

To develop the hurricane model, the Hazus team reviewed existing wind loss estimation modeling techniques. The resulting Hazus hurricane model has five components: a hurricane hazard model, a terrain model, a wind load model, a physical damage model, and a loss model. As illustrated in Fig. 11.1, Hazus has a hierarchical structure that feeds data between models to calculate the building loss and damages associated with the specific wind characteristics of a TC. The model employs empirical data from wind tunnel experiments, and is validated using damage data from post-storm surveys and insurance company data (Vickery et al. 2006b). So, while the publicly available Hazus model does not require validation with claims data or historical observations at the time of use, and many of its methodologies are published, numerous components have been calibrated to proprietary data, e.g., validation with insurance claims data, so it is not entirely open-source.

Hazus is distinguished from the economic approach, described in Sect. 11.3, that empirically regresses losses against wind speed, as it uses physical principles of load and resistance to calculate building response and does not depend purely on historical observations for model development (Schneider and Schauer 2006; Vickery et al. 2006b; Pita et al. 2013). To communicate structural vulnerability, Hazus assigns a damage state for buildings in a region, which the loss model (Fig. 11.1) converts into a loss ratio to quantify structural vulnerability. Here we briefly describe these different model components, with particular attention paid to the wind load model and physical damage model, which contain different aspects of vulnerability as defined in this chapter.

The hurricane hazard model describes the track and wind field of a TC (Schneider and Schauer 2006), and was extended to include a hurricane rainfall model. The terrain model calculates the change in wind speed due to the roughness of the ground surface. As the ground becomes rougher, defined as a function of height and spacing between buildings, the wind speeds near the ground decrease while upper-level winds remain the same (Vickery et al. 2006a). The wind load model uses numerical modeling to estimate the directionally dependent wind-induced pressure acting on

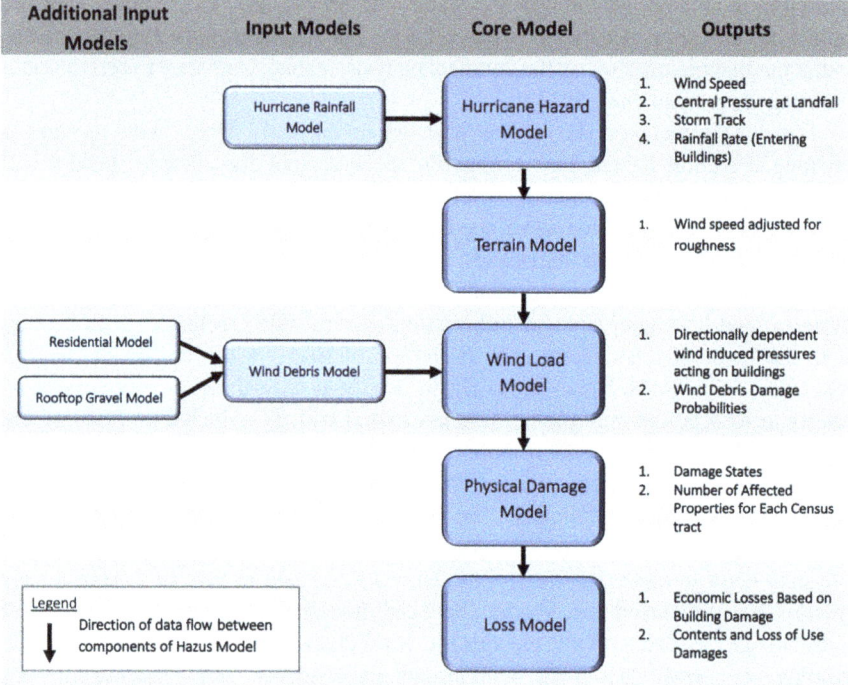

Fig. 11.1 Interconnections of the models used in Hazus as described by Vickery et al. (2006a, b). alongside the outputs from each stage of the process. The Hazus software utilizes multiple models to calculate the damage ratio for the region in question. The input models feed information into the core models, which run consecutively and use the outputs from the prior model. For instance, data from the hurricane rainfall model is used in the hurricane hazard model, the outputs, such as, windspeed and storm track, are input into the terrain model. This highlights the interdependencies between the various components of the Hazus model

the exterior of buildings. The model incorporates pressure coefficient data from wind tunnel tests developed for various roof types, including sloped roofs, flat low-rise buildings, and mid- and high-rise buildings. Validation testing against pressure coefficients obtained from empirical models demonstrated the Hazus model was able to effectively reproduce the variation in load with wind direction found in tunnel tests (Vickery et al. 2006a). The wind load model can also be coupled with a wind debris model to provide the inputs for wind-induced damage and loss. In high wind events, wind debris is a major contributor to damages. Therefore, modeling it is critical to ensure realistic damage representations. Hazus includes two windborne debris models, (1) a residential building type model developed using first principles for estimating hurricane debris impact probabilities and impact energy and (2) a rooftop gravel debris model, developed specifically for Hazus. After calculating the wind conditions associated with a specific TC event, the data from the hurricane hazard, terrain, and wind load models are fed into the loss and damage models (Fig. 11.1).

Physical damages of a building are determined by computing the wind-induced loads on the exterior of a structure. Hazus focuses on damage to the exterior components and casing, including windows, roof cover, joint failures, and wall failures. The model uses a load and resistance methodology to estimate damage of the structure subjected to hurricane winds, as determined by the prior components of the hurricane model. Hazus developed geometric representations of a range of building types to model the resistance of building components using (1) lab test data, (2) engineering analysis, and (3) engineering judgement. For each building type wind loads on all building components are estimated using the direction dependent coefficients from the wind load model and incorporate the wind debris model results. Damage states are developed for each building type. The states are dependent on building performance and range from zero (no damage) to four (destruction). In the case of residential buildings, damage state zero corresponds to limited to no roof failure, water penetration, and no broken windows. As the damage states increase, the occurrence of roof failure, window, and/or door failures increases so that at the highest damage state, there is at least 50% roof failure, extensive interior water damage, and major window and/or door damage (Vickery et al. 2006b). For each building type, the model output estimates the mean number of buildings expected to experience a given damage state for each census tract. Model validation found comparable observed and simulated damages (Vickery et al. 2006b).

Together the wind load and physical damages model account for the vulnerability components of Hazus. To obtain the Hazus equivalent to the damage function (Fig. 11.2a), the loss ratio is computed using exposed value. The loss model estimates losses by applying the building damage states, given by the physical damage model, to exposed value obtained from empirical cost estimation techniques for building repair and replacement (Vickery et al. 2006b). Losses include building damage, contents loss, inventory losses, and loss of building use. Firstly, building loss is estimated by determining the cost of components for each building model. For the given building damage, the rebuild cost is computed for external and internal components. Notably, the model only estimates losses for external building components, thus, empirical functions developed using engineering judgement and insurance company loss data are used to define internal losses. Secondly, contents and inventory loss are related to the performance of buildings during a TC and is largely a function of the volume of water entering a building. The model employs a hurricane rainfall model to estimate the amount of rain entering a building. Rainfall is determined by the storm intensity and, with the inclusion of a calibration factor, provides fair estimates of hurricane rainfall rates. However, Vickery et al. (2006a), acknowledges its limitations to account for all factors affecting hurricane rainfall and states it does not determine inland flooding. Using the output, again, a combination of engineering judgement and insurance loss data is used to estimate contents loss. Thirdly, loss of use is estimated by the time required to rebuild and is proportional to the scale of damages. The three components of the loss model can be summed to give the loss function:

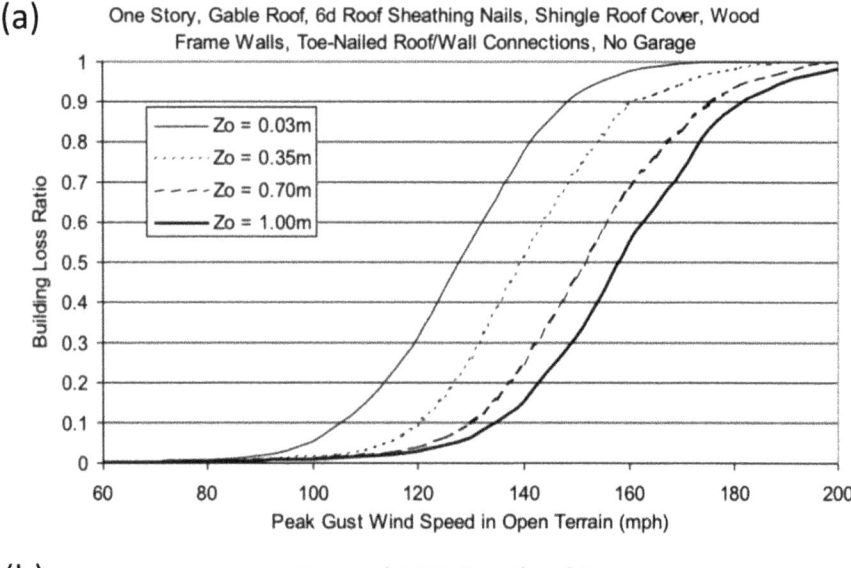

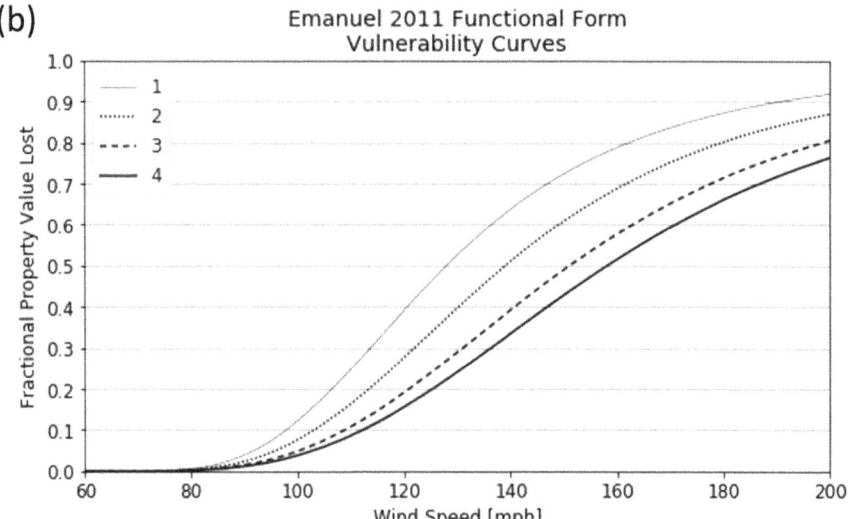

Fig. 11.2 Comparison of damage functions (**a**) Example Hazus loss curves from FEMA (2012). Curves shown are for a Single Family Residential Home (One Story, Gable Roof, 6d Roof Sheathing Nails, Shingle Roof Cover, Wood Frame Walls, Toe-nailed roof/wall connections, no garage); Z_o values represent different levels of surface roughness, where Z_0 varies between open terrain (0.03 m) to treed terrain (1.0 m) (**b**) Emanuel (2011) damage function fitted to V_{half} and V_{thresh} values shown in Hazus curves at left; 1–4 line types in the legend correspond to the 1st to 4th Z_o values in the left panel

$$\text{Loss Ratio} = \frac{Building, Content, and\ Use\ Loss}{Building, Content, and\ Use\ Value} \qquad (11.1)$$

This step in the Hazus model effectively converts the estimated damages captured by the vulnerability component of the model (wind load and physical damages) into exposed value, similar to the damage function approaches. Consequently, the loss ratio is the emergent value of vulnerability graphed in Fig. 11.2a for an example building type. Four scenarios are displayed according to surface roughness, Z_o. A higher value of Z_o corresponds to a rougher surface that slows wind speed more over land. Therefore, as displayed in the graphs, for rougher surfaces, an equivalent level of damage occurs at relatively higher wind speeds.

The output from Hazus differs from economic approaches as it is a process level model that, as this section has explored, calculates the loss according to specific storm and building physics, whereas economic approaches are mostly empirical. The Hazus loss model validation is performed at ZIP code level for single-family residential homes using data obtained from insurance companies. Vickery et al. (2006b) validated the output for four hurricanes and found the model underestimated the losses that occur at low wind speeds (below 100 mph). This is attributed to the model omitting potential sources of damage, such as tree blowdown. Nevertheless, the overall loss and damage estimates well-replicated those observed.

Hazus provides an advanced and detailed model of TC risks based on structural damages. However, Hazus was developed and validated for the USA, leading other studies to build on this work for other, especially less economically developed, countries. For example, the United Nations Disaster Risk Reduction's (UNDRR) Global Assessment Report (GAR 2013) adopts a similar approach to the USA's FEMA Hazus model to quantify earthquake and TC vulnerability, but at a global scale. The approach to estimate the vulnerability functions used in GAR is discussed by Yamin et al. (2014), which proposes a series of vulnerability curves for different building classes. The building classes mirror those used in Hazus, and classify buildings based on construction material, and low/mid/high rise. Additionally, for each building type, the design level, on a scale of poor to high compliance with wind resistance requirements, factors into vulnerability to help reflect variations in building codes across countries. In this study, the vulnerability function models the damage ratio (e.g., the fraction of total exposed value damaged) based on a beta distribution function, as originally proposed in work on seismic vulnerability (Rojahn et al. 1985), pointing to linkage between wind and earthquake vulnerability often cited in this literature. The form is then fit to assess vulnerability in different countries using information on structural vulnerability for different types of buildings. Values for the parameters in this function are key to quantifying vulnerability in the GAR, and these must be estimated for each building class. Yamin et al. (2014) cites 14 studies used to determine these values, including structural laboratory experiment information, wind field models, and observed behaviors from previous hurricanes. At the global scale of the GAR, regional building data is not always readily available. In these cases, the study assigns vulnerability functions to building classes using the Hazus methods. In developing countries, expert opinions are used to propose appropriate parameters for building types not included in Hazus.

Overall, structural approaches enable authors to estimate vulnerability based on a building's performance against TC-hazards using engineering principles. FEMA is a leading open-source model that develops this methodology for the USA through a framework that links multiple models. Similar approaches have been detailed that attempt to apply this approach internationally by considering national variations. These methods can provide highly specific information about TC damages, but suffer from the need for detailed information about a region's building stock still unavailable for many parts of the world.

11.3 Economic Vulnerability Approaches

While structural approaches develop an estimation of TC vulnerability from detailed, building-level modeling, economic approaches take a coarser, top-down approach. Structural approaches are widely the preferred methodology when the desired outcome is a detailed building type analysis and there is sufficient data on the local building stock. However, to gain an appreciation of the wider economic effects, such as future damage projections or cost-benefit analysis, economic methods are often preferable. Additionally, in regions with limited data, economic methods can offer valuable insights into TC risk. In economic studies of TC vulnerability, hazard, exposure, and impacts are considered known quantities from which vulnerability is reverse-engineered via empirical regression. Vulnerability curves are fit to different geographic locations through the function parameters. Additionally, impacts are typically represented by total disaster costs in USD or other currency, as opposed to fatalities or levels of building damage. Here we describe two key aspects of these economic studies: the definition of functional forms to represent vulnerability at different levels of hazard, and the fitting of these functions' parameter values based on historical events. We describe both significant debates that have occurred in this literature and areas that merit further work.

11.3.1 Functional Forms for Vulnerability

In estimating vulnerability, it is often convenient to assume a functional form which relates hazard values to different levels of damage. This simplifies the estimation of vulnerability to fitting a discrete number of parameters in the damage function. Here we describe functional forms that have been used to represent TC vulnerability. Much of the debate over and use of these functional forms has occurred in the economics literature, hence why we discuss them in this economics section. However, functional forms are also sometimes used to represent vulnerability in structural analyses (e.g., Yamin et al. 2014 discussed in Sect. 11.2).

TC vulnerability is commonly modeled as a function that increases with wind speed, or similarly, TC intensity, such that stronger (weaker) winds result in greater (lesser) damages. Such a representation of how TC damages occur is useful but highly idealized as other factors affect TC damages. Most notably, the damages associated with precipitation and storm surge cannot be directly captured by wind speed. Also, wind speed alone ignores other wind-related factors such as variations in wind direction and gustiness. Intuitively, such an idealized function of wind speed should produce zero damages for very low levels of wind speed, and reach a limit of 100% damages for exceedingly high wind speeds. However, understanding how vulnerability varies between these two extremes requires observations, theory, and/or modeling to constrain. As a result, there are a variety of forms for these damage functions (also termed vulnerability functions or impact functions).

As an example, Emanuel (2011) represents vulnerability as a relationship between hazard and fraction of exposed value or property damaged. Emanuel developed this function to quantify risks from modelled storms under present-day and global warming conditions. The equations for the function proposed by Emanuel (2011) are:

$$f = \frac{v_n^3}{1 + v_n^3}, \quad (11.2)$$

where f is the fraction of the property value loss and,

$$v_n = \frac{MAX[(V - V_{thresh}), 0]}{V_{half} - V_{thresh}}, \quad (11.3)$$

where V is the wind speed, V_{thresh} is the wind speed value below which there are zero damages, and V_{half} is the wind speed value at which 50% of exposed value is destroyed. A few different aspects of the Emanuel (2011) damage function are noteworthy. First, an important distinction between damage functions is the power of wind speed assumed to drive damages which we call λ. There have been various attempts to capture this relationship between TC intensities and economic losses (see Sect. 11.3.2.1) below. In the case of Emanuel (2011), λ is 3, whereas in some other studies the power of wind speed is flexibly dependent on empirical parameters (Pielke 2007). Second, this function explicitly includes the behavior that the damage function should only produce damages for wind speeds above a threshold, the value of which is set by a parameter (V_{thresh}). The physical intuitiveness of this parameter, and that of V_{half}, has helped foster the adoption of this functional form in a number of subsequent papers (e.g., Sealy and Strobl 2017; Eberenz et al. 2021).

There are other proposed functional forms beyond Emanuel (2011)'s. The United Nations Global Assessment Report on Disaster Risk Reduction (GAR) utilizes a function detailed in Yamin et al. (2014) that models the damage ratio based on a beta distribution function as originally developed for work on seismic vulnerability (Applied Technology Council 1985). These different functional forms serve similar purposes and, as of now, there is no clear optimal form. Rather, these various vulnerability equations share several common characteristics and each may be appropriate for a given TC risk modeling exercise provided its parameter values are properly constrained for the use case.

11.3.2 Constraining Damage Function Parameters

The damage functions mentioned in the prior section have parameter values that need to be constrained to accurately reflect TC vulnerability. This section describes approaches that reverse engineer these various vulnerability parameter values by combining information about the hazard, exposure, and observed damages. In other words, if damage is a function of hazard, exposure, and vulnerability, vulnerability can be determined by regressing damage against a combination of hazard and exposure.

11.3.2.1 The Debate Over λ

When determining the damage function based on empirical regression methods, an area of great debate within the published literature is the λ parameter (as summarized in Table 11.1). Wind-related damages are modeled as proportional to wind speed raised to the power of λ, so λ effectively represents how strongly damages vary with wind speed. This parameter is important in estimating TC impacts for regions without damage data or for future TCs.

This key damage function parameter, λ, is modeled with observations of damages and wind speed and is estimated with a log-linearized bivariate regression. The general form is as follows:

$$log\left(Damages_{r,t}\right) = \lambda log(V_{r,t}) + \varepsilon_{r,t}, \qquad (11.4)$$

where Damages are dollars lost from a storm, also known as direct damages, and are reported in the dollar value of the year of the storm. V is wind speed which can be measured as maximum wind speed at landfall or maximum wind speed in a geographic area exposed to the storm, such as county or census tract. Both damage and wind speed measures are normalized and aggregated to the storm-level, such as dividing damages by GDP, population, housing units, wealth or property values, and weighting wind speed by population in exposed locations (more on this in the subsequent paragraphs). The subscript r refers to a particular storm, ε is the standard error term, t is the time period of the storm (year or month), and λ is the coefficient from the regression and the parameter of interest. Other versions of this model include time fixed effects. The λ is sensitive to the specification of bivarate regression as well as the sample used.

Table 11.1 Summary of values and approaches to determining λ

Author/s	λ Value	Focus area	Outcome	Treatment	Time	Geographic Level
Emanuel (2005)	3	USA	Theoretical			
Nordhaus (2010)	7.27	Atlantic coastal U.S.	Value: Logarithm of total damages ($) for each storm-year divided by US GDP-year Data Source: Hurricane Data Spreadsheet	Value: Logarithm maximum wind speed Data Source: HURDAT	1900–2005	HURDAT US sub-grid cells at 15×15km
	8.21				1980–2008	
Nordhaus (2010)	9.53			Value: Instrumental variable using logarithm of minimum pressure Data Source: HURDAT	1900–2005	
	10.69				1980–2008	
Strobl (2011)	3.17	Coastal U.S.	Value: Logarithm of monetary damages adjusted for inflation, real wealth per capita, and coastal county population (see Pielke et al. 2008) Data Source: Hurricane Damage figures from Pielke et al. (2008)	Value: Logarithm sum of maximum wind speed weighted by population Data Source: Hazus wind field data	1970–2005	Census Tract
	3.7				1900–2005	County
	3.49				1900–1969	
	4.22				1970–2005	
Strobl (2012)	3.8	Counties Central American and Caribbean	Data Source: Hurricane Damage figures from Pielke et al. (2008)	Data Source: HURDAT and Eastern North Pacific Tracks, paired with Boose et al. (2004) wind field model	1950–2006	County

Focus area is the geographic region the paper calculates and/or applies lambda for. Outcome is the damages data while treatment represents how the hazard was calculated. The time is the period of data used in the calculation. Geographic level represents the scale data is aggregated over. Note that the Hazus wind field data is produced by the Vickery et al. (2000) mean flow model

The academic debate around appropriate λ values was concentrated in the early 2000s–2010s. To start, based on the Intergovernmental Panel on Climate Change's (IPCC) second assessment, Pielke et al. (2000) adopts a linear damage function ($\lambda=1$) relating different TC intensities to economic losses. However, this approach greatly underestimates the wind-damage relationship (Pielke 2007). Emanuel (2005) then presents an index of TC potential destructiveness and proposes that monetary loss is proportional to the cube of wind speed ($\lambda=3$). Two lines of evidence are used to justify this value. First, simple physical arguments common to both wind energy generation and wind damage modeling suggest that the power exerted by wind is proportional to the cube of wind speed (Bister and Emanuel 1998; Emanuel 1999). Second, Emanuel (2005) claims that historical damage data from a study of TCs across many vulnerable regions by Southern (1979) also supports $\lambda=3$. Although Emanuel (2005) is widely cited in subsequent publications, the short letter does not provide calculations to obtain the parameter, which opened the door to subsequent and more detailed attempts to quantify λ.

A few important studies followed Emanuel (2005) in the economics literature, with somewhat divergent conclusions depending on assumptions made in normalizing the damage data across different storms and years. This normalization process amounts to an assumption about the value of assets exposed to storms (hereafter "exposed value"). To accurately quantify λ, Nordhaus (2010) explores the relationship of historical damages in the USA to maximum hurricane wind speeds. Critiquing prior approaches, such as Emanuel (2005), Nordhaus argues such studies where λ is 2 or 3 are based on overly simplified physical constraints that neglect the relationship between wind/water damage and structural design. Using US-based hurricane data from 1900 to 2008, a damage intensity function is calculated. The damage data for the period is normalized using national scale GDP increases, effectively assuming that exposed value is US national GDP. Nordhaus presents a range of potential λ values, with the value varying by the period of years included, but the most likely value being close to 9 (Nordhaus 2010). This is a much higher λ value than that discussed in Emanuel (2005).

Due to the contention between Nordhaus (2010) and Emanuel (2005), subsequent studies conducted additional calculations to determine the value of λ. In particular, Strobl (2011) studies a similar period (1900–2005) to Nordhaus (2010), but instead uses a damage dataset normalized by assets where individual storms make landfall along US coastlines. This normalization method, developed by Pielke and Landsea (1998) and Pielke et al. (2008), adjusts for inflation, wealth, and coastal population. Using the Pielke et al. (2008) damage dataset, Strobl (2011) calculates a range of λ values between 3.18 and 4.22 depending on the period chosen. Strobl (2011) concludes $\lambda= 3.17$ is appropriate to minimise data bias. It is notable that this value is quite close to the cubic power proposed in Emanuel (2005) based primarily on simple physical arguments.

The weight of evidence (both physical and economic) seems to point to a λ value around 3. However, this is not conclusive, partially due to the complexity of the underlying processes, and also because extensive studies have not been conducted outside the USA to determine this value. λ largely reflects structural vulnerability and so may be expected to vary regionally based on local conditions and building standards. For instance, λ may be expected to vary between regions with strong building codes and those with greater susceptibility to physical hazards and/or less resilient building stock (Pita et al. 2013). Therefore, it is possible that λ, alongside other parameters, would need to be fitted regionally and it is unlikely a universally applicable value could fully capture such variations in building stock. Unfortunately, attempts to validate λ in regions other than the USA are, to the authors' knowledge, lacking.

11.3.2.2 Other Damage Function Parameters

Beyond λ, damage functions include other parameters which need to be constrained to accurately estimate vulnerability. These parameters vary depending on the specific form of the damage function. Here, as salient examples, we discuss the two additional parameters present in the Emanuel (2011) damage function -- V_{thresh} and V_{half}-- and attempt to determine their values. V_{thresh} represents the wind speed at and below which no damage occurs. The existence of such a threshold is intuitively understandable as buildings do not exist in still air; instead, they are subject to mild to moderate breezes daily which do not result in damages. V_{half} represents the wind speed at which 50% of exposed assets are expected to be destroyed. Following the suggestion in Emanuel (2011), studies employing this particular damage function have generally left V_{thresh} constant at 50 kts, and varied V_{half} to capture differences in vulnerability across different regions caused by, for example, differences in building construction standards.

While Emanuel (2011) proposed some reasonable guesses for these parameter values, follow-up work has attempted to constrain these values more rigorously for different regions via comparing simulated to historical damages (Elliott et al 2015; Eberenz et al. 2020). Elliott et al. (2015) uses changes in nightlights data as a proxy for reductions in economic activity and historical damages over China. In this framework, V_{thresh} of 50 kts can effectively simulate damages, but does not provide clear constraints on V_{half}. In contrast, Eberenz et al. (2020) uses reported country-scale TC damage estimates from the EM-DAT database (EM-DAT 2020) to represent historical damages, specifically to fit V_{half}. This study finds large uncertainty in optimal V_{half} values depending on the method of fit. For example, V_{half} is estimated to be 188.4 m/s to accurately simulate total damages across all storms, and 85.7 m/s to accurately simulate damages for any given storm (equally weighting damages from more and less destructive storms). Reasons for the great uncertainty in optimized vulnerability may include variations in vulnerability between urban versus rural regions not captured when using one damage function across an entire country, the lack of surface roughness modeling in this particular study, hazards beyond wind that were not simulated (e.g., rainfall, flooding, storm surge), and limitations of the employed population-based exposed value layer in capturing the spatial distribution of agriculture.

While these prior studies seeking to constrain V_{thresh} and V_{half} take steps forward in determining these parameters at country and multi-country-scales, we believe further consideration of these parameter values is merited. The large uncertainties highlighted in Eberenz et al. (2020) speak to significant limitations in empirically determining vulnerability based only on wind, exposed value, and total economic impacts, especially when the impacts are aggregated across large regions (as in EM-DAT). More work could be done to fit these parameters for individual provinces/regions, especially urban versus rural locales. Additionally, the common approach of holding V_{thresh} and λ constant and modifying only V_{half} seems poorly justified. Intuitively, V_{thresh} and λ might vary with construction standards, correlate with V_{half}, and also depend on whether only directly wind-caused damages are being modeled or if wind is being used as a proxy for all damages. For this reason, future work might explore which parameters should be constrained to optimally model damages, in addition to seeking to constrain those parameters

11.4 Multi-hazard Approaches

In this review, we overall focus on vulnerability to wind speed, both as a hazard in isolation and as an approximate proxy for the total combination of hazards inflicted by a storm. However, TC damages result from a number of different hazards, including intense rainfall and storm surge of surrounding seas, which both cause flooding, and landslides. These hazards are not always directly related to wind speed. For example, when Hurricane Harvey hit Texas in August 2017, although the storm had significant wind speed and size, the damages were predominantly from the unprecedented rainfall caused by the storm remaining stationary over the Houston area for about four days (van Oldenborgh et al. 2017). The study suggests this may be part of a positive trend in the intensity of precipitation since the 1880s, partly due to increased moisture holding capacity of the atmosphere. Ideally, in risk analyses these various hazards and their related vulnerabilities might each be explicitly modeled, both to more accurately model present risks and to account for different trends of the sub-perils under global warming. Here we briefly highlight studies that take strides in the direction of multi-hazard modeling of TC risk and vulnerability.

In our discussion of Hazus, we noted the model's incorporation of rainfall into its wind damage assessment, necessary since a large proportion of wind-induced structural damage is mediated through rainfall. Hazus also has the ability to incorporate storm surge and flood hazards; pointing to the relative ease of simultaneously analyzing vulnerability to multiple physical hazards when using structural approaches. However, it is noted that studies using Hazus tend to focus on a single hazard, likely due to the challenges of using Hazus multi-hazard desktop software (Casamassina 2019). Other commonly used flood loss curves are available from a book known as the Multi-Colored Manual, or MCM, which assessed vulnerability to flood for many different buildings based on a large number of interviews (Penning-Rowsell et al. 2005). An example flood damage function is shown in Fig. 11.3,

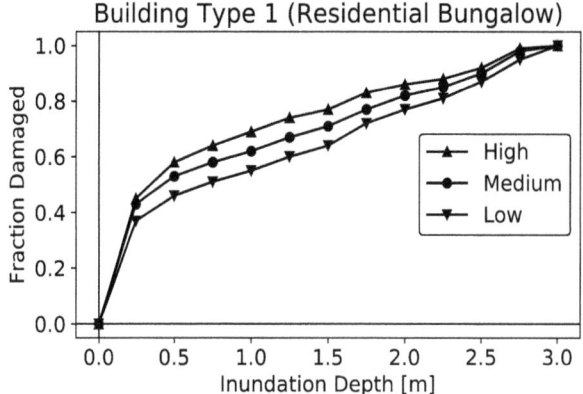

Fig. 11.3 Example flood/storm surge damage function
Flood inundation depth loss curves for a residential bungalow type building provided in the Multi-Colored Manual (MCM; Penning-Rowsell et al. 2005) and used in Aerts et al. (2013) in determining TC storm surge impacts. High, medium, and low curves represent different possible levels of damage. Note that the MCM provides loss curves for many other building types as well, and this is just one

where the most notable difference from wind speed damage functions (e.g., Fig. 11.2) is the substantial increase of damage for even low levels of the flood hazard.

Studies have begun to synthesize TC wind and surge related risks. Surge return levels are modeled using various methods, including historical data paired with extreme value statistics (Hallegatte et al. 2011) and hydrodynamic modeling (Lin et al. 2012). Damages are then estimated by pairing such surge analyses with flood vulnerability curves. For example, Aerts et al. (2013) model return periods for storm surge damages for each of the five boroughs of NYC. Damage is calculated by using storm surge height to determine an inundation map, which is then combined with storm surge damage functions from the MCM and asset maps to determine the cost of flood damage for different storms and exceedance probability loss curves. Sealy and Strobl (2017) build on these studies further, using a composite approach to TC risk modeling that calculates damages in regard to wind speed and storm surge. The authors utilize Emanuel (2011)'s vulnerability curve for wind speed-related damages (see Sect. 11.3) fit to Hazus structural vulnerability data, but combine it with proprietary flood vulnerability curves to include the effects of storm surges, and conduct a risk assessment of coastal property losses in the Bahamas. Notably, they find that for this low-lying island nation, a damaging hurricane on average leads to 24% of property value lost from storm surge and 0.2% directly from strong winds. These findings are context specific to this island, and may not be broadly applicable. Nevertheless, this study highlights the potentially significant impact of other non-wind TC hazards and underlines the great need for further work on multi-hazard integration in TC vulnerability studies.

These existing multi-hazard studies tend to be at least partially grounded in structural methods of assessing flood and wind vulnerability. Economic methods, as they presently exist, pose some challenges for multi-hazard integration. While there are a variety of open-source estimates for total damages from TCs, to the authors' knowledge, few if any open-sources of data split TC damages up by sub-peril. As a result, it is unclear how to constrain vulnerability curves for wind and flooding separately. TC damage estimates available by sub-peril could open up significant opportunities for economic approaches to multi-hazard risk modeling. More broadly, multi-hazard TC risk modeling is a relatively nascent area, and represents a critical target for future TC vulnerability studies.

11.5 Social Vulnerability Approaches

In Sects.11.2 and 11.3 we described methods of assessing vulnerability based on structural engineering or top-down analyses of economic impacts, which, according to the O'Brien et al. (2007) framework, are outcome-oriented approaches. In this section, we consider vulnerability as related to social characteristics of local populations, which is more closely aligned with contextual approaches. We emphasize that the contextual analysis of disaster vulnerability is an extensive field of study, which we only briefly examine with a few relevant examples.

Analysis of social vulnerability is founded on different types of inquiry and information than the previous outcome-oriented methods we have discussed. Rather than just quantitative information on populations or structures, the qualitative experience of individuals during disasters is considered in detail. Post-disaster interviews with people impacted by particular events are employed to capture the complex dynamics of how social systems are impacted by and respond to TC disasters. As an example of this type of inquiry, Adams et al. (2011) interviewed 57 first responders from the New Orleans Police Department following Hurricane Katrina to determine the factors that promote resilience in the face of this devastating event. These types of interviews can highlight chaotic chains of events during disasters which influence impacts but are difficult to understand through quantitative data alone (Adams and Stewart 2015). In this way, contextual analyses of vulnerability such as these are distinct from, but also complementary to, more outcome-oriented approaches.

When social vulnerability is quantified, it tends to be presented as indices that vary across space and provide a proxy of different regions' relative vulnerabilities (De Sherbinin et al. 2019). Social vulnerability indices have generally been used to explore associations with natural disaster damages, rather than directly estimate economic or other quantifiable outcomes from events. The study of social vulnerability is most developed for multi-hazard sets of disasters and disasters other than TCs. In the USA, a prominent index is FEMA's Social Vulnerability Index (or SoVI) (Cutter et al. 2003), which is a composite index available based on county and census tract level information, intended to be relevant to a range of natural disaster risks across the USA. This index was developed by using Principal Component

Analysis to reduce a set of 42 census variables to 11 that explain most of the variance in the local population characteristics, where the top three explanatory variables are personal wealth, age, and density of the built environment. Recent work has shown that for specific hazards, improved social vulnerability indices can be constructed from census data by comparing observed fatalities or property damages for rigorous variable selection and validation (as demonstrated for inland floods, Tellman et al. 2020). Another broad-scale social vulnerability map is that of the Global Earthquake Model (GEM), which supplies the Global Earthquake Social Vulnerability Map (Burton and Toquica 2021). This is a composite country-scale index that reflects the quality of social systems that support disaster resilience. The composite index combines many publicly available variables purported to be relevant to social vulnerability in the published literature. A machine learning technique was then used to create a combined index that can most accurately model past adverse impacts of earthquakes. Ongoing work is seeking to provide analogous maps at sub-national province scales in the Asia-Pacific region, such as municipalities in Japan and the Philippines.

There are limited studies that map social vulnerability to TCs and integrate it with risk analysis, but those that do exist suggest social vulnerability is highly influential in TC risk. In a study of TC risk in Mexico, Dominguez et al. (2021) use two different social vulnerability indices available across Mexico: the "local marginalization", a measure of access to resources such as housing, education, and appropriately compensated labor, and the "local social gap", a measure of social deprivations such as lagged education, poor quality housing, and lack of access to adequate healthcare. They find that combining TC hazards with a normalized version of these indices exhibits strong spatial correlation with maps of historical TC damages during El Niño versus La Niña years, suggesting that these indices accurately reflect spatial variation in TC vulnerability across Mexico. Further studies that examine social vulnerability indices' relevance to quantitative TC damages represents one possible way forward in integrating economic and social approaches to improve TC vulnerability assessment.

11.6 Discussion and Conclusions

An overview of approaches to quantifying TC vulnerability has been presented, with a focus on methods to model vulnerability to strong winds. The variation in data and modeling techniques demonstrates there is no single approach or agreed definition of what components must be used to determine vulnerability. Instead, the appropriate approach depends on the spatial scale of the region being modeled, impacts of interest and the data available for the region of interest. The open-source approaches outlined in this chapter stand in contrast to the proprietary methods used across the insurance industry. These methods provide transparent insight into the techniques risk modelers are using to capture vulnerability, and are open for scrutiny and debate through peer-review.

Vulnerability assessment described in this chapter can be categorized into three broad approaches: structural, economic, and social. The structural approach relies on engineering models and information about local building characteristics to determine vulnerability. This method is generally preferable to economic approaches when researchers are seeking an intricate understanding of the damages at a building type level, as it calculates a damage ratio considering the local building stock specificities of the region in question. As a representative example of this type of approach, we discussed the wind damage module from FEMA's Hazus software. This module is composed of five models that operate in a hierarchy to simulate wind-related damages and economic losses. In the first three models, using a user-created or real storm, Hazus determines wind speed, direction, and the wind load buildings are subjected to. The outputs are input into the damage model to generate damage states that enable the 'loss ratio' of building, content, and use loss to be calculated. While the wind load acting on buildings and damage states determines the vulnerability, the loss model converts the damages into an economic value to compute the loss ratio—Hazus's damage function. This approach is driven by numerical modeling of different types of idealized buildings under different wind conditions, parameterized via empirical information from wind tunnel experiments and validated for some historical TCs. Hazus-type approaches also form the backbone for an international analysis of TC risk in the United Nations Office for Disaster Risk Reduction (UNDRR) Global Assessment Report.

Economic methods differ from structural approaches in that they focus on studying aggregate regional economic effects over the detailed building-levels. This means economic methods are well suited to decision-making applications that are informed by economic analysis such as cost-benefit or future damage projections. Additionally, while economic approaches have been used across the US where there is sufficient data to use engineering-based methods, it is also particularly useful in regions with limited data on the local building stock. For instance, Elliot et al. (2015) employed economic methodologies to calculate TC vulnerability in China by using nightlight data in the absence of detailed building data. Therefore, while the economic and structural approaches both seek to quantify TC vulnerability, they do so in complementary ways, with different purposes related to the intended application of results or data availability.

In the economic approach, functional forms are used to represent the relationship between wind speed and damages, and then the parameters of these functional forms are fit to successfully replicate observed damages when paired with layers for exposed value and hazard. There are a few different functional forms that have been proposed, though a number of studies we discuss use that proposed by Emanuel (2011) (Fig. 11.2b). The majority of this top-down work has tended to use wind speed as a proxy for all damages—a useful idealization of the many different hazards associated with TCs. There has been significant debate over what power of wind speed (or λ value) drives damages. Regression-based analyses find λ-values ranging from 3 to 9 depending on how damages from different storms are normalized to reflect differences in exposed assets. At this point, we believe the weight of evidence points to $\lambda = \sim 3$, as this value is consistent with both basic physical arguments and analyses that normalize for the coastal assets actually exposed to storms. Examples of other parameters critical to determining the relationship between wind and damage are V_{half} and V_{thresh} from Emanuel (2011). V_{half} is the wind speed at which 50% of assets are destroyed and V_{thresh} is the wind speed below which no damage occurs, as it is assumed no damage occurs at low wind speeds. A few studies have sought to rigorously constrain these parameters, though further analysis is merited especially for V_{thresh}. The optimal value for each parameter seems to vary depending on both the region of interest and the metrics of the statistical fit.

Moving beyond just wind-related hazards, we summarized ongoing work to estimate multi-hazard TC risks. This requires vulnerability information for multiple different sub-perils of TCs, such as wind and flooding. For structural approaches, there are existing approaches to consider several physical hazards simultaneously, including Hazus's additional models to include impacts from rain and storm surge, and building-type-specific flood loss curves based on post-disaster surveys from the Multi-Colored Manual. However, multi-hazard methodologies create challenges when applied to economic-based methods, as the historical data that underlies these methods tends not to categorize damages by specific hazard. Therefore, there are no or highly limited examples of multi-hazard economic-based vulnerability analyses. Provision of economic damage data divided by sub-peril could provide many opportunities for further work to perform multi-hazard TC risk analysis.

While structural and economic vulnerability approaches are outcome-oriented, and focus on quantification of risk, social vulnerability approaches are more contextual in nature and rooted in qualitative inquiry into the human experience of disaster. Interviews of people post major TCs have provided context on how individuals and social systems are affected by and affect the outcomes of these disasters. Inspired by this qualitative work, quantitative indices have been developed that map regions of relatively higher or lower social vulnerability based on population characteristics. In the US, these indices are largely based on census data, and incorporate information on age, population density, wealth, race, etc. Well-developed social vulnerability indices exist for a cross-section of 18 hazards in the USA, and earthquakes across the world. Measures of inequality and access to public resources have shown promise in capturing areas experiencing greater TC-related losses in Mexico. However, to the authors' knowledge, there does not exist a commonly used TC-specific index of social vulnerability. Future work might evaluate the efficacy of existing multi-hazard social vulnerability indices in capturing TC impacts.

A key question for this review was whether these different disciplinary approaches to constraining TC vulnerability might be usefully integrated. As discussed previously, social vulnerability approaches which focus on disaster contexts are in some ways fundamentally distinct from the other outcome-oriented approaches discussed here (O'Brien et al. 2007). However, structural and economic approaches, both focused on quantitative impacts, can more readily inform one another. Some papers, such as Elliot et al. (2015) and Sealy and Strobl (2017), have sought to combine information from these two approaches using a particular functional form (e.g., that in Emanuel 2011) and deriving or checking parameter values based on Hazus loss curves. We explore this type of integration as follows: using the loss curves for a particular Hazus building type in Fig. 11.2a as an example, we approximate V_{half} as the wind speed at which a 0.5 loss ratio is reached and V_{thresh} as the wind speed when the loss ratio first becomes greater than 0; Fig. 11.2b shows the Emanuel (2011) vulnerability curves that result from this approximation. While the curve shapes are similar, there are some distinct differences. Most notably, the Emanuel (2011) damage function reaches 100% damages at higher wind speeds than the Hazus curves. Disagreement between the damage functions may partially result from their somewhat different intended purposes: the Emanuel (2011) curve is an aggregate for all building types, while the selected Hazus curve indicates damages for a specific building type (single family home). However, it is still noteworthy that the curves produced from these different methods are not entirely interchangeable, as even when fit to be as similar as possible they produce different damage states for events with comparable wind characteristics.

11.7 Directions for Future Work

Across all approaches, there are multiple opportunities to better assess the human and physical systems that the TC vulnerability models seek to represent. One significant area for improvement is in availability of open-source data. Firstly, there is a clear data and research deficit for TC vulnerability outside of the USA. When studies do attempt to determine TC risks in other regions, they often fall back on parameters developed from US damages and building codes. To confidently apply such techniques internationally, research is required to determine (a) country-level, regional, and local damage function parameter values, such as λ, and (b) how effectively US-developed structural approaches, such as Hazus, can be applied internationally. Such work requires high-quality, localized data about building types and socioeconomic characteristics which is often unavailable, especially for lower- and middle-income countries (LMICs) where impacts of TCs on wellbeing are most devastating (Hallegatte et al. 2017). Ideally, a global database containing information on local building stock and damages from historical TCs would exist to help the open-source community build a more comprehensive understanding of TC vulnerability. Projects such as Open Cities (N.D.) are seeking to meet this data demand via mapping of infrastructure and its vulnerability at the municipality/ neighborhood level in LMICs. However, the gap between the scale of Open Cities data and top-down regional or country-scale analyses of vulnerability (e.g., Eberenz et al. 2021) remains vast. To bridge these scales, future research might seek to create larger scale building maps using machine learning of satellite or Google Street View data (Ayush et al. 2020; Kang et al. 2018). Secondly, standard formatting of infrastructure and socioeconomic vulnerability data would greatly increase the comparability of data sources, making it easier to incorporate diverse sources of information into modeling techniques. Organizations such as OASIS

(2021) are presently attempting to create standards to make vulnerability and other natural hazard risk modeling data more accessible and comparable.

In addition to needs for improved data and data integration, there are various areas in which TC risk model structure can be improved to more accurately capture vulnerability and resulting damages. First, there are many random components of vulnerability that are not clearly accounted for in existing open-source frameworks. Models already include epistemic uncertainties due to data and knowledge limitations, but there are also less appreciated random components of TC vulnerability which are not explicitly accounted for (aleatory uncertainty). For example, the particular orientation of individual structures with respect to a given storm's wind field or human activities on the day of a storm are somewhat random processes that can be very influential in a given storm's impacts. Research should be conducted to estimate how much can be known about TC vulnerability versus how much of vulnerability is irreducibly uncertain and may be better represented by techniques that account for this, such as Monte Carlo simulations. Second, while this review has focused on studies quantifying wind-related damages, further work is needed to develop multi-hazard TC risk frameworks that account for storm surge and precipitation in addition to wind. By seeking to quantify vulnerability to more specific hazards rather than using wind speed as a proxy for all TC damages, uncertainty in TC vulnerability and risk modeling should decrease, or at least be better understood. However, it is worth noting that wind speed-based TC risk modeling has proliferated partially due to its simplicity; multi-hazard approaches add complexity which may decrease usability of models. Finally, research might seek to merge different methods of quantifying vulnerability for more holistic vulnerability assessment. For example, while prior work has treated social and structural vulnerability as distinct, local information on characteristics of both buildings and populations might be integrated to develop vulnerability maps for disaster preparedness and response. To ensure accuracy, new vulnerability metrics ought to be validated against quantitative TC impacts (such as losses in dollars, fatalities, human migration, etc.) in risk modeling frameworks that combine hazard, exposure, and vulnerability. However, such quantitative risk analysis should always keep in mind the lessons learned from qualitative social vulnerability research: that vulnerability reflects complex dynamics of human systems that are necessarily, but nonetheless, simplified in indices, loss curves, and damage functions.

Acknowledgements K.M.W. was supported by funding from the Climate & Society Graduate Research Scheme at Columbia University. J.W.B was supported by the Lamont-Doherty Earth Observatory Postdoctoral fellowship at Columbia University. R.M.Y. was supported by the Science, Technology, and Environmental Policy (STEP) Program at Princeton University's School of Public and International Affairs. We thank Suzana Camargo, Adam Sobel, Chia-Ying Lee, Deborah Coen, and two anonymous reviewers for providing constructive feedback on earlier versions of this chapter. Additionally, we appreciate the insightful discussions we had with Yuki Miura, Brian Walsh, Kyle Mandli, George Deodatis, Dail Rowe, Stuart Fraser, Peter Vickery, Michael Oppenheimer, and Gabriel Vecchi that supported the development of this chapter.

References

Adams TM, Stewart LD (2015) Chaos theory and organizational crisis: a theoretical analysis of the challenges faced by the New Orleans Police Department during Hurricane Katrina. Public Organ Rev 15:415–431. https://doi.org/10.1007/s11115-014-0284-9

Adams T, Anderson L, Turner M, Armstrong J (2011) Coping through a disaster: lessons from Hurricane Katrina. J Homel Secur Emerg Manag 8. https://doi.org/10.2202/1547-7355.1836

Aerts JCJH, Lin N, Botzen W et al (2013) Low-probability flood risk modeling for New York City. Risk Anals 33:772–788. https://doi.org/10.1111/risa.12008

Ayush K, Uzkent B, Burke M, et al (2020) Generating interpretable poverty maps using object detection in satellite images. In: IJCAI international joint conference on artificial intelligence. International joint conferences on artificial intelligence, pp 4410–4416

Bister M, Emanuel KA (1998) Dissipative heating and hurricane intensity. Meteorol Atmos Phys 65:233–240. https://doi.org/10.1007/BF01030791

Boose ER, Serrano MI, Foster DR (2004) Landscape and regional impacts of hurricanes in Puerto Rico. Ecol Monogr 74:335–352. https://doi.org/10.1890/02-4057

Burton C, Toquica M (2021) Global Earthquake Social Vulnerability Map (Version 2020.1). In: GEM. https://www.globalquakemodel.org/gem-maps/global-earthquake-social-vulnerability-map. Accessed 25 May 2021

Casamassina NA (2019) Estimating losses from Hurricane Harvey using FEMA's HAZUS-MH model. Texas A&M Univ, pp 89–120

Crowell M, Coulton K, Johnson C, et al. (2010) An estimate of the U.S. population living in 100-year coastal flood hazard areas. https://doi.org/10.2112/JCOASTRES-D-09-00076.1

Cutter SL, Boruff BJ, Shirley WL (2003) Social vulnerability to environmental hazards. Soc Sci Q 84:242–261. https://doi.org/10.1111/1540-6237.8402002

de Sherbinin A, Bukvic A, Rohat G et al (2019) Climate vulnerability mapping: a systematic review and future prospects. Wiley Interdiscip Rev Clim Change 10:e600

Dominguez C, Jaramillo A, Cuéllar P (2021) Are the socioeconomic impacts associated with tropical cyclones in Mexico exacerbated by local vulnerability and ENSO conditions? Int J Climatol 41:E3307–E3324. https://doi.org/10.1002/joc.6927

Eberenz S, Lüthi S, Bresch DN (2020) Regional tropical cyclone impact functions for globally consistent risk assessments. Nat Hazards Earth Syst Sci Discussions:1–29. https://doi.org/10.5194/nhess-2020-229

Eberenz S, Lüthi S, Bresch DN (2021) Regional tropical cyclone impact functions for globally consistent risk assessments. Nat Hazards Earth Syst Sci 21:393–415. https://doi.org/10.5194/nhess-21-393-2021

Elliott et al (2015) The local impact of typhoons on economic activity in China: a view from outer space. J Urban Econ 88:50–66. https://doi.org/10.1016/j.jue.2015.05.001

Emanuel KA (1999) The power of a hurricane: An example of reckless driving on the information superhighway. Weather 54:107–108. https://doi.org/10.1002/j.1477-8696.1999.tb06435.x

Emanuel K (2005) Increasing destructiveness of tropical cyclones over the past 30 years. Nature 436:686–688. https://doi.org/10.1038/nature03906

Emanuel K (2011) Global warming effects on US hurricane damage. Weather Clim Soc 3:261–268. https://doi.org/10.1175/WCAS-D-11-00007.1

EM-DAT (2020) EM-DAT: the international disaster database. https://public.emdat.be/

Federal Emergency Management Agency (FEMA) (2012) Hazus multi-hazard loss estimation methodology: Hurricane Model (Hazus MH 2.1. technical manual). Mitigation Division, Department of Homeland Security, Federal Emergency Management Agency, Washington, DC

Global Assessment Report on Disaster Risk Reduction (GAR) (2013) Probabilistic modelling of natural risks at the global level: global risk model. United Nations Office for Disaster Risk Reduction, Geneva

Hallegatte S, Ranger N, Mestre O et al (2011) Assessing climate change impacts, sea level rise and storm surge risk in port cities: a case study on Copenhagen. Clim Change 104:113–137. https://doi.org/10.1007/s10584-010-9978-3

Hallegatte S, Vogt-Schilb A, Bangalore M, Rozenberg J (2017) Unbreakable: building the resilience of the poor in the face of natural disasters. World Bank, Washington, DC

Holland G, Bruyère CL (2014) Recent intense hurricane response to global climate change. Clim Dyn 42:617–627. https://doi.org/10.1007/s00382-013-1713-0

Iman RL, Johnson ME, Schroeder TA (2002a) Assessing hurricane effects. Part 2 – Uncertainty analysis. Reliab Eng Syst Saf 78:147–155. https://doi.org/10.1016/S0951-8320(02)00134-5

Iman RL, Johnson ME, Schroeder TA (2002b) Assessing hurricane effects. Part 1 – sensitivity analysis. Reliab Eng Syst Saf 78:131–145. https://doi.org/10.1016/S0951-8320(02)00133-3

Iman RL, Johnson ME, Watson CC (2005a) Sensitivity analysis for computer model projections of hurricane losses. Risk Anal 25:1277–1297. https://doi.org/10.1111/j.1539-6924.2005.00673.x

Iman RL, Johnson ME, Watson CC (2005b) Uncertainty analysis for computer model projections of hurricane losses. Risk Anal 25:1299–1312. https://doi.org/10.1111/j.1539-6924.2005.00674.x

IPCC (2014) Climate Change 2014: Synthesis Report. Contribution of Working Groups I, II and III to the Fifth Assessment Report of the Intergovernmental Panel on Climate Change, Geneva, Switzerland, 151 pp

IPCC (2021) Summary for policymakers. In: Climate Change 2021: The Physical Science Basis. Contribution of Working Group I to the Sixth Assessment Report of the Intergovernmental Panel on Climate Change. Cambridge University Press, Cambridge, United Kingdom and New York, NY, USA, pp. 3−32. https://doi.org/10.1017/9781009157896.001

Kang J, Körner M, Wang Y et al (2018) Building instance classification using street view images. ISPRS J Photogramm Remote Sens 145:44–59. https://doi.org/10.1016/j.isprsjprs.2018.02.006

Khanduri AC, Morrow GC (2003) Vulnerability of buildings to windstorms and insurance loss estimation. J Wind Eng Ind Aerodyn 91:455–467. https://doi.org/10.1016/S0167-6105(02)00408-7

Li K, Li GS (2013) Risk assessment on storm surges in the coastal area of Guangdong Province. Nat Hazards 68:1129–1139. https://doi.org/10.1007/s11069-013-0682-2

Lin N, Emanuel K, Oppenheimer M, Vanmarcke E (2012) Physically based assessment of hurricane surge threat under climate change. Nat Clim Change 2:462–467. https://doi.org/10.1038/nclimate1389

Little CM, Horton RM, Kopp RE et al (2015) Joint projections of US East Coast sea level and storm surge. Nat Clim Change 5:1114–1120. https://doi.org/10.1038/nclimate2801

Nordhaus WD (2010) The economics of hurricanes and implications of global warming. Source. Clim Change Econ 1:1–20. https://doi.org/10.2307/climchanecon.1.1.1

O'Brien K, Eriksen S, Nygaard LP, Schjolden A (2007) Why different interpretations of vulnerability matter in climate change discourses. Clim Policy 7:73–88. https://doi.org/10.1080/14693062.2007.9685639

OASIS (2021) Oasis loss modelling framework: open source catastrophe modelling platform. https://oasislmf.org/. Accessed 29 May 2021

Open Cities (N.D.) Open Cities Africa. In: Open cities. https://opencitiesproject.org/. Accessed 29 May 2021

Penning-Rowsell E, Johnson C, Tunstall S et al (2005) The benefits of flood and coastal risk management: a handbook of assessment techniques. Middlesex University Press, London

Pielke RA (2007) Future economic damage from tropical cyclones: sensitivities to societal and climate changes. Philos Trans R Soc A Math Phys Eng Sci 365:2717–2729. https://doi.org/10.1098/rsta.2007.2086

Pielke Jr R, Landsea C (1998) Normalized hurricane damages in the United States: 1925–95. Weather Forecast 13(3): 621–631. https://doi.org/10.1175/1520-0434(1998)0132.0.CO;2

Pielke RA, Klein R, Sarewitz D (2000) Turning the big knob: an evaluation of the use of energy policy to modulate future climate impacts. Energy Environ 11(2):255–275. https://doi.org/10.1260/0958305001500121

Pielke RA, Gratz J, Landsea CW et al (2008) Normalized hurricane damage in the United States: 1900–2005. Nat Hazards Rev 9:29–42. https://doi.org/10.1061/(asce)1527-6988(2008)9:1(29)

Pita GL, Pinelli JP, Gurley KR, Hamid S (2013) Hurricane vulnerability modeling: development and future trends. J Wind Eng Ind Aerodyn 114:96–105. https://doi.org/10.1016/j.jweia.2012.12.004

Pita G, Pinelli J-P, Gurley K, Mitrani-Reiser J (2015) State of the art of hurricane vulnerability estimation methods: a review. Nat Hazards Rev 16:04014022. https://doi.org/10.1061/(asce)nh.1527-6996.0000153

Rojahn C, Abel MA, Ayres JM et al (1985) ATC-13 earthquake damage evaluation data for California. Applied Technology Council, Redwood City

Schneider PJ, Schauer BA (2006) HAZUS—its development and its future. Nat Hazards Rev 7:40–44. https://doi.org/10.1061/(asce)1527-6988(2006)7:2(40)

Sealy KS, Strobl E (2017) A hurricane loss risk assessment of coastal properties in the Caribbean: evidence from the Bahamas. Ocean Coast Manag 149:42–51. https://doi.org/10.1016/j.ocecoaman.2017.09.013

Shultz J, Russell J, Espinel Z (2005) Epidemiology of tropical cyclones: the dynamics of disaster, disease, and development. Epidemiol Rev 27(1):21–35. https://doi.org/10.1093/epirev/mxi011

Soares MB, Gagnon AS, Doherty RM (2012) Conceptual elements of climate change vulnerability assessments: a review Article in International Journal of Climate Change Strategies and Management · Asia Regional Resilience to a Changing Climate (ARRCC) View project CONFER: Co-production of Climate Services for East Africa View project Conceptual elements of climate change vulnerability assessments: a review. https://doi.org/10.1108/17568691211200191

Southern (1979) The global socio-economic impact of tropical cyclones. Aust Meteorol Mag 27:176–195

Strobl E (2011) The economic growth impact of hurricanes: evidence from U.S. Coastal counties. Rev Econ Stat 93:575–589

Strobl E (2012) The economic growth impact of natural disasters in developing countries: Evidence from hurricane strikes in the Central American and Caribbean regions. J Dev Econ 97:130–141. https://doi.org/10.1016/j.jdeveco.2010.12.002

Tellman B, Schank C, Schwarz B et al (2020) Using disaster outcomes to validate components of social vulnerability to floods: flood deaths and property damage across the USA. Sustainability 12:6006. https://doi.org/10.3390/su12156006

United Nations (2017) Factsheet: people and oceans. In: The ocean conference United Nations. UN, New York

van Oldenborgh GJ, van der Wiel K, Sebastian A et al (2017) Attribution of extreme rainfall from Hurricane Harvey, August 2017. Environ Res Lett 12:124009. https://doi.org/10.1088/1748-9326/aa9ef2

Vickery et al (2000) Hurricane wind field model for use in hurricane simulations. J Struct Eng 10:126. https://doi.org/10.1061/(ASCE)0733-9445(2000)126:10(1203)

Vickery PJ, Lin J, Skerlj PF et al (2006a) HAZUS-MH hurricane model methodology. I: Hurricane hazard, terrain, and wind load modeling. Nat Hazards Rev 7:82–93. https://doi.org/10.1061/(ASCE)1527-6988(2006)7:2(82)

Vickery PJ, Skerlj PF, Lin J et al (2006b) HAZUS-MH hurricane model methodology. II: Damage and loss estimation. Nat Hazards Rev 7:94–103. https://doi.org/10.1061/(asce)1527-6988

Ward PJ, Blauhut V, Bloemendaal N et al (2020) Review article: natural hazard risk assessments at the global scale. Nat Hazard Earth Syst Sci 20:1069–1096. https://doi.org/10.5194/nhess-20-1069-2020

Watson CC, Johnson ME (2004) Hurricane loss estimation models: opportunities for improving the state of the art. Bull Am Meteorol Soc 85:1713–1726. https://doi.org/10.1175/BAMS-85-11-1713

Yamin LE, Hurtado AI, Barbat AH, Cardona OD (2014) Seismic and wind vulnerability assessment for the GAR-13 global risk assessment. Int J Disaster Risk Reduct 10:452–460. https://doi.org/10.1016/j.ijdrr.2014.05.007

Open Access This chapter is licensed under the terms of the Creative Commons Attribution 4.0 International License (http://creativecommons.org/licenses/by/4.0/), which permits use, sharing, adaptation, distribution and reproduction in any medium or format, as long as you give appropriate credit to the original author(s) and the source, provide a link to the Creative Commons license and indicate if changes were made.

The images or other third party material in this chapter are included in the chapter's Creative Commons license, unless indicated otherwise in a credit line to the material. If material is not included in the chapter's Creative Commons license and your intended use is not permitted by statutory regulation or exceeds the permitted use, you will need to obtain permission directly from the copyright holder.

Chapter 12
Assessing the Drivers of Intrinsically Complex Hurricane Insurance Purchases: Lessons Learned from Survey Data in Florida

Juan Zhang, Jeffrey Czajkowski, W. J. Wouter Botzen, Peter J. Robinson, and Max Tesselaar

Abstract In the United States (U.S.), there is no one base policy for property insurance that can cover all disaster perils such as floods and windstorms. Hurricane-based insurance is intrinsically complex because the disaster peril may be excluded from a regular insurance policy and thus homeowners need to purchase a separate policy for that risk. Besides, the coverage for disaster perils often comes with separate deductibles and coverage limits. As a result, homeowners need to acquire a significant amount of information and knowledge to understand the insurance policies and make informed decisions about their coverage choices. This study utilizes decision trees to provide a comprehensive overview of flood and wind insurance purchase outcomes in the state of Florida. We also examine the behavioral, personal, and socio-demographic factors that influence the decision to obtain natural disaster insurance coverage for the various identified types of insurance purchases.

J. Zhang
College of Business, Eastern Kentucky University, Richmond, KY, USA
e-mail: juan.zhang@eku.edu

J. Czajkowski (✉)
Center for Insurance Policy and Research, National Association of Insurance Commissioners, Kansas City, MO, USA
e-mail: jczajkowski@naic.org

W. J. W. Botzen
Department of Environmental Economics, Institute for Environmental Studies, VU University Amsterdam, Amsterdam, The Netherlands

Utrecht University School of Economics (U.S.E.), Utrecht University, Utrecht, The Netherlands

Risk Management and Decision Processes Center, The Wharton School, University of Pennsylvania, Philadelphia, PA, USA
e-mail: wouter.botzen@vu.nl

P. J. Robinson · M. Tesselaar
Department of Environmental Economics, Institute for Environmental Studies, VU University Amsterdam, Amsterdam, The Netherlands
e-mail: peter.robinson@vu.nl; max.tesselaar@vu.nl

Our empirical analyses are based on homeowner survey data collected from coastal residents in Florida. We find that different types of flood and wind insurance purchases are related to unique factors, which highlights the importance of distinguishing insurance purchase outcomes. We also provide policy implications that focus on specific targets to improve insurance uptake.

Keywords Natural disaster insurance · Insurance purchase · Hurricane · Flood · Real-time surveys · Protection gap

12.1 Introduction

Hurricanes are a continued threat to the physical and financial well-being of coastal residents throughout the United States (U.S.). From 1980 to 2021, the U.S. experienced 56 tropical cyclones classified as billion-dollar disasters totaling over $1.1 trillion dollars in damage and resulting in 6697 fatalities.[1] In particular, for Florida and other U.S. states bordering the Gulf of Mexico, tropical storms and hurricanes are the costliest natural disasters, with Florida experiencing 29 billion-dollar tropical cyclone events totaling $220 billion in damage from 1980 to 2021, and all Gulf Coast States combined experiencing 43 billion-dollar tropical cyclone events totaling $740 billion in damage over the same period.[2] Climate change may increase the severity of these storms in the future (Marsooli et al. 2019), which highlights the need to improve preparedness for hurricanes to mitigate future damages.

Most damage during a hurricane or tropical storm occurs as a result of powerful winds, as well as from flooding due to large amounts of rainfall and/or high storm surges. Households can reduce potential property damage caused by flooding and windstorms by implementing risk reduction measures. Such measures range from structural alterations to the property, for example home-elevation or flood-proofing, to emergency preparation measures taken during an immediate threat of a hurricane. These latter measures include boarding windows, applying sandbags, and moving belongings to higher floors. Besides physical preparation for storms and hurricanes, households can choose to purchase natural hazard insurance to financially protect against flood and windstorm losses. However, the uptake of natural hazard insurance in the U.S. has been notoriously low, causing a significant insurance protection gap (Lingle and Kousky 2018). For example, of the $7 billion in expected annual flood losses to single-family homes in the U.S., more than 87% are uninsured by the National Flood Insurance Program (NFIP), which is the largest flood insurance provider in the country (Milliman 2021). The insurance gap was exposed by

---

[1] https://www.ncdc.noaa.gov/billions/summary-stats.

[2] Cost numbers were extracted from https://www.ncdc.noaa.gov/billions/summary-stats/FL/1 980-2021 and https://www.ncdc.noaa.gov/billions/summary-stats/GCS/1980-2021. Both costs were CPI-adjusted and did not include Hurricane Nicholas in September 2021.

Hurricane Harvey (the second largest hurricane flood loss in the U.S.) in Texas where only 15% of impacted homeowners in the area had flood insurance (Munich Re 2020).

It is plausible that part of the U.S. insurance gap is due to the intrinsic complexity of the natural disaster insurance purchase itself. In the U.S., there is no one base policy for property insurance that can cover all disaster perils.[3] Instead, policyholders need to purchase an additional endorsement or even a separate insurance policy to cover certain natural disasters, such as floods and windstorms. Additionally, in areas at high risk of these perils, insurance coverage may be mandatory (for example, required for a federally backed mortgage), or is otherwise voluntary. Finally, the coverage that is purchased for natural disasters typically comes with separate deductibles and coverage limits. Consequently, homeowners need to acquire a significant amount of information and knowledge to understand their homeowners' insurance policies and make informed decisions about their coverage options.

The goal of this study is to first demonstrate the intrinsic complexity of natural disaster property insurance coverage in the U.S. by examining the types of flood and wind insurance purchases in the state of Florida. Accordingly, we use decision trees to illustrate the conditions and choices that Florida homeowners need to consider when purchasing natural disaster property insurance coverage. Based on these choices and conditions, homeowners will end up with different choice sets for insurance coverage. As a result, we identify various types of disaster insurance purchases. We then attempt to shed light on behavioral, personal, and socio-demographic factors that influence the decision to obtain disaster insurance coverage for the various identified types of disaster insurance purchases. To do this, we apply empirical analyses of homeowner survey data collected as part of a multi-year research effort on hurricane preparedness by coastal residents in Florida.

Designing policies to improve disaster preparedness, including insurance coverage purchase, needs a better understanding of individual decision-making at different points in time, since certain impacting factors, such as subjective risk or social norms, may be more important at times of high risk compared to low-risk situations. Most studies of individual natural disaster risk perceptions and their relation to risk reduction activities rely on cross-sectional data that are collected at one point in time after the disaster has occurred (e.g., Botzen et al. 2019; Mol et al. 2020), but both risk perceptions and preparedness activities may evolve over time (e.g., Bubeck et al. 2020; Mondino et al. 2020). In our study, we address the notion of evolving risk perceptions and preparedness by applying survey data collected at different times during the hurricane season. One survey was collected during a high threat level of flood and wind damage conducted at the end of the 2020 hurricane season when Hurricane Eta approached Florida, and another was collected during a low-threat situation at the beginning of the 2021 hurricane season.

---

[3]There may be comprehensive policies that include added-on endorsements to cover disaster perils.

We conduct empirical analyses using the data from the two surveys separately and compare their results because the two surveys were given at different time points to different populations and covered different geographical areas. Our previous survey outcomes have shown that individuals are likely to change their insurance policy between the end and the beginning of the hurricane seasons (Botzen et al. 2020b). Besides accounting for risk levels at different points in time, we also differentiate between types of flood and wind insurance uptake to test for different factors driving uptake of varying insurance policies.

The most important ways in which flood insurance policies differ are in the level of coverage and whether coverage is optional or mandatory. Individuals in the U.S. may be mandated to purchase flood insurance when they live in high-risk flood areas and have federally backed mortgages. Theoretically, we expect that the socio-demographic factors, house characteristics, and individual risk perceptions are more likely to affect voluntary purchase than mandatory purchase. The major underwriter of flood insurance is the NFIP. In Florida, there are several private insurers that provide private flood insurance products. Homeowners may purchase private flood insurance products for more comprehensive flood coverage. We identify four types of flood insurance purchases – mandatory purchase of an NFIP policy, mandatory purchase of a private product, voluntary purchase of an NFIP policy, and voluntary purchase of a private product.

For wind insurance, most individuals can obtain coverage through their standard homeowners' insurance. However, if an individual lives in a coastal area where the wind peril is widely excluded from the standard homeowners' insurance policy, that person only has the choice set of a wind-only policy from state-run programs. In certain areas in Florida, insurers may choose to exclude the windstorm peril and the policyholders must purchase a wind endorsement or a separate wind-only policy to obtain the wind coverage. The wind-only policy is provided mainly through the state-run program – the Florida Citizens Property Insurance Corporation (Citizens). Some private insurers also offer a wind-only policy or a wind-endorsement. We identify four types of wind insurance purchases – homeowners' insurance from a private insurer, homeowners' insurance from Citizens, wind-only coverage from a private insurer, and wind-only policy from Citizens.

We find that the determinants of insurance uptake vary across different types of insurance purchases. For example, regarding flood insurance, we find that the value of contents and home buildings is only positively related to the voluntary purchase and not related to the mandatory purchase and that, in general, the mandatory purchase is much less related to covariates than voluntary purchase. For wind insurance, we find that being a homeowner increases the probability of having coverage through homeowners' insurance, while being a homeowner does not positively relate to the uptake of wind-only policies. Homeowners more frequently have insurance coverage than renters and thus are more likely to have windstorm coverage through their homeowners' insurance.

The remainder of the chapter is organized as follows. First, Sect. 12.2 provides the insurance market context and details the decisions that Florida homeowners need to consider when purchasing natural disaster property insurance coverage for floods and windstorms. Section 12.3 outlines the survey instrument and its implementation in the field during the two different time periods. Section 12.4 provides the insurance purchase types identified and the regression methodology deployed to assess the behavioral, personal, and socio-demographic factors that relate to the decision to obtain disaster insurance coverage for each type. In Sect. 12.5, we present the empirical results for flood insurance coverage purchases and then wind insurance coverage purchases. Section 12.6 provides our concluding discussion and policy implications.

12.2 Decision Trees and Insurance Purchase Types

We utilize decision trees to demonstrate the complexity of purchasing flood insurance and windstorm coverage, thereby illustrating the process that leads to different types of insurance purchases. The root and branch nodes represent a decision, and the end nodes show the outcomes (i.e., choice sets). We note that consumers often use an agent to purchase insurance. Consumers may not consciously make the decisions shown in the decision tree because the agent can collect their information and help them determine the appropriate coverage. However, the usage of a decision tree can still explain the underlying decision-making process by an agent or a consumer. The complexity of the process also indicates the opacity of the property insurance market and the substantial knowledge that consumers need to acquire to understand their insurance policy. Also, we do not distinguish admitted and non-admitted carriers because sourcing insurance purchases often depends on the agent and is not a policyholder's choice. If the agent can write in both the standard and the non-admitted market, the policyholder is likely to choose the insurer based on the price.

12.2.1 Flood Insurance Decision Tree

Flood risk is not covered by a standard homeowners insurance policy in the U.-S. Homeowners need to purchase a separate flood insurance policy to obtain coverage. Moreover, homeowners located in high-risk flood areas with mortgages from government-backed lenders are required to have flood insurance (FEMA n.d.).[4] The high-risk flood area is also called the Special Flood Hazard Area (SFHA) that

---

[4]Federal banking regulators have allowed for either a NFIP policy or a private flood insurance. A rule on acceptance of private flood insurance was finalized in 2019. However, Federal Housing Administration (FHA) regulations currently do not allow FHA-insured properties to purchase private flood insurance to fulfill the mandatory requirement.

has a 1-in-100 year flood probability.[5] Homeowners without the mandatory requirement, such as those living in moderate- to low-risk flood areas, can purchase flood insurance on a voluntary basis.

Most flood insurance is provided through the NFIP, which is a federal program administered by the Federal Emergency Management Agency (FEMA). Only property owners in the participating communities can purchase an NFIP policy. There are approximately 23,000 NFIP participating communities nationwide; only 9 communities in Florida do not participate.[6] Homeowners living in non-participating communities must purchase a private flood insurance product that is designed and underwritten by a private insurance company. The private flood insurance product can be a stand-alone flood policy or an endorsement of the homeowners' insurance.[7]

Owners of high-value homes may want to purchase a private flood insurance product to obtain additional coverage beyond the NFIP policy. The NFIP policy has coverage limits of $250,000 for the building and $100,000 for the building contents. In comparison, private flood insurance products have much higher coverage limits, along with some additional benefits, such as more deductible choices, a shorter waiting period, and fewer underwriting questions. In sum, for a mandated flood insurance purchase (top branch of the decision tree), homeowners may have three options to obtain flood coverage: an NFIP policy plus a private flood insurance policy for additional coverage, a sole NFIP policy,[8] or a sole private flood insurance policy (Fig. 12.1).

For a voluntary flood insurance purchase (bottom branch of the decision tree), homeowners living in the NFIP participating communities can buy an NFIP policy. Homeowners located in moderate- to low-risk flood areas (non-SFHA areas, Zone B, C, or X) are eligible for a Preferred Risk Policy (PRP), which has the same coverage as a standard-rated NFIP policy but charges a lower cost.[9] Similarly, policyholders with a voluntary purchase can add a private flood insurance policy for additional coverage. In sum, there are five options for homeowners who purchase flood insurance voluntarily: an NFIP PRP plus a private flood insurance policy, an NFIP standard-rated policy plus a private flood insurance policy, a sole NFIP PRP policy, a sole NFIP standard-rated policy, and a sole private flood insurance policy.

---

[5] High-risk flood areas begin with the letters A or V on the FEMA flood maps. Moderate- to low-risk flood areas are designated with the letters B, C, and X on the FEMA flood maps. More than 40% of all NFIP flood claims came from outside of high-risk flood areas between 2015 and 2019 (FEMA 2020).

[6] The numbers for Florida are obtained from the Community Status Book, retrieved October 5, 2020 from https://www.fema.gov/national-floodinsurance-program-community-status-book.

[7] Alternatively, homeowners may have a flood endorsement onto a dwelling fire policy.

[8] Private insurance companies can underwrite the NFIP standard insurance policy through the Write Your Own (WYO) Program. But the financial liabilities of these NFIP policies are fully on federal government.

[9] More policy information regarding PRPs is available at https://www.fema.gov/pdf/nfip/manual201105/content/09_prp.pdf.

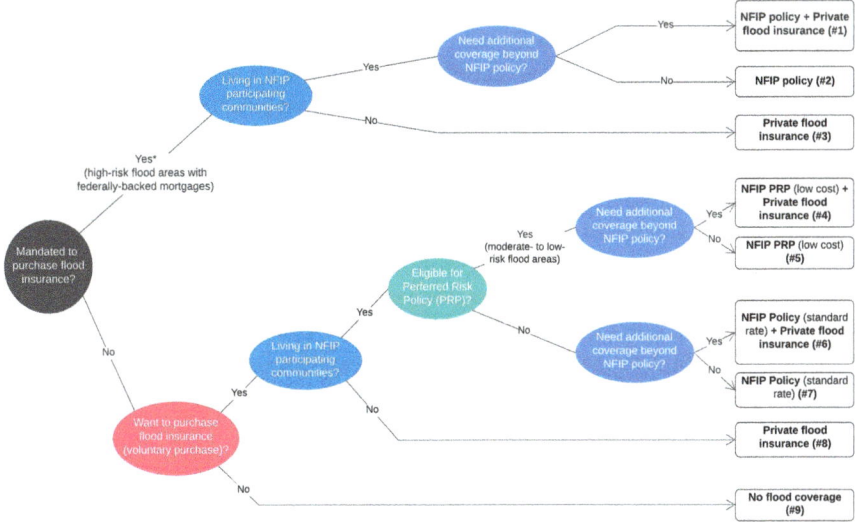

Fig. 12.1 Flood insurance decision tree in Florida. **Note**: *In practice, it is possible that properties that are mandated to have flood insurance do not purchase flood insurance. FHA-insured properties that are mandated to have flood insurance but not located in the NFIP participating communities cannot purchase private flood insurance to fulfill the requirement at this stage. Insurance regulators have been working to change this rule

12.2.2 Windstorm Coverage Decision Tree

Windstorm peril is typically covered by the standard homeowners' (HO) multi-peril insurance policy except in some wind-prone areas. The windstorm loss in some coastal states is subject to a separate deductible. There are three types of wind deductibles:

- Hurricane deductible applied to windstorm damages caused by a named hurricane.
- Named-storm deductible is less restrictive than Hurricane deductible and additionally applies to damages caused by named tropical storms that are not a hurricane at landfall.
- Windstorm deductible is the broadest type and applies to windstorm damages from any source.

Based on the III (2020) and the National Association of Insurance Commissioner's Center for Insurance Policy and Research (CIPR) website, 19 states and the District of Columbia currently have a hurricane or named storm deductible in place.[10] Unlike the NFIP flood policy, homeowners' insurance is specific to states because different

---

[10] More detailed information is available at https://www.agordon.com/blog/bid/163479/wind-deductible-vs-hurricane-vs-named-storm-deductibles and the CIPR website https://content.naic.org/cipr_topics/topic_hurricane_deductibles.htm.

states can employ different triggers and amounts for windstorm deductibles. The residual markets in different states may also have different eligibility requirements, policies, programs, and management rules. We focus on the decision-making process in Florida because our survey data do not cover other states.

In Florida, homeowners living in coastal areas may not find a standard HO multi-peril insurance policy from a private insurer or may only find coverage with extremely high premiums. In this case, homeowners can turn to the Citizens Property Insurance Corporation (Citizens), the residual market and last resort of high-risk homeowners.[11] A homeowner is eligible for a Citizens policy if one of the following criteria is met: (1) no comparable private-market offers of coverage are received, or (2) comparable private-market offers of coverage are received, but the premiums are more than 15% higher than a comparable Citizens policy.

Another issue is that, in some high-risk regions, private insurers can exclude the windstorm peril from the coverage. According to the Florida Statute s. http://www.leg.state.fl.us/statutes/index.cfm?App_mode=Display_Statute&URL=0600-0699/0627/Sections/0627.712.html, admitted insurers are required to offer windstorm coverage in the base policy except in areas covered by the Citizens Coastal Account. The Citizens Coastal Account, formerly known as High Risk Account, is for wind-only and multi-peril policies for personal residential, commercial residential, and commercial nonresidential risks located in eligible coastal high-risk areas, i.e., in areas that were defined on January 1, 2002 to be eligible for coverage by the Florida Windstorm Underwriting Association. Surplus line underwriters are not subject to s. http://www.leg.state.fl.us/statutes/index.cfm?App_mode=Display_Statute&URL=0600-0699/0627/Sections/0627.712.html.

When the windstorm peril is excluded, homeowners can purchase a wind-only policy from Citizens covering only damages from hail and windstorms. Only properties in areas within the boundaries of the Citizens Coastal Account are eligible for Citizens wind-only policies. On the demand side, policyholders may voluntarily exclude the windstorm peril and purchase a wind-only policy if the latter is a cheaper option.

High-value properties are ineligible to obtain coverage from Citizens and thus, must purchase coverage from private insurers. Effective January 1, 2017, housing units with a replacement cost of $0.7 million or over are not eligible for any coverage by Citizens (the replacement cost limit is $1 million in Miami-Dade and Monroe counties).[12] In Florida, a few private insurers provide a wind-only policy or an

---

[11] Citizens was established by the Florida Legislature in 2002 when the state combined two separate high-risk insurance pools – the Florida Windstorm Underwriting Association and the Florida Residential Property & Casualty Joint Underwriting Association. The company website is https://www.citizensfla.com/insurance-101.

[12] Based on s. 627.351(6)(a)3.d., effective January 1, 2017, a structure that has a dwelling replacement cost of $700,000 or more, or a single condominium unit that has a combined dwelling and contents replacement cost of $700,000 or more, is not eligible for coverage by the corporation. Such dwellings insured by the corporation on December 31, 2016 may continue to be covered by the corporation until the end of the policy term. Rules and processes were revised in December 2019; see details at Citizens website.

endorsement for windstorm damages.[13] When the windstorm peril is excluded from the base policy, the high-value houses must find wind-only coverage from these private insurers

In Florida, both HO multi-peril insurance and wind-only insurance have a hurricane deductible that applies to wind damages caused by a named hurricane. A hurricane deductible can be either a flat amount of $500 or 2%, 5%, or 10% of the home's total insured value. The $500 flat deductible is only available for certain types of policies, such as homes with a total insured value of less than $100,000. The hurricane deductible applies only once during a hurricane season.[14] In sum, policyholders in Florida may have four options to obtain windstorm coverage: a standard homeowners' multi-peril policy from a private insurer, a homeowners' multi-peril policy from Citizens, and a private homeowners' multi-peril policy excluding windstorm peril plus a Citizens wind-only policy, and a private homeowners' multi-peril policy excluding windstorm peril plus a private wind-only policy (Fig. 12.2).

In Florida, there are several programs to help homeowners access the state's increasingly expanding insurance market. The Clearinghouse program established by Citizens helps policyholders with no option other than Citizens to shop around and find better property coverage from private insurers. Policyholders are not eligible for Citizens if a comparable offer of coverage is received through the Clearinghouse with a premium less than 15% higher than the Citizens premium. The Homeowners Rate Comparison Tool (CHOICES) on the Florida Office of Insurance Regulation (OIR) website[15] provides users the average rate quotes for three coverage examples and the user's county from various insurance companies (including Citizens). The quotes reflect the most recent rate filings approved by the OIR office. The Florida Market Assistance Plan (FMAP), run by Citizens, is a free and online referral service that matches property owners with agents who can help the property owners find private-market coverage.

Under Florida law, policyholders can obtain premium discounts for implementing certain types of mitigation measures. The first layer discount is for new building codes and eligible for houses built after 2001 or houses built before 2001 but with an updated roof construction that meets the 2001 Florida Building Code. The second layer discount only applies to the hurricane-wind portion of the premium. It includes two types of wind mitigation measures – securing the roof and protecting windows from flying debris.[16]

---

[13] There are around 10 private insurers that sell wind-only policies in Florida, according to our communications with the Florida Office of Insurance Regulation (OIR) staff.

[14] When homeowners incur wind losses under the second hurricane, the deductible of the second claim will be either the remainder of the unused hurricane deductible or the AOP deductible, whichever is greater. See more details at the Florida's Chief Financial Officer's website.

[15] The CHOICES system is for four types of insurance including homeowners at https://www.floir.com/choices.aspx.

[16] To learn more about the wind mitigation discount, see Form OIR-B1-1655 from Florida OIR available at https://www.floir.com/siteDocuments/OIR-B1-1655.pdf.

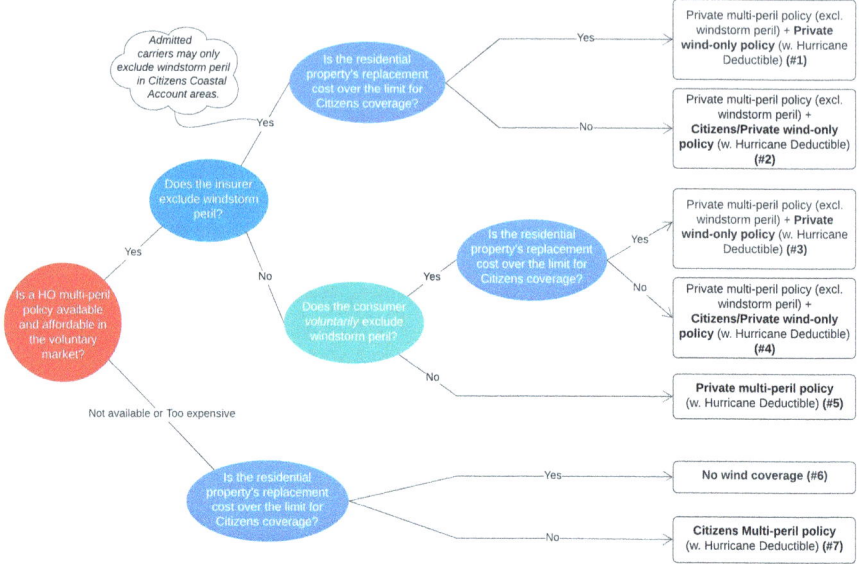

Fig. 12.2 Windstorm coverage decision tree in Florida. **Note**: The voluntary/private market includes admitted carriers and surplus line (i.e., non-admitted) underwriters. The replacement cost limit for the Citizens coverage is $0.7 million in Florida except in Miami-Dade and Monroe counties, where the limit rises to $1 million. A private wind-only policy at the end nodes can be a separate wind-only policy or an endorsement for windstorm damages onto the base policy

Our analysis focuses on the state of Florida. However, the situations in other coastal states can be different. The NFIP program provides federal flood insurance, but private flood insurance products may not be a choice for homeowners in all coastal states. For wind insurance coverage, the deductibles of wind coverage vary across states. For example, Louisiana has three types of windstorm deductibles in different areas. Homeowners in wind-prone areas may have a hurricane or named-storm deductible, whereas those in low-risk wind areas may have a windstorm & hail deductible of a lower amount. As a result, the wind coverage in Louisiana can vary by the types of deductibles. Besides, the state-run program is different among states in terms of eligibility requirements and other regulations rules. Therefore, the decision trees for other coastal states should be tailored to state-specific conditions and regulations.

12.3 Survey Instrument and Field Implementation Summary

12.3.1 Real-Time, Repeated Surveys

This study is part of a multi-year research effort on hurricane preparedness of coastal residents in Florida. We conducted five surveys from 2019 to 2021, some of which were distributed during a hurricane threat (real-time survey). We aim to better understand individual decision-making during the threat of a disaster as well as in its aftermath.

The analysis of this chapter focuses on the last two surveys (i.e., survey 4 and survey 5) that we conducted during the 2020 hurricane season and at the beginning of the 2021 hurricane season. The 2020 hurricane season produced a record-breaking 30 named storms, including six major hurricanes (Blackwell 2020). Survey 4 was distributed just before Hurricane Eta hit Florida. Hurricane Eta approached Florida from Central America but decreased in power before landfall on the Florida Keys on November 8 as a tropical storm with maximum sustained winds of 100 km/h (65 mph) (Insurancejournal.com 2020). After reentering the Gulf of Mexico, it regained power, becoming a category 1 hurricane, and veered back towards Tampa Bay on November 11, where heavy rains and a powerful storm surge caused significant damage. On November 12, Eta was reduced to a tropical storm with maximum sustained winds of 85 km/h (50 mph) and made landfall for the second time in Florida in Cedar Key. Although no deaths were reported because of the storm, estimated direct damage to structures exceeded $1.1 billion, of which insurance firms covered approximately half (AON 2020). Figure 12.3 exhibits the track of Hurricane Eta and its development in terms of strength classifications.

Survey 4 was conducted as an online survey on November 10 and 11, 2020. It was given to households living along the Gulf coast of Florida. As Fig. 12.4 shows, most survey respondents were located close to the Tampa Bay area where Hurricane Eta was expected to make landfall on November 12. In total, the survey received 844 responses.

Survey 5, also conducted as an online survey, was given to households living along the Gulf and Atlantic coasts of Florida. The survey was sent out between May 26–June 7, 2021, to examine individual hurricane preparedness before a hurricane

Fig. 12.3 Final track of hurricane/storm Eta. (Source: weather.com 2020)

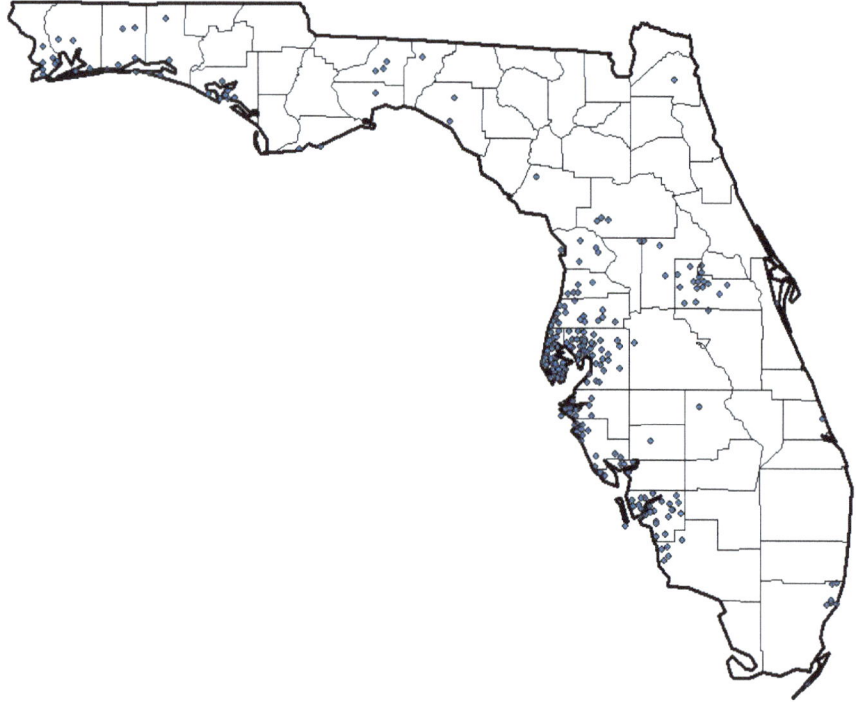

Fig. 12.4 Location of respondents to the survey conducted in November 2020

season. In total, 1245 respondents completed the survey, and their locations are shown in Fig. 12.5.

12.3.2 Sampling and Variable Coding

This chapter focuses on findings from Surveys 4 and 5 since these surveys contained detailed questions regarding the types of respondents' insurance purchases. Table 12.1 summarizes the information of Survey 4 and Survey 5. Both survey 4 and survey 5 were implemented via the market research company Downs & St. Germain Research. Their panel was randomly sampled from the population in the geographic regions of interest with the overall aim to obtain a representative sample based on socio-demographic characteristics.

Table 12.2 provides information about the way in which the survey questions were asked as well as their variable coding for the statistical analysis.

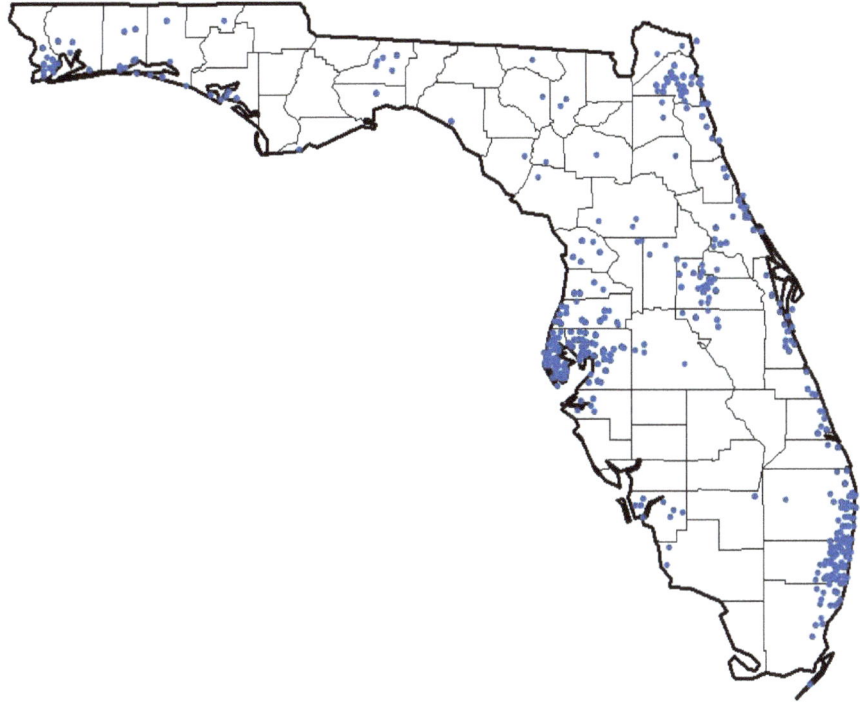

Fig. 12.5 Location of respondents to the survey conducted in May and June 2021

Table 12.1 Information of two surveys

	Survey 4 (n = 844)	Survey 5 (n = 1245)
Dates	November 10–11, 2020	May 26–June 7, 2021
Real-time?	Yes (Hurricane Eta)	No, beginning of the season
Location	Florida, Gulf of Mexico	Florida, Gulf & Atlantic coasts
Survey method	Online	Online
Sampling method	Random sampling	Random sampling

12.4 Measures and Method

12.4.1 Measures of Insurance Purchase Types

Based on the survey questions, we can separate mandatory versus voluntary purchase of flood insurance. The mandatory and voluntary flood insurance purchase can

Table 12.2 Coding of variables

Variable	Coding
Number of ex ante risk reduction measures	The sum of ex ante risk reduction measures applied by a respondent, including home elevation; flood-proof paint or coating; a sump pump and/or a drainage system; flood-resistant building materials; water-resistant floor; and the installation of electrical and central heating systems above potential flood levels.
Premium discount for flood risk mitigation[a]	"Did you receive a premium discount on your flood insurance for taking any of these (the above flood risk mitigation) measures?" 1 = Yes, 0 = No
Window protection	"Did you implement the following measures to reduce the windstorm damages to your home? Window protection such as shutters, plywood panels, or hurricane proof glass." 1 = Yes (in or before 2021), 0 = No (plan to do in 2021 or not plan to do)
Roof retrofit[a]	"Did you implement the following measures to reduce the windstorm damages to your home? Roof construction that meets the 2001 Florida Building Code such as roof covering, roof-deck attachment, and roof-to-wall connection." 1 = Yes (in or before 2021), 0 = No (plan to do in 2021 or not plan to do)
Hip roof[a]	"Did you implement the following measures to reduce the windstorm damages to your home? Hip roof, i.e., roof sloping down to meet all your outside walls (like a pyramid)." 1 = Yes (in or before 2021), 0 = No (plan to do in 2021 or not plan to do)
Premium discount for wind risk mitigation[a]	"Did you receive a premium discount on your windstorm insurance coverage for taking any of these measures?" 1 = Yes, 0 = No
Worry about flooding	"I am worried about the danger of a flood at my current residence." 1 = Strongly disagree to 5 = Strongly agree
Perceived flood impact	"What would it cost to repair the damage to your home and its contents if your home did flood?" 1 = Less than $10,000 to 7 = $200,000 or more
Perceived flood probability	"What is your best estimate of how often a flood will occur at your home?" 1 = Less often than 1 in 1000 years to 7 = More often than 1 in 10 years
Worry about windstorm[a]	"I am worried about the danger of a windstorm at my current residence." 1 = Strongly disagree to 5 = Strongly agree
Perceived wind impact[a]	"What would it cost to repair the damage to your home and its contents if your home did suffer a windstorm?" 1 = Less than $10,000 to 7 = $200,000 or more
Windonly_territory	Identify whether the respondent's home is located within the Florida Citizens Coastal Account Area (wind-only policy eligible area) based on the latitude and longitude of their home. 1 = Yes, 0 = No

(continued)

Table 12.2 (continued)

Variable	Coding
Trust in government flood policies	"How much do you trust the ability of government officials to limit flood risk where you live, for example by maintaining levees and enforcing building codes? Do you:" 1 = Not trust them at all to 4 = Trust them completely
Risk taking/risk aversion	"Using a 10-point scale, where 0 means you are not willing to take any risks and 10 means you are very willing to take risks, what number reflects how much risk you are willing to take?" For risk aversion, the inverse is taken, i.e., 0 = Very willing to take risk (response = 10) to 10 = Not willing to take risk (response = 0)
Internal locus of control	"Using a 10-point scale, where 0 means you have no control and 10 means you have complete control, what number reflects how much control you think you have over how your life turns out?" Scale from 0 to 10
Social norm for insurance uptake	"Most people who are important to me would think that someone in my situation ought to purchase flood insurance." 1 = Strongly disagree to 5 = Strongly agree
Regret of no insurance	"I would regret not purchasing flood insurance coverage if a flood were to occur next year." 1 = Strongly disagree to 5 = Strongly agree
Regret of having insurance	"I would regret purchasing flood insurance coverage if no flood were to occur next year." 1 = Strongly disagree to 5 = Strongly agree
House owner	0 if the respondent rents his/her house; 1 if the respondent is a property owner
Value of home building	"What is approximately the current market value of your home?" 1 = Less than $100k to 8 = $800k or more
Value of home content	"What is approximately the value of your home contents?" 1 = Less than $5000 to 8 = $75,000 or more
Length of residence	"How long have you lived in your home (in years)?"
Underfloor basement	"Does your home have a basement, cellar or crawlspace?" 1 = Yes for basement, 0 = No basement
Age	"How old are you?" in years
Education	"What is your highest completed level of education?" 1 = Some high school to 5 = Post graduate
Income	"Which of the following describes your total household income for 2019 before taxes?" 1 = Less than $10,000 to 6 = $125,000 or more
Female	Was the respondent male of female? female = 1, male = 0
Financial difficulty due to COVID-19	"Did you experience any financial difficulties as a result of the coronavirus that prevented you from purchasing insurance for your home?" 1 = Yes, 0 = No
Trouble purchasing flood insurance[a]	"Have you had trouble getting or renewing your flood insurance because of natural disasters in the past?" 1 = Yes, 0 = No
Trouble purchasing homeowners insurance[a]	"Have you had trouble getting or renewing your homeowners insurance because of natural disasters in the past?" 1 = Yes, 0 = No

[a]These questions were only included in Survey 5

Table 12.3 Distribution of flood insurance purchase types

Flood PH types	Definition	Survey 4		Survey 5	
		Freq.	Percent	Freq.	Percent
Mandatory NFIP	Mandated to purchase flood insurance, and only purchased the NFIP policy	46	6%	79	8%
Mandatory private insurer	Mandated to purchase flood insurance, and purchased a private flood product	28	4%	60	6%
Voluntary NFIP	Voluntarily chose to purchase flood insurance, and only purchased the NFIP policy	62	8%	70	7%
Voluntary private insurer	Voluntarily chose to purchase flood insurance, and purchased a private flood product	41	6%	108	11%
None	No flood insurance	567	76%	690	69%
Subtotal		**744**	**100%**	**1007**	**100%**
Don't know[a]	Don't know if have a flood policy	91		171	
	Having flood insurance but don't know insurance type (if mandatory or insurance provider)	9		67	
Total		**844**		**1245**	

[a]The data of respondents who don't know any needed information are not used in the analysis due to lack of information

be further distinguished by where the individuals obtained the insurance – NFIP policy versus private flood insurance. The flood policyholders are primarily measured by four *mutually exclusive* types – *mandatory NFIP, mandatory private insurer, voluntary NFIP, and voluntary private insurer.*

Table 12.3 shows the frequency and percentage of different types of flood insurance purchases. For both Surveys 4 and 5, we have slightly more voluntary purchase of flood insurance than mandatory purchase. About 10% of respondents in Survey 4 and 17% of respondents in Survey 5 purchased a private flood insurance product. The NAIC report[17] showed that the private flood premiums written was 8% of the total flood premiums written in Florida in 2018. We have a slightly higher percentage of private flood insurance buyers possibly because our sample is mostly limited to coastal counties.

Based on our survey questions, in Survey 4, the wind coverage purchases can be divided into two types – obtaining wind coverage from the homeowners' insurance (*homeowners' insurance*) or from a wind-only policy (*wind-only policy*). There are 331 respondents whose homeowners insurance covered the windstorm peril and 34 respondents who had to obtain windstorm coverage through a wind-only policy.

In Survey 5, we asked the policy type question separately for individuals who purchased wind coverage from private insurers and from the Florida Citizens. Therefore, there are four types of wind insurance purchases for Survey 5 –

---

[17] National Association of Insurance Commissioners (NAIC), December 2019, "Considerations for State Insurance Regulators in Building the Private Flood Insurance Market."

Table 12.4 Distribution of wind insurance purchase types

Wind PH types	Definition	Survey 4 Freq.	Survey 4 Percent	Survey 5 Freq.	Survey 5 Percent
Homeowners' insurance from a private insurer	Coverage through a homeowners' insurance policy from private insurers	331	47%	309	33%
Homeowners' insurance from Citizens	Coverage through a homeowners' insurance policy from Florida Citizens			112	12%
Wind-only coverage from a private insurer	Coverage through a wind-only policy (or wind endorsement) from private insurers	34	5%	73	8%
Wind-only policy from Citizens	Coverage through a wind-only policy from Florida Citizens			15	2%
None	No wind coverage	341	48%	435	46%
Subtotal		**706**	**100%**	**944**	**100%**
Don't know[a]	Don't know if have wind coverage	138		253	
	Having wind insurance but don't know insurance type (insurance provider or policy type)			48	
Total		**844**		**1245**	

[a]The data of respondents who don't know any needed information are not used in the analysis due to lack of information

homeowners' insurance from a private insurer, homeowners' insurance from Citizens, wind-only coverage from a private insurer, and wind-only policy from Citizens.

Table 12.4 shows the frequency and percentage of different types of wind insurance purchases. The majority of respondents (47% in Survey 4 and 45% in Survey 5) obtained their wind coverage through their homeowners' insurance policy. Only 10% or fewer respondents purchased a wind-only policy.

12.4.2 Other Survey Variables

To study insurance purchase behavior, we examine the influence of sociodemographic factors, house characteristics, flood risk perceptions, wind risk perceptions (only in Survey 5), the regret of having or not having flood insurance, the social norm for flood insurance, the mitigation measures, the premium discount for implementing measures (only in Survey 5), the trouble of obtaining insurance due to disaster activities (only in Survey 5), the financial difficulty of purchasing insurance due to COVID-19, and the trust in the government's ability to limit flood risk.

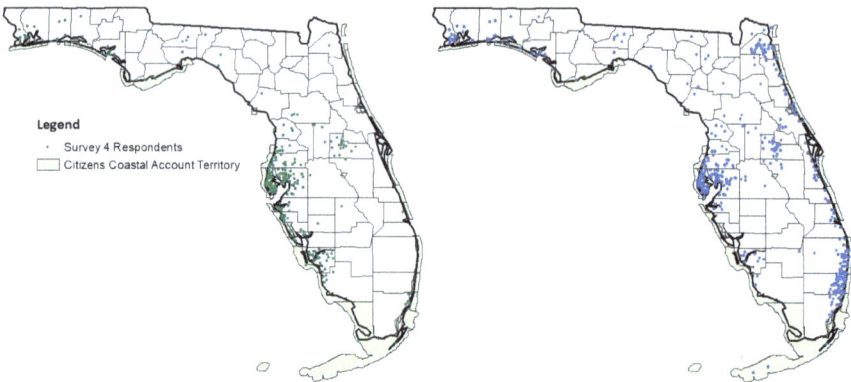

Fig. 12.6 Respondents in citizens coastal account territories

In addition to using individuals' wind risk perceptions, we also include a proxy for objective wind risk to study how windstorm coverage uptake is affected by objective risk. The variable, *Windonly_territory*, is based on whether a respondent lives within the boundaries of Florida Citizens Coastal Account Territories. The two maps below show the overlap of our respondents with the Coastal Account Territories for Survey 4 and Survey 5, respectively (Fig. 12.6). Essentially, the Coastal Account Territory covers the areas along the coastal lines of the Florida state. These areas are deemed as high-risk wind areas by Florida Citizens; only residents located within these areas are eligible for Citizens' wind-only policies. The variable *Windonly_territory* takes a value of 1 for the respondents located within the Coastal Account Territories and 0 otherwise.

The summary statistics of variables are displayed in Table 12.5. The average flood insurance uptake rate was 25% in Survey 4 and 36% in Survey 5. The average wind insurance uptake rate was 52% in Survey 4 and 56% in Survey 5, both higher than the flood insurance uptake rate of the same survey. Surveys 4 and 5 have different populations, but the two populations have similar mean values for the factors that may influence the insurance purchase.

12.4.3 Regression Methods

We use different models to examine the overall insurance uptake and the uptake of different insurance purchase outcomes. Our method to examine the overall insurance uptake is the fixed-effects Logit regression model. The dependent variable, *insurance purchase*, is a dummy variable equal to 1 if the respondent has purchased

Table 12.5 Summary statistics of variables

	Survey 4			Survey 5		
	N	Mean	Std. dev	N	Mean	Std. dev
Insurance uptake:						
Flood policy	753	0.247	0.432	1074	0.358	0.479
Wind policy	706	0.517	0.500	992	0.561	0.496
Influencing factors:						
Worry about flooding	839	2.712	1.268	1210	2.756	1.199
Perceived flood probability	676	0.105	0.207	974	0.090	0.183
Perceived flood impact	674	3.714	1.969	971	3.821	1.975
Worry about windstorm				1208	3.475	1.118
Perceived wind impact				980	3.789	1.959
Trust in government flood policies	796	2.734	0.780	1137	2.553	0.810
Number of ex ante risk reduction measures	844	1.315	1.524	1245	1.530	1.935
Window protection	794	0.486	0.500	1245	0.516	0.500
Roof retrofit				1245	0.561	0.497
Hip roof				1245	0.425	0.495
Premium discount for flood risk mitigation				739	0.108	0.311
Premium discount for wind risk mitigation				804	0.267	0.443
Trouble purchasing flood insurance				1245	0.142	0.349
Trouble purchasing homeowners' insurance				1245	0.153	0.361
Internal locus of control	835	7.404	2.258	1229	7.206	2.346
Risk taking	838	5.827	2.527	1228	5.818	2.575
Regret of no insurance	795	3.551	1.134	1142	3.613	1.083
Regret of having insurance	807	2.984	1.168	1152	2.941	1.186
Social norm for insurance uptake	790	3.108	1.176	1140	3.111	1.118
Financial difficulty due to COVID-19	844	0.217	0.412	1060	0.208	0.406
Age	832	46.73	17.49	1245	49.66	19.20
Education	839	3.222	1.055	1229	3.265	1.095
Income	813	3.480	1.405	1194	3.626	1.373
Female	837	0.687	0.464	1245	0.640	0.480
Value of home building	692	3.251	1.814	1031	3.596	1.856
Value of home content	716	4.581	2.279	1058	4.757	2.179
House owner	814	0.649	0.478	1190	0.703	0.457
Length of residence	814	9.026	9.723	1178	11.09	10.73
Underfloor basement	844	0.040	0.197	1245	0.074	0.262
Windonly_territory	803	0.052	0.223	1181	0.094	0.292

Note: The length of residence is winsorized at the 99% level because the maximum value is extremely large

insurance coverage and 0 if they have not purchased insurance. The fixed-effects Logit regression model is given by:

$$Insurance\ purchase_i = f(\beta X_i + \gamma_c + \epsilon_i)$$

where i indicates survey respondents and γ_c is county-fixed effects. On the right-hand side, we include independent variables X_i for demographic factors (e.g., age, female, education, income), house characteristics (e.g., home value, contents value, length of residence, underfloor basement), risk perceptions (e.g., worry, damage estimated), the mitigation measures implemented and the associated premium discount, psychology factors (e.g., internal locus of control, risk-taking, regret, social norm), and other factors (e.g., trust in government's ability to deal with flood risk, financial difficulty due to COVID-19). ϵ_i is the zero-mean error term.

To examine the uptake of different types of purchase outcomes, we use a Multinomial Probit regression model because the dependent variable for purchase outcome has different categories. The outcome of not buying insurance is always specified as the base outcome so that each insurance policy type is compared against the no coverage category. The Multinomial Probit regression model is given by:

$$Insurance\ purchase\ outcome_i = f(\beta X_i + \epsilon_i)$$

where $insurance\ purchase\ outcome_i$ has multiple categories because households can have different insurance purchase outcomes for flood and wind. The flood insurance purchase has five categories. Each category can be linked to the decision tree (Fig. 12.1) as follows:

- Outcome 1: no insurance = no flood coverage (end node 9); base outcome
- Outcome 2: mandated purchase and buying only NFIP policies (end node 2)
- Outcome 3: mandated purchase and buying private flood product (end nodes 1 & 3)
- Outcome 4: voluntary purchase and buying only NFIP policies (end nodes 5 & 7)
- Outcome 5: voluntary purchase and buying private flood product (end nodes 4, 6, 8)

For the windstorm coverage, the dependent variable for Survey 4 has three outcomes and is linked to the decision tree (Fig. 12.2) as follows:

- Outcome 1: None = no wind coverage (end node 6); base outcome
- Outcome 2: wind coverage through homeowners' multi-peril insurance (end nodes 5 & 7)
- Outcome 3: wind coverage through a wind-only policy (end nodes 1–4)

The dependent variable for Survey 5 has five outcomes and relates to the decision tree (Fig. 12.2) as follows:

- Outcome 1: None = no wind coverage (end node 6); base outcome
- Outcome 2: wind coverage through homeowners' multi-peril insurance from private insurers (end node 5)
- Outcome 3: wind coverage through a wind endorsement or a wind-only policy from private insurers (end nodes 1–4)
- Outcome 4: wind coverage through homeowners' multi-peril insurance from Florida Citizens (end node 7)
- Outcome 3: wind coverage through a wind-only policy from Florida Citizens (end nodes 2 & 4)

Mitigation measures and insurance purchases may be jointly determined by factors such as an individual's risk aversion level and the public mitigation measures implemented at the county level. We asked about the individual's risk aversion (or risk taking) level in the surveys and have controlled this factor in the regression. We also incorporated county-fixed effects in the Logit model for the overall insurance coverage uptake. The fixed effects capture unobservable factors that are the same within a county.

The mitigation measure variables may be endogenous because of reverse causality. Individuals may decide whether to implement mitigation measures depending on their insurance coverage level. The timing of insurance purchases and mitigation measure implementation cannot be established based on the survey questions. We cautiously interpret our results as correlations rather than causations.

12.5 Empirical Results

12.5.1 Results of Flood Insurance

12.5.1.1 Survey 4 Results

For all regression models, we report marginal effects at the means.[18] The sample size of the regressions is based on non-missing observations. The respondents could answer "Not sure" or "Don't know" to most of the survey questions.

Table 12.6 shows the regression results for the overall flood insurance uptake. We find the value of home contents and the worry about flooding positively relate to purchasing flood insurance. For example, when the value of home contents increases by one level, the probability of purchasing flood insurance increases by 0.23 on average, with other things equal. Anticipating regret for not having insurance if a flood were to occur next year is associated with an 0.35 increase in the probability of purchasing flood insurance. A stronger social norm for insurance uptake also

---

[18]This is the marginal effect of x variable on the y variable when holding other covariates at their mean values.

Table 12.6 Overall flood insurance uptake (Survey 4)

	Flood insurance uptake
Age	0.005
	(0.008)
Education	0.078
	(0.143)
Income	−0.239
	(0.143)
Female	−0.377
	(0.278)
House owner	0.021
	(0.211)
Value of home building	−0.122
	(0.168)
Value of home content	0.230**
	(0.072)
Length of residence	−0.015
	(0.019)
Underfloor basement	1.038*
	(0.407)
Worry about flooding	0.281*
	(0.124)
Perceived flood probability	−0.796
	(0.727)
Perceived flood impact	0.103
	(0.071)
Risk taking	−0.008
	(0.049)
Internal locus of control	−0.058
	(0.042)
Regret of no insurance	0.353*
	(0.171)
Regret of having insurance	−0.179
	(0.101)
Social norm for insurance uptake	0.683**
	(0.161)
Number of ex ante risk reduction measures	0.205
	(0.124)
Financial difficulty due to COVID-19	−1.067**
	(0.345)
Trust in government flood policies	−0.161
	(0.207)
County fixed-effects	Yes
Observations	402
Log-likelihood	−158.5

Note: The table reports the marginal effect at the mean of the fixed-effects Logit Regression model. Standard errors in parentheses are clustered by county
**$p < 0.01$, *$p < 0.05$

positively relates to the flood insurance purchase probability and the average increase is 0.68.

Having a basement is positively related to purchasing flood insurance and having financial difficulty due to Covid-19 has a negative relationship with the flood insurance purchase. The magnitudes of the marginal effects for both variables are greater than one. The marginal effect being greater than one is possible because the derivative at a point is the tangent line of the curve at that point, which could be steeper than one.[19]

The flood insurance purchase can be further divided into five outcomes. Theoretically, we expect that different types of insurance purchases may be related to different factors. For example, the demographic factors, house characteristics, and risk perception variables may affect voluntary purchase more than mandatory purchase because policyholders are not supposed to make decisions based on these factors when they are mandated to purchase insurance. The purchase of a private flood insurance product may be related to factors that reflect the households' needs for more comprehensive coverage.

The regression results regarding the flood insurance purchases across various types are reported in Table 12.7. Our results show that different types of flood insurance purchases (in different columns) are associated with different factors. The mandatory purchase of an NFIP policy is not significantly associated with any factors, and the mandatory purchase of a private product is only significantly and positively related to the worry about flooding. One-level increase in the worry about flooding increases the probability of the mandatory purchase of a private product by 0.005.

In comparison, voluntary purchase of flood insurance is associated with more factors. The voluntary purchase of an NFIP policy is positively related to the value of contents, the worry about flooding, the regret of having no insurance when a flood occurs next year, and the social norm for flood insurance. The financial difficulty due to Covid-19 and the trust in the government's ability to deal with flood risk negatively relate to the voluntary purchase of an NFIP policy.

The voluntary purchase of a private flood insurance product is positively related to the dummy variable measure of respondents' homes having a basement. This may be explained by the fact that the NFIP policy does not cover the contents in the basement so individuals must obtain the basement coverage from a private flood insurance policy.

Comparing the results in Tables 12.6 and 12.7, we find that some factors that do not have a significant relationship with the overall insurance uptake may significantly relate to certain insurance purchase outcomes. For example, the trust in the government's ability to deal with flood risk negatively relates to the voluntary purchase of an NFIP policy. Individuals may have a lower incentive to purchase flood insurance if they believe the local government is effectively dealing with the

---

[19] See more at the Stata website https://www.stata.com/support/faqs/statistics/marginal-effect-greater-than-1/.

Table 12.7 Uptake of various flood insurance purchase types (Survey 4)

	No insurance (base)	Mandatory purchase of NFIP	Mandatory purchase of a private product	Voluntary purchase of NFIP	Voluntary purchase of a private product
Age	−0.002	−0.000	−0.000	0.001	0.001
	(0.002)	(0.000)	(0.000)	(0.001)	(0.001)
Education	−0.004	0.000	−0.000	0.011	−0.007
	(0.018)	(0.000)	(0.003)	(0.011)	(0.013)
Income	0.027	−0.000	−0.002	−0.013	−0.011
	(0.025)	(0.000)	(0.005)	(0.013)	(0.015)
Female	0.087*	0.000*	−0.000	−0.047	−0.040
	(0.035)	(0.000)	(0.004)	(0.030)	(0.021)
House owner	−0.019	0.000	0.006	0.040	−0.028
	(0.031)	(0.000)	(0.004)	(0.032)	(0.027)
Value of home building	0.023	0.000	−0.005	−0.019	0.001
	(0.020)	(0.000)	(0.005)	(0.010)	(0.010)
Value of home content	−0.045**	−0.000	0.005	0.025**	0.015
	(0.010)	(0.000)	(0.005)	(0.007)	(0.008)
Length of residence	0.004	0.000	−0.001	−0.002	−0.001
	(0.002)	(0.000)	(0.001)	(0.001)	(0.001)
Underfloor basement	−0.172**	−0.000	0.000	0.076	0.114*
	(0.064)	(0.000)	(0.006)	(0.052)	(0.051)
Worry about flooding	−0.045*	−0.000	0.005*	0.029*	0.011
	(0.019)	(0.000)	(0.002)	(0.012)	(0.011)
Perceived flood probability	0.161*	0.000	0.004	−0.114	−0.051*
	(0.070)	(0.000)	(0.010)	(0.065)	(0.025)
Perceived flood impact	−0.010	−0.000	0.003	−0.002	0.009
	(0.012)	(0.000)	(0.002)	(0.007)	(0.007)
Risk taking	−0.003	0.000	−0.001	0.004	0.000
	(0.010)	(0.000)	(0.001)	(0.007)	(0.004)
Internal locus of control	0.011	0.000	0.001	0.000	−0.013*
	(0.007)	(0.000)	(0.002)	(0.006)	(0.006)
Regret of no insurance	−0.047*	0.000	0.002	0.039**	0.006
	(0.024)	(0.000)	(0.004)	(0.011)	(0.017)
Regret of having insurance	0.029*	−0.000	0.006	−0.012	−0.023*

(continued)

Table 12.7 (continued)

	No insurance (base)	Mandatory purchase of NFIP	Mandatory purchase of a private product	Voluntary purchase of NFIP	Voluntary purchase of a private product
	(0.012)	(0.000)	(0.004)	(0.010)	(0.009)
Social norm for insurance uptake	−0.090**	0.000	0.005	0.049**	0.036*
	(0.026)	(0.000)	(0.005)	(0.015)	(0.016)
Number of ex ante risk reduction measures	−0.001	0.000	0.011	−0.010	−0.000
	(0.023)	(0.000)	(0.008)	(0.008)	(0.011)
Financial difficulty due to COVID-19	0.118**	−0.000	−0.013	−0.100**	−0.005
	(0.043)	(0.000)	(0.013)	(0.024)	(0.019)
Trust in government flood policies	0.039	0.000	−0.002	−0.042*	0.004
	(0.026)	(0.000)	(0.004)	(0.020)	(0.014)
County fixed-effects	No				
Observations	413				
Log-likelihood	−295.2				

Note: This table reports the marginal effect at the mean of the Multinomial Probit Regression model. Standard errors in parentheses are clustered by county
**$p < 0.01$, *$p < 0.05$

flood events. Anticipating regret if one were to hold insurance when there was no flood has a negative relationship with the voluntary purchase of a private flood insurance product. This means that individuals are less likely to purchase the private flood insurance product for additional coverage beyond the NFIP policy if they anticipate regretting this decision in the case no flood occurs.

Moreover, some factors that have a significant relationship with the overall insurance uptake may not hold the same relationship with the different types of insurance purchases. For example, the dummy variable for having a basement only positively relates to the voluntary purchase of a private flood insurance product, perhaps because an NFIP policy does not provide coverage for contents in the basement. The value of home contents is only positively related to the voluntary purchase of an NFIP product and not related to mandatory purchases.

12.5.1.2 Survey 5 Results

We conduct the same regressions in Tables 12.6 and 12.7 again using the responses from Survey 5 because our previous surveys have shown that households may change the insurance purchase decision (add a policy or drop a policy) during the period between the end of last year's hurricane season and the beginning of this year's hurricane season. Therefore, we examine whether the findings from Survey 4 (the end of 2020 hurricane season) are still observed in Survey 5 (the beginning of 2021 hurricane season).

Table 12.8 reports the results for the overall flood insurance uptake. Anticipated regret for having no insurance if a flood were to occur next year and the social norm for buying flood insurance consistently have a positive relationship with flood insurance uptake; the financial difficulty due to Covid-19 consistently relates to reduced flood insurance uptake.

The dummy variable for respondents' homes having a basement is negatively related to flood insurance uptake, which is opposite to the Survey 4 finding. This relationship will be further explored when we examine the insurance purchase outcomes by type because the dummy variable for a basement does not have the same relationship with all types of insurance purchase outcomes.

The value of the home building is similar to the value of home contents and positively relates to the flood insurance uptake. The risk-taking variable is a new factor in Survey 5 that positively relates to the probability of purchasing insurance. This is not consistent with the extant literature that insurance purchase is positively associated with an individual's risk aversion (Robinson et al. 2021). However, the significant relationship disappears when we examine the insurance purchase outcomes by type.

Table 12.9 reports the regression results regarding various types of flood insurance purchases from Survey 5. For Survey 5, we find more factors related to the mandatory flood purchase than Survey 4. The difference may be due to the difference in samples or the difference in survey time. In practice, the mandatory purchase requirement is not well-enforced (Lingle and Kousky 2018), which leaves some room for individuals to make their own choice.

We observe a similar pattern that some factors that do not have a significant relationship with the overall insurance uptake may significantly relate to certain insurance purchase outcomes. For example, the length of residence and the number of ex-ante risk reduction measures taken do not relate to the overall flood insurance purchase but are positively associated with the mandatory purchase of an NFIP policy. Age and the trust in the government's ability to deal with flood risk have insignificant relationships with overall flood insurance uptake but are negatively related to the mandatory purchase of a private flood insurance product.

The dummy variable for whether the household has ever experienced trouble obtaining flood insurance is not related to the overall flood insurance uptake; however, it has different relationships with different types of insurance purchases. It is positively related to the mandatory purchase of an NFIP policy and negatively related to the voluntary purchase of a private flood insurance product. Based on the decision tree in Fig. 12.1, individuals may have trouble obtaining flood insurance if their communities do not participate in NFIP. In this case, individuals need to purchase a private flood insurance product. We expect a positive relationship between this variable and the uptake of an NFIP policy because individuals who

Table 12.8 Overall flood insurance uptake (Survey 5)

	Flood insurance uptake
Age	−0.006
	(0.009)
Education	0.222
	(0.116)
Income	−0.149
	(0.123)
Female	0.161
	(0.175)
House owner	0.048
	(0.323)
Value of home building	0.205**
	(0.071)
Value of home content	−0.035
	(0.062)
Length of residence	0.008
	(0.011)
Underfloor basement	−0.801**
	(0.278)
Worry about flooding	0.088
	(0.121)
Perceived flood probability	1.316
	(0.745)
Perceived flood impact	0.011
	(0.093)
Risk taking	0.098**
	(0.036)
Internal locus of control	0.055
	(0.056)
Regret of no insurance	0.274*
	(0.109)
Regret of having insurance	−0.059
	(0.063)
Social norm for insurance uptake	0.589**
	(0.100)
Number of ex ante risk reduction measures	0.108
	(0.085)
Premium discount for flood risk mitigation	0.545
	(0.363)
Trouble purchasing flood insurance	0.327
	(0.411)
Financial difficulty due to COVID-19	−0.985**

(continued)

Table 12.8 (continued)

	Flood insurance uptake
	(0.183)
Trust in government flood policies	0.099
	(0.142)
County fixed-effects	Yes
Observations	433
Log-likelihood	−204.6

Note: The table reports the marginal effect at the mean of the fixed-effects Logit Regression model. Standard errors in parentheses are clustered by county
**$p < 0.01$, *$p < 0.05$

used to live in non-participating communities but sought flood insurance might make efforts to make their community join the NFIP program and increase the availability of the NFIP insurance coverage.

The pattern that some factors that have a significant relationship with the overall insurance uptake may not hold the significant relationship with all types of insurance purchases also exists in Survey 5. For example, the value of buildings, similar to the value of contents in Survey 4, is only positively related to the voluntary purchase of an NFIP product. The dummy variable for having a basement has a negative relationship with the overall flood insurance uptake in Survey 5, which is contrary to the finding in Survey 4 because the basement indicator is only negatively related to the voluntary purchase of an NFIP policy. The underlying explanation is likely to be the same. The basement indicator is either positively related to the purchase of a private flood insurance product or negatively related to just buying an NFIP policy because an NFIP policy does not provide coverage for contents in the basement. Such coverage needs to be obtained through private flood insurance products.

12.5.2 Results of Wind Coverage

12.5.2.1 Survey 4 Results

Table 12.10 shows the regression results for the overall uptake of windstorm coverage. Owning a home significantly increases the probability of having windstorm coverage by 0.85, possibly because most homeowners buy property insurance for their homes, but few tenants purchase renters' insurance. Implementing window protection against wind damages positively relates to the purchase of windstorm coverage. This is possibly due to the premium discount for implementing window protection, which is mandated in Florida.

Table 12.11 reports the results regarding different types of windstorm insurance purchases. Buying coverage through the homeowners' insurance and buying coverage through the wind-only policy are associated with different factors. The

Table 12.9 Uptake of various flood insurance purchase types (Survey 5)

	No insurance (base)	Mandatory purchase of NFIP	Mandatory purchase of a private product	Voluntary purchase of NFIP	Voluntary purchase of a private product
Age	0.003	0.000	−0.002*	−0.001	−0.000
	(0.002)	(0.001)	(0.001)	(0.001)	(0.002)
Education	−0.046	0.016	−0.001	−0.006	0.038*
	(0.026)	(0.012)	(0.016)	(0.008)	(0.018)
Income	0.013	0.005	−0.004	0.002	−0.015
	(0.030)	(0.014)	(0.018)	(0.014)	(0.021)
Female	0.001	0.007	−0.010	−0.007	0.010
	(0.042)	(0.028)	(0.033)	(0.027)	(0.032)
House owner	0.012	−0.008	0.046	0.042	−0.092
	(0.069)	(0.021)	(0.038)	(0.057)	(0.056)
Value of home building	−0.044*	0.010	0.011	0.034**	−0.010
	(0.018)	(0.009)	(0.009)	(0.013)	(0.009)
Value of home content	0.013	−0.016	−0.010	−0.006	0.019
	(0.014)	(0.008)	(0.008)	(0.011)	(0.010)
Length of residence	−0.002	0.003*	−0.001	−0.001	0.000
	(0.003)	(0.001)	(0.001)	(0.001)	(0.002)
Underfloor basement	0.219**	−0.040	−0.069	−0.055**	−0.055
	(0.081)	(0.033)	(0.046)	(0.017)	(0.077)
Worry about flooding	−0.033	0.036**	0.015	0.013	−0.031
	(0.027)	(0.011)	(0.017)	(0.008)	(0.021)
Perceived flood probability	−0.206	0.016	0.034	0.133*	0.023
	(0.135)	(0.041)	(0.074)	(0.066)	(0.113)
Perceived flood impact	−0.009	−0.013	0.001	0.009	0.013
	(0.018)	(0.010)	(0.011)	(0.007)	(0.014)
Risk taking	−0.012	0.008	−0.003	−0.009	0.016
	(0.008)	(0.004)	(0.004)	(0.005)	(0.009)
Internal locus of control	−0.015	0.005	0.005	0.021**	−0.017
	(0.013)	(0.006)	(0.006)	(0.006)	(0.009)
Regret of no insurance	−0.062**	0.060**	0.008	0.029**	−0.035*
	(0.022)	(0.016)	(0.015)	(0.009)	(0.017)
Regret of having insurance	0.024	0.006	−0.015	−0.004	−0.012

(continued)

Table 12.9 (continued)

	No insurance (base)	Mandatory purchase of NFIP	Mandatory purchase of a private product	Voluntary purchase of NFIP	Voluntary purchase of a private product
	(0.014)	(0.011)	(0.011)	(0.010)	(0.012)
Social norm for insurance uptake	−0.137**	0.029*	0.051**	0.013	0.044**
	(0.025)	(0.013)	(0.015)	(0.012)	(0.010)
Number of ex ante risk reduction measures	−0.032*	0.024**	0.006	0.000	0.002
	(0.016)	(0.007)	(0.008)	(0.005)	(0.008)
Premium discount for flood risk mitigation	−0.054	0.019	0.074	−0.066	0.026
	(0.081)	(0.027)	(0.040)	(0.034)	(0.043)
Trouble purchasing flood insurance	0.013	0.052*	0.051	0.038	−0.155**
	(0.092)	(0.027)	(0.044)	(0.037)	(0.054)
Financial difficulty due to COVID-19	0.256**	0.011	−0.072	−0.037	−0.159*
	(0.049)	(0.037)	(0.037)	(0.046)	(0.073)
Trust in government flood policies	−0.011	0.031	−0.025*	0.007	−0.002
	(0.036)	(0.021)	(0.011)	(0.013)	(0.027)
County fixed-effects	No				
Observations	428				
Log-likelihood	−428.0				

Note: This table reports the marginal effect at the mean of the Multinomial Probit Regression model. Standard errors in parentheses are clustered by county
**p < 0.01, *p < 0.05

homeowner indicator only has a positive relationship with the coverage through the homeowners' insurance. The wind-only policy uptake is higher for tenants than homeowners. The window protection variable also has a positive relationship solely with the purchase of homeowners' insurance. This is possible because the state-level mandated premium discount for wind mitigation measures primarily applies to the standard homeowners' insurance policies from private insurers.

The variable, *Windonly_territory*, is an indicator for whether the property is located within the Florida Citizens coastal account territories. It increases the probability of wind-only policy purchase as only properties within these territories are qualified for purchasing a wind-only policy from Florida Citizens.

Table 12.10 Overall wind coverage uptake (Survey 4)

	Wind coverage uptake
Age	0.012
	(0.008)
Education	−0.131
	(0.114)
Income	0.136
	(0.088)
Female	−0.133
	(0.261)
House owner	0.852**
	(0.166)
Value of home building	0.110
	(0.093)
Value of home content	0.035
	(0.052)
Length of residence	−0.009
	(0.010)
Windonly_territory	1.125
	(0.695)
Risk taking	0.016
	(0.043)
Internal locus of control	0.065
	(0.062)
Window protection	0.423**
	(0.159)
Financial difficulty due to COVID-19	−0.003
	(0.144)
County fixed-effects	Yes
Observations	491
Log-likelihood	−261.6

Note: The table reports the marginal effect at the mean of the fixed-effects Logit Regression model. Standard errors in parentheses are clustered by county
**p < 0.01, *p < 0.05

Overall, we find that different types of wind insurance purchases relate to different factors. Some factors may only affect certain insurance purchase types (e.g., house owner and window protection for coverage through homeowners' insurance).

12.5.2.2 Survey 5 Results

We conduct similar regressions using Survey 5 to compare the findings of the two surveys. We include more variables for Survey 5 because it covered more survey

Table 12.11 Uptake of various wind coverage purchase types (Survey 4)

	No coverage (base)	Homeowners' insurance	Wind-only policy
Age	−0.003	0.004*	−0.001
	(0.002)	(0.002)	(0.001)
Education	0.038	−0.030	−0.008
	(0.031)	(0.034)	(0.006)
Income	−0.025	0.008	0.017
	(0.025)	(0.029)	(0.013)
Female	0.029	−0.024	−0.005
	(0.066)	(0.073)	(0.014)
House owner	−0.258**	0.289**	−0.031**
	(0.040)	(0.037)	(0.011)
Value of home building	−0.014	0.006	0.008
	(0.020)	(0.018)	(0.006)
Value of home content	−0.016	0.023*	−0.007*
	(0.012)	(0.012)	(0.003)
Length of residence	0.004	−0.005	0.001
	(0.002)	(0.002)	(0.001)
Windonly_territory	−0.271*	0.197	0.074**
	(0.120)	(0.123)	(0.020)
Risk taking	−0.007	0.005	0.002
	(0.011)	(0.012)	(0.003)
Internal locus of control	−0.020	0.023	−0.003
	(0.014)	(0.012)	(0.003)
Window protection	−0.095*	0.093*	0.002
	(0.040)	(0.040)	(0.015)
Financial difficulty due to COVID-19	0.073*	−0.130**	0.057**
	(0.035)	(0.031)	(0.013)
County fixed-effects	No		
Observations	494		
Log-likelihood	−357.7		

Note: This table reports the marginal effect at the mean of the Multinomial Probit Regression model. Standard errors in parentheses are clustered by county
**$p < 0.01$, *$p < 0.05$

questions than Survey 4. The results for the overall wind coverage uptake are reported in Table 12.12. In addition to the house owner dummy variable, the worry about windstorms and the dummy for being located in wind-only territories both increase the probability of having wind coverage. The properties located within the Florida Citizens coastal account territories (wind-only territories) tend to have higher objective wind risk; the property owners may be more concerned about windstorm damage and are more willing to obtain insurance than those not located in high-risk wind areas.

In addition to window protection, we also included two other questions about whether respondents have implemented protection against wind damage in Survey 5. All wind damage mitigation measures we included are qualified for a premium discount based on Florida law. The variable of premium discount for implementing wind risk mitigation measures increases the probability of having windstorm coverage. The financial difficulty due to Covid-19 reduces the overall uptake of wind coverage.

For Survey 5, we divide the insurance purchase outcomes by where the individuals purchased the coverage and have four types of insurance purchase outcomes. The results for the uptake of different types of windstorm insurance purchases are reported in Table 12.13. Similar to Survey 4, the positive relationship between being a homeowner and the wind coverage uptake is only for coverage through homeowners' insurance. The premium discount for implementing mitigation measures only positively relates to the homeowners' insurance policies from private insurers.

The worry about windstorms is only positively related to the homeowners' insurance from Citizens. We failed to find a significant effect of *Windonly_territory* on wind-only policies because the last column reports insignificant marginal effects. This may result from the small sample problem – the Citizens wind-only policy category only has 15 respondents.

In Table 12.12, we find that individuals who have had trouble obtaining wind coverage are more likely to have wind coverage at the survey time. But this positive relationship disappears when we examine the various types of insurance purchases. Based on the decision tree in Fig. 12.2, the private insurers can only exclude the windstorm peril from the standard homeowners' insurance policy in some high-wind risk areas; in this case, households must purchase a wind-only policy. Thus, individuals who have had trouble obtaining wind coverage from their standard homeowners' insurance policy may have a better understanding of the wind-only policies. However, we fail to find significant relationships in Table 12.13 possibly because the sample size of the Citizens wind-only policy category is small.

12.6 Discussion and Conclusion

Inadequate insurance coverage and disaster preparation are major obstacles for society to deal with the increasing risk posed by hurricanes. Surrounded by water and regularly impacted by hurricanes, Florida is extremely vulnerable to flood and wind damage. Although it has the highest flood insurance market penetration rate in the U.S. (35% NFIP and 3% private sector policies in 2018), there is still a considerable coverage gap (Lingle and Kousky 2018).

By conducting and analyzing two surveys of households in Florida, in this chapter, we sought to explore motives and characteristics of households with regard to the uptake of flood insurance during the direct threat of Hurricane Eta in November 2020 and in June 2021 at the start of the hurricane season. Moreover, the unique factors that only drive the insurance purchase of a specific type of policy

Table 12.12 Overall wind coverage uptake (Survey 5)

	Wind coverage uptake
Age	0.023**
	(0.006)
Education	0.023
	(0.080)
Income	−0.126
	(0.113)
Female	−0.104
	(0.350)
House owner	0.493*
	(0.240)
Value of home building	0.070
	(0.094)
Value of home content	−0.052
	(0.096)
Length of residence	−0.009
	(0.005)
Worry about windstorm	0.325**
	(0.113)
Perceived wind impact	0.092
	(0.082)
Windonly_territory	0.587*
	(0.277)
Risk taking	−0.031
	(0.044)
Internal locus of control	0.057
	(0.064)
Window protection	−0.221
	(0.169)
Roof retrofit	0.409
	(0.260)
Hip roof	0.378
	(0.278)
Premium discount for wind risk mitigation	1.058**
	(0.233)
Trouble purchasing homeowners insurance	0.812**
	(0.290)
Financial difficulty due to COVID-19	−0.951*
	(0.474)
County fixed-effects	Yes
Observations	493
Log-likelihood	−201.9

Note: The table reports the marginal effect at the mean of the fixed-effects Logit Regression model. Standard errors in parentheses are clustered by county
**$p < 0.01$, *$p < 0.05$

Table 12.13 Uptake of various wind coverage purchase types (Survey 5)

	No coverage (base)	Homeowners' insurance from a private insurer	Wind-only coverage from a private insurer	Homeowners' insurance from Citizens	Wind-only policy from Citizens
Age	−0.005**	0.004*	0.003*	−0.001	−0.000
	(0.001)	(0.002)	(0.001)	(0.001)	(0.000)
Education	0.001	−0.030	0.022	0.004	0.000
	(0.015)	(0.024)	(0.023)	(0.016)	(0.000)
Income	0.025	−0.025	0.001	0.001	−0.000
	(0.021)	(0.023)	(0.016)	(0.021)	(0.000)
Female	0.020	−0.114*	0.029	0.057*	0.000
	(0.061)	(0.046)	(0.032)	(0.028)	(0.001)
House owner	−0.120**	0.203**	−0.091	0.006	−0.000
	(0.046)	(0.075)	(0.067)	(0.043)	(0.000)
Value of home building	−0.010	0.016	−0.003	−0.003	−0.000
	(0.016)	(0.013)	(0.013)	(0.013)	(0.000)
Value of home content	0.009	−0.003	−0.002	−0.004	−0.000
	(0.017)	(0.021)	(0.014)	(0.009)	(0.000)
Length of residence	0.002	−0.006*	0.002	0.002	−0.000
	(0.001)	(0.003)	(0.002)	(0.002)	(0.000)
Worry about windstorm	−0.071**	0.030	0.004	0.036**	−0.000
	(0.023)	(0.024)	(0.018)	(0.014)	(0.000)
Perceived wind impact	−0.017	0.005	0.014	−0.001	−0.000
	(0.015)	(0.017)	(0.011)	(0.006)	(0.000)
Windonly_territory	−0.142**	−0.022	0.122	0.039	0.000
	(0.045)	(0.079)	(0.066)	(0.043)	(0.000)
Risk taking	0.006	−0.006	−0.009	0.008*	0.000
	(0.009)	(0.009)	(0.006)	(0.004)	(0.000)
Internal locus of control	−0.026*	0.043**	−0.013	−0.004	−0.000
	(0.012)	(0.013)	(0.010)	(0.005)	(0.000)
Window protection	0.026	−0.036	0.009	0.004	−0.000
	(0.044)	(0.050)	(0.054)	(0.031)	(0.000)
Roof retrofit	−0.076	0.079	0.056	−0.055	−0.000
	(0.042)	(0.054)	(0.045)	(0.037)	(0.000)
Hip roof	−0.046	−0.019	0.013	0.041	0.000
	(0.055)	(0.043)	(0.035)	(0.038)	(0.001)
Premium discount for wind risk mitigation	−0.193**	0.121**	0.052	0.012	0.000

(continued)

Table 12.13 (continued)

	No coverage (base)	Homeowners' insurance from a private insurer	Wind-only coverage from a private insurer	Homeowners' insurance from Citizens	Wind-only policy from Citizens
	(0.042)	(0.038)	(0.033)	(0.027)	(0.000)
Trouble purchasing homeowners insurance	−0.098	−0.067	0.097	0.058	0.000
	(0.072)	(0.072)	(0.059)	(0.037)	(0.001)
Financial difficulty due to COVID-19	0.163*	−0.083	−0.036	−0.041	0.000
	(0.065)	(0.072)	(0.053)	(0.043)	(0.000)
County fixed-effects	No				
Observations	476				
Log-likelihood	−538.0				

Note: This table reports the marginal effect at the mean of the Multinomial Probit Regression model. Standard errors in parentheses are clustered by county
**$p < 0.01$, *$p < 0.05$

are assessed. We demonstrate that various types of insurance purchases can exist for flood and windstorm insurance, and they can have a unique decision-making process as they may have a different choice set.

We use a decision tree to illustrate the complex insurance purchase process and show how individuals can end up having different insurance purchase outcomes. We also conducted regression analyses to assess the drivers of various types of insurance purchase outcomes. In general, we find that different types of insurance purchases relate to unique factors. With flood insurance, we find that mandatory purchase is related to fewer explanatory variables than voluntary purchase. For example, only the voluntary purchase of an NFIP policy is positively related to the value of possession contents, the anticipated regret of having no insurance when a flood occurs next year, and the social norm for flood insurance. Regarding wind coverage, we find that being a homeowner increases the probability of purchasing homeowners' insurance policies but does not increase the uptake of wind-only policies. Homeowners more frequently have insurance coverage than renters and thus are more likely to have windstorm coverage through their homeowners' insurance.

This research contributes to the limited existing literature that distinguishes types of natural disaster insurance purchases in understanding insurance uptake decisions (Brody et al. 2017; Botzen et al. 2019; Petrolia et al. 2015). Botzen et al. (2019) make the distinction between mandatory and voluntary flood insurance coverage in their study. Brody et al. (2017) specifically focus on the voluntary purchase of NFIP policies by properties located outside the 100-year floodplain. We provide a comprehensive overview of insurance policy types. For flood insurance, in addition to the distinction between mandatory and voluntary purchase, we also distinguish the insurance purchase by the underwriter. As the private flood insurance products provide coverage beyond the NFIP coverage limits, the purchase of a private flood insurance product is positively associated with the homeowner's demand for basement coverage and negatively related to the financial difficulty due to Covid-19. With the development of the private flood insurance market, the distinction based on the underwriter will become more important in the future.

Regarding wind insurance coverage, Petrolia et al. (2015) study wind insurance in coastal states. They look at the overall wind coverage uptake but include a dummy variable to indicate whether the wind peril is excluded from the regular homeowners' insurance policy and to represent wind-only policies. In our analysis of drivers of wind insurance purchases, we explicitly distinguish regular homeowners' insurance and wind-only policies to examine the unique factors associated with each type. Since a few insurance companies in Florida can offer wind-only policies, we also separate the wind insurance purchase based on the underwriter to provide a comprehensive view of the wind insurance purchase outcomes in Florida. We find that factors such as wind damage mitigation measures and the premium discount for wind mitigation measures are positively associated with only homeowners' insurance purchases.

Our results also highlight the importance of distinguishing different types of insurance purchases when studying the drivers of natural disaster insurance purchases. We find factors that do not have a significant relationship with the overall insurance uptake but are significantly related to certain insurance purchase types. For example, trust in the government's ability to deal with flood risk does not relate to the overall flood insurance uptake but is negatively related to the voluntary purchase of an NFIP policy. Individuals may have a lower incentive to purchase flood insurance if they believe the local government is effectively dealing with the flood events.

We also find factors that have different relationships with different types of insurance purchases. The indicator for having a basement has a positive relationship with the overall flood insurance uptake in Survey 4 and a negative relationship with the overall flood insurance uptake in Survey 5. The two results are contrary because the basement indicator has opposite relationships with two insurance purchase types. When we look at the flood insurance purchase at a more granular level, the basement indicator is positively related to buying a private flood insurance product in Survey 4 and negatively related to buying an NFIP policy in Survey 5. As the NFIP policy does not cover the contents in the basement, such coverage needs to be obtained through a private flood insurance product. Therefore, if we do not distinguish insurance purchase types, we may fail to understand the underlying reasons for observing different relationships between having a basement and the overall insurance uptake across the two surveys.

Given the increased costs from natural disasters, insurance against hurricane perils (e.g., flood, windstorm) is vital to individuals to cover their property damages and reduce their financial vulnerability to damage caused by natural disasters.

Acknowledging the drivers of flood and wind insurance uptake can inform policy design related to insurance uptake. For example, the perception of flood risk may be low for individuals who lack knowledge about flood risk, causing a low uptake of coverage within this group. This may require campaigns to raise risk awareness, such as the NFIP Community Rating System (CRS) seeks to do (Li and Landry 2018). Consistent with Robinson and Botzen et al. (2019), our result shows the psychological factor – the anticipated regret of having no insurance when a flood occurs next year, increases flood insurance demand, especially the demand for voluntary purchase. This suggests that policies can promote communication to enhance insurance uptake and overcome the feeling of regret.

Another type of policy is through the norm-nudge (Mol et al. 2021), where individuals are made aware of insurance uptake and risk-reduction effort in their neighborhood. Previous papers have found that insurance and mitigation measures are complements for flood and wind risks (Botzen et al. 2019; Petrolia et al. 2015). Our results also show such a positive relationship for mandatory flood insurance purchase (mandatory requirement is not well-forced in practice) and for wind coverage via homeowners' insurance. Therefore, policies may be designed to apply a degree of social pressure on individuals that have not taken mitigation measures.

In this chapter, we seek to uncover what is deterring certain types of individuals from purchasing natural disaster insurance and contribute to the policy debate related to the low demand for natural disaster insurance. Although our data and analysis are limited to Florida, our method of distinguishing the types of natural disaster insurance purchases may apply to other coastal states as well. The decision trees and the specific insurance purchase outcomes should be tailored to state-specific regulations and conditions.

Bibliography

AON (2020) Global catastrophe recap: November 2020. Available online at: http://thoughtleadership.aon.com/documents/20201210_analytics-if-november-global-recap.pdf

Blackwell J (2020) Record-breaking Atlantic hurricane season draws to an end. National Oceanic and Atmospheric Administration. Available at: https://www.noaa.gov/media-release/record-breaking-atlantic-hurricane-season-draws-to-end

Botzen WJW (2021) Economics of insurance against natural disaster risks. In: Oxford research encyclopedia of environmental science.

Botzen WJW, Kunreuther H, Michel-Kerjan E (2019) Protecting against disaster risks: why insurance and prevention may be complements. J Risk Uncertain 59(2):151–169

Botzen WJW, Mol JM, Robinson PJ, Zhang J, Czajkowski J (2020a) Individual hurricane preparedness during the COVID-19 pandemic: insights for risk communication and emergency management policies. Available at SSRN 3699277

Botzen WJW, Robinson PJ, Mol JM, Czajkowski J (2020b) Improving individual preparedness for natural disasters: lessons learned from longitudinal survey data collected from Florida during and after hurricane Dorian. Institute for Environmental Studies – Vrije Universiteit, Amsterdam

Brody SD, Highfield WE, Wilson M, Lindell MK, Blessing R (2017) Understanding the motivations of coastal residents to voluntarily purchase federal flood insurance. J Risk Res 20(6): 760–775

Bubeck P, Berghäuser L, Hudson P, Thieken AH (2020) Using panel data to understand the dynamics of human behavior in response to flooding. Risk Anal 40(11):2340–2359

Federal Emergency Management Agency (FEMA) (2020) Special flood hazard area (SFHA). Available at: https://www.fema.gov/glossary/special-flood-hazard-area-sfha

Federal Emergency Management Agency (FEMA) (n.d.) National Flood Insurance Program. https://www.fema.gov/flood-insurance. Retrieved 24 Feb 2021

Insurance Information Institute (III) (2020) Background on: hurricane and windstorm deductibles. Available at: https://www.iii.org/article/background-on-hurricane-and-windstorm-deductibles

Insurance Journal (2020). With Economic Damages of $9B, Hurricanes Eta and Iota Topped Off Record Year. Available online at: https://www.insurancejournal.com/news/international/2020/12/14/593841.html

Kunreuther H, Pauly M (2004) Neglecting disaster: why don't people insure against large losses? J Risk Uncertain 28(1):5–21

Li J, Landry CE (2018) Flood risk, local hazard mitigation, and the community rating system of the National Flood Insurance Program. Land Econ 94(2):175–198

Lingle B, Kousky C (2018) Florida's private residential flood insurance market. Risk Management and Decision Processes Center, University of Pennsylvania. https://riskcenter.wharton.upenn.edu/wp-content/uploads/2018/09/Florida-Private-Flood-Issue-Brief.pdf

Marsooli R, Lin N, Emanuel K, Feng K (2019) Climate change exacerbates hurricane flood hazards along US Atlantic and Gulf Coasts in spatially varying patterns. Nat Commun 10(1):1–9

Milliman (2021) Case study: Wright Flood teams with Milliman to close the flood protection gap with private flood insurance. Available at: https://f.hubspotusercontent20.net/hubfs/9214628/Resources/wright-flood-case-study.pdf?hsCtaTracking=be10bc30-f411-4d39-9d13-df036157ec23%7C033d7e40-bcb6-4ea9-a8eb-26f3862100a6

Mol JM, Botzen WJW, Blasch JE (2020) Behavioral motivations for self-insurance under different disaster risk insurance schemes. J Econ Behav Organ 180:967–991

Mol JM, Botzen WJW, Blasch JE, Kranzler EC, Kunreuther HC (2021) All by myself? Testing descriptive social norm-nudges to increase flood preparedness among homeowners. Behav Public Policy 1–33. https://doi.org/10.1017/bpp.2021.17

Mondino E, Scolobig A, Borga M, Albrecht F, Mård J, Weyrich P, Di Baldassarre G (2020) Exploring changes in hydrogeological risk awareness and preparedness over time: a case study in northeastern Italy. Hydrol Sci 65(7):1049–1059

Munich Re (2020) The flood insurance gap in the United States. Available at: https://www.munichre.com/topics-online/en/climate-change-and-natural-disasters/natural-disasters/floods/the-flood-insurance-gap-in-the-us.html

Petrolia DR, Hwang J, Landry CE, Coble KH (2015) Wind insurance and mitigation in the coastal zone. Land Econ 91(2):272–295

Robinson PJ, Botzen WJW (2019) Determinants of probability neglect and risk attitudes for disaster risk: an online experimental study of flood insurance demand among homeowners. Risk Anal 39(11):2514–2527

Robinson P, Botzen WJW, Kunreuther H, Chaudhry S (2021) Default options and insurance demand. J Econ Behav Organ 183:39–56

Weather.com (2020) Hurricane Eta drenches Central America, then measurers toward Florida. Available at: https://weather.com/storms/hurricane/news/2020-11-12-tropical-storm-eta-landfall-florida-southeast-flooding-rain

Chapter 13
Exploring the Role of Social Networks in Hurricane Preparedness Planning: A Study of Public Housing Residents

Robin Ersing, Beverly Ward, and Jennifer M. Collins

Abstract Situated on the eastern end of the Gulf Coast, Florida is often ground zero for hurricanes and tropical storms 6 months of the year. In 2017 Hurricane Irma made landfall causing widespread destruction in the Florida Keys before impacting the Tampa Bay region. Weather related hazards threaten communities of all types yet little is known about the impact on public housing developments. This exploratory descriptive study engaged adult residents in public housing in Tampa, Florida to understand the role of social networks in preparedness planning and evacuation decision-making. Surveys were conducted to learn about the dependability and diversity of social networks and their value in disaster preparedness. Findings from the study suggest the integration and mobilization of social connections have important consequences for women of color and individuals with disabilities living in public housing. Furthermore, relationships with family, friends and neighbors may influence both disaster preparedness behavior and evacuation decision-making. Although family members living nearby were perceived as a positive social support, the strongest social connections were with neighbors. Results from this exploratory study are intended to assist Public Housing Authority (PHA) leaders and those in local emergency management to consider policies and practices to promote the use of strong social connections in disaster planning and evacuation decision-making. Recommendations include ways to improve communication and influence evacuation behavior to promote safety and reduce loss of life within public housing developments.

Keywords Public housing · Disasters · Social vulnerability · Social connections

R. Ersing (✉)
University of South Florida, School of Public Affairs, Tampa, FL, USA
e-mail: rersing@usf.edu

B. Ward
BG Ward and Associates, LLC, Tampa, FL, USA

J. M. Collins
School of Geosciences, University of South Florida, Tampa, FL, USA

13.1 Introduction

As the Federal Emergency Management Agency (FEMA) and other government led organizations request communities to do more with less, self-reliance in preparing for and recovering from a natural disaster becomes increasingly important. The utility of social ties and networks among particularly vulnerable populations has gained attention as an integral factor in building capacity for resilience. Residents in public housing are considered a sub-population that tends to become hidden in the emergency preparedness literature. While research in this area is still emerging, one untapped aspect is knowledge about the role social connections play in decision-making about being storm ready including the need to evacuate.

In 2017, Hurricane Irma was forecasted as a Category 4 storm as it tracked toward the Gulf of Mexico. On September 10th, Irma eventually made landfall along the southwest coast of Florida near Marco Island. As a low-end Category 4 storm, it left a trail of significant damage to property, flooded many coastal towns and cities, and knocked out power to nearly 7 million people (Bousquet and Klas 2017). As Hurricane Irma neared the Tampa Bay area, residents in some public housing developments were ordered by emergency officials to evacuate. The order came despite potential barriers stemming from a number of social, economic, and health factors making the decision to leave their homes challenging (O'Donnell 2017).In the classic hazards literature the concept of disaster vulnerability is expressed both through personal attributes as well as social systems that might interject harm (Cutter et al. 2008). Wisner et al. (2003) define vulnerability to a hazard as the "characteristics of a person or group and their situation that influence their capacity to anticipate, cope with, resist and recover from the impact of a natural hazard" (p. 11). Indeed, challenges associated with low socioeconomic status (SES) have been found to correlate with and perhaps even influence one's decision-making with regard to hurricane evacuation orders (Fothergill and Peek 2004; SAMHSA 2017). To understand the plight of individuals and families residing in public housing from a disaster vulnerability perspective, one should consider factors within the social, economic, and physical environments (ISDR 2004). Together, these forces may contribute to the further marginalization of residents in public housing during a hazard event.

13.1.1 Factors of Social Vulnerability

Studies examining social vulnerability and disasters have identified several key factors related to risk. Among these are age, gender, race, income and disability. Disparities across these socio-demographic attributes have long been correlated to increased susceptibility and risk during a natural hazard (Bergstrand et al. 2015; Morrow 1999; Wisner et al. 2003). At each end of the age spectrum, both children and older adults are viewed as more vulnerable to the impact of a weather related disaster (Zoraster 2010). As individuals age some become more isolated and therefore may have fewer opportunities to connect socially with family and friends. This may result in reduced access to resources to mitigate exposure to a natural disaster (Bergstrand et al. 2015; Ngo 2001).

Gendered inequalities place women at continued risk of harm and burden due to exposure to a natural disaster (Cutter 2017). In many households females remain the dominant care provider for both children and aging adults (Peek and Fothergill 2008; Pickering et al. 2021). Vulnerability further increases among single female headed households where women have the added responsibility of providing financially for those in their care (David and Enarson 2012; Fothergill and Peek 2004) Disparities relevant to gender have been attributed to more women dying as a result of a disaster and often at an earlier age compared to men (Neumayer and Plümper 2007).

After the catastrophic events of Hurricane Katrina in 2005, race, ethnicity, and social vulnerability gained significant attention in the United States (David and Enarson 2012; Dyson 2007; Squires and Hartman 2006). Studies suggest minorities are more likely to live in hazard-prone or more hazard vulnerable areas (Davies et al. 2018) For example, Blacks and Latinos are more likely to live in high-risk flood zones (Bakkensen and Ma 2020). African American and Latino populations are generally thought to experience greater hardships both preparing for and recovering from a disaster resulting from longstanding issues of racial and cultural inequality (Cutter and Emrich 2006; Perry and Lindell 1991). Those with limited English proficiency face additional obstacles (Drolet et al. 2018). Households with Spanish as the first language are often the least prepared during a natural disaster (Bethel et al. 2013). This may be due in part to a lack of access to risk reduction information in a language other than English.

Income is another factor associated with disaster vulnerability both in terms of financial ability and social ties. Individuals with limited economic means are often less likely to afford costs related to preparedness planning, protection coverage, and evacuation efforts (Bergstrand et al. 2015; Fothergill 2004). This has been found in the case of older adults, particularly those with fixed or limited incomes. Lack of financial support can restrict options for sufficient preparedness such as access to transportation for an evacuation and money to cover expenses to shelter away from home (Al-rousan et al. 2014; Vatsa 2004). Sometimes individuals of lower socioeconomic status also experience greater social marginalization which can result in smaller social networks and networks with access to fewer resources (Vatsa 2004).

Individuals with disabilities are an often overlooked population in the natural hazards literature. According to Bethel et al. (2011), persons with health or mental health disabilities are at greater likelihood to be vulnerable to a disaster, and often experience obstacles preventing them from effectively preparing to reduce their risk of harm (Alexander et al. 2012). While each individual with a disability is unique, when encountering a natural disaster, many face challenges with evacuation routes and transportation, accessible public shelters, and adequate notification to prepare effectively (Twigg et al. 2018). In addition to environmental barriers, the disabled community may experience challenges in establishing and maintaining strong and diverse social networks (Simplican et al. 2015). Detachment from employment, location of housing, ability to interact and engage socially (Stough and Kang 2015) may also heighten social vulnerability.

13.1.2 Social Vulnerability and Public Housing

In 2005, Hurricane Katrina devastated the U.S. Gulf Coast region leaving behind significant casualties and significant damage to property (Vigdor 2008). This weather-related disaster also raised awareness of the social vulnerabilities and inequalities that spanned race, class and gender as people attempted to prepare for and recover from the storm (Dyson 2007). Indeed, the literature on threats to public housing residents resulting from a natural hazard note that low-income and minority households tend to experience greater loss and face a longer recovery period (Hamideh and Rongerude 2018; Peacock et al. 2014; Peek et al. 2014) compared with those living in the broader community.

An area of particular interest to the current study was the impact of Hurricane Katrina on public housing residents in New Orleans who were predominantly women of color and their families (David and Enarson 2012). Where one lives can influence the social ties and networks we develop. In the case of public housing developments, some studies suggest there may be a limit to the diversity of social connections available which in turn may affect the amount and types of support, both emotional and tangible, found in those relationships (Curley 2009; Kleit 2001). Others have focused on the strength of social ties among those who share a common sense of place and have established trust (Laakso 2013). The latter suggests that although there may be a gap in having access to a range of relationships, those who we can relate to best may prove more dependable when there is a need.

Twelve years after Katrina, Hurricane Irma became one of the largest mass evacuation events in U.S. history with millions of Floridians leaving their homes for safer shelter (Bousquet and Klas 2017). Collins et al. (2018) examined the effect of social connections in the general public comparing those who evacuated to those who stayed at home for Hurricane Irma and found that those who evacuated tended to have more social connections. Still little is known about the role of social connections and the effect on disaster preparedness planning and evacuation behavior of individuals residing in public housing.

Public housing in the United States has traditionally been funded by the federal Office of Housing and Urban Development (HUD) through the 1937 U.S. Housing Act. Government owned and subsidized public housing is one of the longest running and most recognized programs to assist lower-income households in need of affordable housing (Davlasheridze and Miao 2021). A local Public Housing Authority (PHA) generally contracts with HUD to administer the program within their respective community. The Tampa Housing Authority, incorporated in 1939, serves as a PHA to provide affordable housing opportunities to income eligible residents in the Tampa Bay, Florida region. The Tampa PHA currently lists 17 public development sites for a total of 1740 housing units (Tampa Housing Authority 2021). In 2017 the HUD Public Housing Assessment System reported 61% of the Tampa PHA households make below 30% of area median income (ProPublica 2019).

13.2 Methods

This exploratory descriptive study used a cross-sectional design to collect data from residents in several public housing developments in the city of Tampa. The data were gathered in summer 2019, 2 years after Hurricane Irma impacted the area. The overarching research question asked how the types of social connections (diversity), and reliability of connections (dependability) influenced preparation and evacuation behavior during Hurricane Irma. Survey data were analyzed using a statistical software package (SPSS) to record descriptive statistics of variables.

13.2.1 Study Sites

The Tampa PHA manages 17 public housing developments in the city. Using the Hurricane Evacuation Assessment Tool (HEAT), we selected four sites based on their geographic vulnerability to a hurricane. HEAT is an instrument provided by the local county emergency management system to assist residents in determining whether they reside in an evacuation zone. The tool also provides information on the location of shelters and other useful information to prepare for a storm related evacuation. Each of the four sites chosen for the study was geographically situated within a declared county hurricane evacuation zone in Tampa, Florida including Level A (Hillsborough County Board of County Commissioners 2019). This means emergency management officials may issue an evacuation order starting with those in Level A zones, based on the threat of wind speed (74 mph and higher) and storm surge.

Once the four sites were selected, surveys were administered on-site at each public housing development. Each development contained a mix of individual and family units. One location was designated for seniors and individuals with disabilities. Figure 13.1 shows the location of the survey sites.

13.2.2 Sample

A convenience sample of adult residents living in public housing managed by the PHA was used in the study. Flyers created by the researchers were distributed by the PHA to recruit eligible participants. Inclusion criteria required individuals to be 18 years of age or older and currently living in the housing unit they lived in during Hurricane Irma. Bilingual members of the research team were available to collect data in English and Spanish. The Institutional Review Board (IRB) for human subject protection at the University of South Florida determined the study was minimal risk and qualified for exempt status. Participation in the study was voluntary and individuals were each given a $5 gift card to use at a local supermarket as an

Tampa Public Housing Study Areas

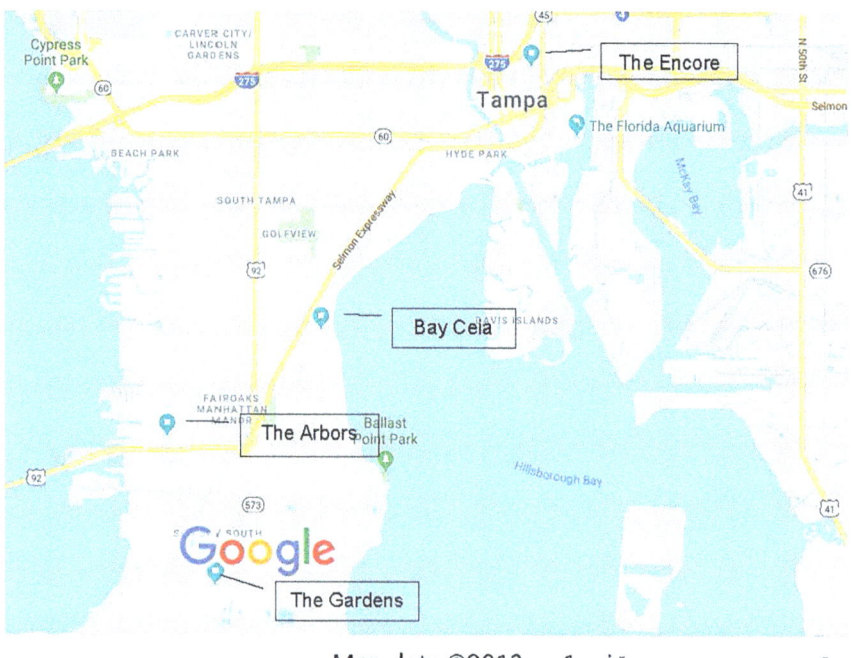

Fig. 13.1 Tampa public housing study areas. As shown in Fig. 13.1, four public housing complexes in the City of Tampa were selected as study areas: (i) Arbors at Rubin Padgett Estate. (ii) Bay Ceia Apartments. (iii) Gardens at South Bay. (iv) The Encore District. (Source: Hillsborough County Board of County Commissioners (2019))

incentive for their time to complete the survey. The research team was deliberate to not collect any personally identifying information given the sample population resides within a government managed public housing authority and to avoid potential discomfort among residents if they believed someone might connect their responses. Given that the four sites used in the study are located in hazard evacuation zones, this increased the potential to recognize which developments were included in the study. Therefore it was determined that data from the survey would be analyzed and reported in aggregate form to further protect confidentiality of participants. Research team members acknowledged consent to participate after reading from an IRB approved form indicating the study was confidential, voluntary, could be stopped at any time, and that results from the survey would only be presented in aggregated form.

Demographics of the sample (N = 90) show the average age to be 67 years old. Relationship status varied across the group with 38% being separated or divorced, 23% never married, 21% widowed, and 16% currently married. More than three-fourths (77%) of those surveyed were female. The race and ethnicity of participants was 42% African American, 37% Latino, and 13% White non-Hispanic and 8% other race. Overall 48% of those surveyed had at least some college or a college degree, while 32% reported a high school diploma or GED. Twenty percent had not graduated high school. Over half the participants spoke English (55%), while 22% spoke only Spanish and another 22% lived in bilingual households. Forty percent of the sample (n = 36) acknowledged having some type of disability.

13.2.3 Instrument

An 18 item survey was developed to assess social ties and access to resources resulting from these connections. The purpose was to understand whether the diversity and dependability of these social relationships influenced hurricane preparedness and evacuation decision-making during Hurricane Irma. The survey took approximately 20 minutes to complete and was administered face-to-face with responses recorded by a trained member of the research team. The instrument consisted of mostly categorical and Likert-type items with a single open-ended question that allowed participants to share any additional information about their experience. Questions were asked on the following topics:

- Household composition and ties to the local community
- Experience with hurricane preparedness and evacuation challenges
- Resources depended on in the event of a hurricane
- Frequency of being in touch with family, friends and neighbors
- Perceptions of quality of life in the local community
- Perceptions of personal social connections
- Demographic information

Social factors were measured in terms of personal social connections using nine items that were a sub-scale excerpted from a larger instrument designed to assess perceptions of individual social networks (Van der Gaag and Snijders 2005). Each item was scored on a Likert scale of 1–4 with 1 representing "Strongly disagree" and 4 representing "Strongly agree". The first dimension of the scale contained four items to represent integration of *different types* of relationships in one's network (i.e., diversity). Examples of items include "most of my friends know each other"; "my good friends also know my family members". The remaining five items represented the dimension of reliability and *propensity to mobilize* those relationships to access social resources (i.e., dependability). Examples of items included "I do not easily ask for help when I need it"; "I can't expect my neighbors to help me with serious problems"; "before I trust someone I have to be sure of his/her intentions". The sub-scale was scored in accordance with the original instrument by tallying Likert responses for each dimension (Van der Gaag and Snijders 2005). A total score of 12 or higher on the first dimension (diversity), suggested having more diverse social connections in one's life, while a total score of 15 or higher on the second dimension (dependability), suggested having greater ability to put those social resources into action.

13.3 Results

This study explored how the types of social connections (diversity), and reliability of those connections (dependability) may have influenced preparation and evacuation behavior during Hurricane Irma among residents in four public housing developments. The housing sites were located in county designated evacuation zones. The PHA had pre-warned all residents to prepare for the storm and to heed the warning to evacuate. Table 13.1 provides an overview of descriptive data derived from variables used in the study.

13.3.1 Hurricane Evacuation Planning

To understand evacuation planning, participants were asked about preparing for Hurricane Irma in 2017 as it approached the region. Despite being located within a county hurricane evacuation zone, an overwhelming majority of residents in the study (78%) said they did not believe they had received a warning notice to prepare for an evacuation. In contrast, 20% of the sample did recall being notified. Of the latter group, 67% (n = 13) stated it was their desire to evacuate if necessary.

When asked about actual evacuation behavior at the time of Hurricane Irma, 22% of residents did evacuate. Of that group, 7% (n = 6) acknowledged seeking safety by going to a public shelter outside of the evacuation zone. Another 15% (n = 14) said they evacuated their homes but did not seek out a public shelter. In comparison 74% (n = 67) reported not evacuating but instead sheltering in place at home. Several participants either did not recall taking action or did not believe they actually lived in an evacuation zone.

Given that residents experienced Hurricane Irma 2 years prior to the study we wanted to understand possible changes in evacuation strategy and their use of social connections. Specifically we asked, in the event of a hurricane evacuation now, do you have family or friends outside the Tampa area you could stay with until allowed to return back home? Over one-third of residents (37%) responded they have no family or friends that they could stay with during a hurricane. Another 47% reported being able to stay with family or friends that were located less than 50 miles away.

13.3.2 Integration and Mobilization of Social Connections

As discussed previously, a subset of nine questions within the survey were used to assess the diversity and dependability of personal social connections with family, friends and neighbors. More diverse types of social ties are believed to provide greater access to more resources that can be used to prepare for and evacuate from a natural hazard (Hawkins and Maurer 2010). Likewise, highly dependable social ties

Table 13.1 Descriptive survey data[a]

Variables	Percent	(Frequency)	Mean	(Range)
Sex:				
Male	23	(21)		
Female	77	(69)		
Age (years):			67.0	21.0–97.0
Marital status:				
Married	16	(14)		
Separated/divorced	38	(34)		
Widowed	21	(19)		
Never married	23	(21)		
Race/ethnicity:				
White non-Hispanic	13	(12)		
African American	42	(38)		
Latino	37	(33)		
Education:				
Some college/college degree	48	(43)		
High school diploma/GED	32	(29)		
Less than high school	20	(18)		
Language spoken at home:				
English	55	(50)		
Spanish	22	(20)		
Bilingual	22	(20)		
Disabled:	40	(36)		
Type of disability[b]				
Mobility	60	(22)		
Vision	16	(6)		
Hearing	9	(3)		
Cognitive	7	(3)		
Actual evacuation behavior (Hurricane Irma):				
Evacuated to public shelter	7	(6)		
Evacuated to non-shelter	15	(14)		
Did not evacuate	74	(67)		
Anticipated evacuation behavior:				
Shelter with family/friends nearby	47	(42)		
Shelter alone	37	(33)		
Social connections:				
Integration of relationships				
(Low diversity)	75	(68)		
Mobilization of relationships				
(High utility)	85	(77)		

Notes:
[a]Sample size N = 90
[b]Based on participants reporting a disability (N = 36)

are linked to one's trust and reliance on the utility of available resources (Atilano Pena-Lopez and Sanchez-Santos 2017). In other words, diverse social connections may give access to a broader array of resources, while dependable social connections ensures those resources will be there when needed.

Results were analyzed along two dimensions: Integration of different types of relationships in one's social network (i.e., diversity of relationships), and mobilization of those social resources (i.e., utility of relationships). Among residents sampled from the public housing developments in this study, 75% were found to have low integration of social relationships. This means their social connections were less diverse with regard to the types of relationships they have through family, friends, and neighbors. Integration of relationships also includes having a variety of ways to socialize with those in your life. The opposite was found with regard to mobilization of resources. An overwhelming majority of residents (85%) revealed high mobilization within their social relationships. This means the trust capital gained through ties with family, friends, and neighbors has dividends by feeling able to ask for help when needed. In other words, the gains from these social connections can be put to use when necessary.

Understanding integration and mobilization of social connections among the sample may factor into results that revealed 37% (n = 33) of residents claim to have no friends or family that they could shelter with during a hurricane. Another 30% (n = 27) believe they would be able to stay with family or friends within 50 miles. When asked how often they stay in touch with those in their social network, 61% (n = 55) indicated they talk with neighbors at least once per day compared to family or other friends. This may account for 79% (n = 71) of residents saying they want to shelter within their own homes during a hurricane. Another factor might be the need for transportation to access family or friends a distance from the public housing development.

13.3.3 Gender, Race and Social Connections

Women accounted for 77% (n = 69) of the sample for this study. Specifically, in terms of race, 42% (n = 29) of women identified as African American, 37% (n = 26) identified as Latino, and 13% (n = 9) reported being Caucasian. Overall 80% of all women in the study reported living in single-female headed households due to separation or divorce, having never married, or being widowed. Each of these factors (e.g. gender, race, and household composition) raise important implications for social connections and natural hazards planning. Participants were asked how often they talk with family, friends and neighbors in a typical month. Among white female residents, 15% acknowledged being in touch with family members either nearby (living within the county) or a distance away (outside the county or state) at least once per month. Less than 5% of African American and Latino women reported talking with any family members or friends at least monthly.

When asked about resources they could depend on for shelter in the event of a hurricane, Latino women would rely on family members, African Americans would rely on family or the government, and Whites would rely on family or no one. When asked what evacuation action they took in Hurricane Irma, 30% African American women and 25% Latino women said they chose to stay home and ride out the storm. In comparison, less than 10% of White women reported remaining home during the hurricane.

13.3.4 Individuals with Disabilities and Social Connections

Hurricane evacuation can pose additional challenges when someone has a disability. In this study 40% (n = 36) of participants reported having a disability. Of that subgroup, 60% (n = 22) identified a mobile disability, 16% (n = 7) a vision disability, 9% (n = 4) a hearing disability and 7% (n = 3) a cognitive disability. To understand types of social connections, we asked participants how often they are in touch with friends and neighbors. Seventy percent (n = 25) said they talk at least daily with friends living nearby (within the county), while 60% (n = 22) say they communicate daily with neighbors. Slightly more than one-third (35%) of those with a disability reported having daily contact with a family member including those that live nearby, live a distance more than 50 miles away, or live out of state. In addition to suggesting the types of social connections, it appears that those with a disability maintain closer connections by having daily interactions with those in their social networks.

The utility of social connections was analyzed through responses to a set of questions about dependability of resources in the event of a hurricane. Specifically, when asked whom they could depend on for emotional support in a hurricane, 20% (n = 8) indicated relatives living within the county. However, one's reliance on government resources was identified most often with regard to seeking shelter during a hurricane (20%), needing financial help (21%), or cleaning up and getting repairs made after a hurricane (26%). In the event of needing temporary housing post-hurricane, 17% said they could rely on family members nearby.

13.4 Discussion

The focus of this exploratory study was to understand how types of social connections (diversity), and reliability of those connections (dependability) influenced preparation and evacuation behavior during Hurricane Irma. Specifically we looked at the diversity and dependability of social connections among individuals living in four urban public housing developments in Florida during Hurricane Irma to understand decisions for disaster planning and actual evacuation behavior. Despite each development being situated in a county designated hurricane evacuation zone, three out of four residents chose to ride out the storm by sheltering in place. This raised serious implications for the Public Housing Authority (PHA) as well as government emergency management officials as they attempted to prevent casualties and mitigate suffering.

Public communication to inform individuals about a natural hazard as it approaches the area is critical to effective disaster preparedness planning (Sorensen and Sorensen 2007). Indeed, sending out warnings to evacuate and encouraging people to seek safe shelter at public sites or through other resources including family and friends is integral to prevent loss of life and reduce suffering due to structural damage, exposure to flood water, loss of electricity, and other storm related conditions (Sorensen and Mileti 1991). In this study we found that only a small number of residents in public housing recalled receiving emergency notice to evacuate, but of that group an overwhelming majority did heed the warning and sought shelter elsewhere. This is consistent with other studies on the influence of evacuation messaging on actual evacuation behavior (Collins et al. 2018; Ersing et al. 2020). The challenge for emergency management is to find messaging strategies that are effective in public housing developments. Indeed, our results revealed that although the diversity of social connections was low among those in the study, the level of dependability among those ties was high. This may aid in hazard planning by recognizing the strength in shared communications even among a smaller close knit group of individuals. Furthermore, demographics from our study suggest vulnerability factors such as age, gender, race, and physical ability may require communication outlets that will resonate with those in public housing. It is notable that neighbors seem to be an important source of support within a development, which might further influence the decision to shelter-in-place so that strong social connections are not broken, thus increasing personal anxiety. The influence of neighbors on people who decided not to evacuate, by providing dependable social connections, was also noted in Collins et al. (2017).

With this in mind, the PHA becomes a critical component for effective communication to pass along storm related information to residents sheltering in place. This might include instructions on storm readiness including how to prepare one's unit. It may be important for the PHA to ensure generators are available to address any disruption to power. This is vital for those needing to keep medications such as insulin refrigerated, or those requiring power for medical devices. The PHA should consider an effective and efficient transportation plan in case conditions deteriorate to the extent that residents must be physically evacuated to special needs medical shelters within the county. Finally, each development might become a "point of distribution" site.

for residents to obtain water, meals, first aid, comfort care, and other necessities until living conditions return to normal. Disaster service organizations including local nonprofits, faith-based groups, and community health agencies are potential partners with the PHA to provide for residents that shelter-in-place.

Our research question also realized that a lack of diversity or integration of social connections was found to influence evacuation behavior. Previous research suggests that lack of social ties with people outside an evacuation zone may result in not evacuating despite a warning (Collins et al. 2017). Low integration of social relationships was found in our study. In other words, those who chose to shelter at home reported fewer diverse social connections, which in turn resulted in reduced access to resources that could aid in an evacuation. This included having a place to stay temporarily, help with finances, and emotional support. Low integration of social connections may be due to limited opportunities to socialize outside the housing development. Constraints such as access to transportation, age, being single, or health status likely contribute to fewer opportunities to build connections. In contrast to the diversity of social connections, study participants revealed that mobilization of the social connections that do exist are highly dependable. This means for those in public housing, it might not be the size and variety of relationships that matters, but instead having a few highly reliable connections that are available when needed. This fits with our results suggesting residents communicate more frequently with their neighbors compared to family and friends. This might be contextualized as a "neighbor helping neighbor" safety net for those in public housing developments.

Females comprised a vast majority of participants in our sample, with most identifying as single due to divorce, separation, loss of spouse, or never being married. Social connections among these women suggested they were more likely to rely on family for safe shelter in the event of a hurricane however, most reported being in communication with those family members on less than a monthly basis. This weak social connection may be reflected in actual evacuation behavior during Hurricane Irma with women of color particularly deciding to shelter at home. This suggests for women in public housing, the integration and mobilization of social relationships is more challenging. The implication might mean they either do not have the means or resources to leave the hazard area, or they must rely on government assistance to do so. The latter might raise a concern for personal safety among single females considering the use of a public evacuation shelter. The PHA may be able to provide support to residents in this situation by helping to establish a "buddy" system for residents to link up and seek safe shelter together. The availability of transportation arranged by the PHA to public shelter sites may also influence a woman's decision to evacuate when a warning is issued.

Public housing residents with disabilities are another population vulnerable to the threat of a natural hazard. As such, the role of social connections in disaster planning and evacuation decision-making is critical. We discovered that although family members were perceived as a positive social tie for emotional support, the amount of communication with these individuals varied from weekly to monthly to not at all. In comparison, most in the study reported daily communication with nearby friends and neighbors suggesting stronger connections. What is less clear is the dependability of resources these closer social relationships might provide. Indeed, residents with disabilities in the study indicated a reliance on government supports for shelter during hurricane, providing financial assistance, and repairing any damage caused by the hurricane. Some studies suggest older individuals with disabilities may already be linked to various government or social service programs and resources. If so, this connection likely serves as a source of information and assistance particularly for those with limited means and fixed incomes, especially in the event of a hurricane.

In general, those with a disability have an added vulnerability when preparing for a hurricane and possible evacuation. Building social connections with more diverse and dependable support systems may help to mitigate obstacles other community members may not encounter.

This is an area for additional study to discover specifically what people with disabilities find most difficult in coping with a hurricane, and what public housing or government sources can do to better facilitate preparation and evacuation.

Several limitations to the study should be noted. First, the sample demographics for this study revealed age, gender and race to be less distributed. Specifically participants were older and predominantly women of color. This is likely attributed to the use of convenience sampling methodology to recruit participants. Convenience sampling in disaster research is often used, particularly to document and study those who survived the hazard event (Norris 2006). An assumption of convenience sampling is homogeneity of the target population, making it impossible to generalize results beyond the sample (Etikan et al. 2016). Given the focused lens of this study on public housing residents in the path of Hurricane Irma, the sampling method seemed warranted. A second factor relevant to sampling is the context of urban public housing. Dating back to research conducted in 2003 on Hurricane Andrew (Sanders et al. 2004), to studies on Hurricane Katrina in 2005 (Henrici et al. 2010, 2015), sample demographics drawn from public housing developments tend to reflect higher concentrations of predominantly women of color. Likewise some housing developments are targeted for older residents which may skew the age variable. Future research in this area should consider a sampling method to provide greater diversity among the participants.

Furthermore, the overlap of factors that contribute to social vulnerability and characteristics of the population found in public housing developments presents another limitation. The concentration of variables relevant to race, gender, disability and age often found in subsidized housing may confound results. It would seem a control group from outside the PHA may be warranted to address this issue.

Although not a significant limitation, native language was a consideration during the administration of surveys. Some participants had limited comfort in speaking English. We did have several research team members who were fluent in Spanish and therefore, were able to give the survey in that language. Future consideration should be given to developing a Spanish language version of the survey to improve the comfort level of participants in responding to questions.

Finally the artifact of memory decay is important to note. According to Stallings (2002), a gap between when a disaster event occurs and when data are actually collected can contribute to faulty recall of the decision-making process. Our survey was administered in 2019 and participants were being asked to recall events, behaviors, and decisions made two years prior in 2017 during the onset of Hurricane Irma. Conclusions drawn from this study should have a caveat recognizing an individual's recall of their social connections and perceptions of behavior at the time of an event may have been altered with time (Brewer 2000; Lindell et al. 2005).Despite limitations to the study, the results presented here offer an important opportunity to better understand an often overlooked and hidden population in the disaster management literature. As noted above, individuals residing in public housing often experience social and economic characteristics that contribute to their personal vulnerability making it difficult to test whether the driving force relevant to evacuation preparedness is the person or the collective social environment of the housing development. While it is important to note these variables might confound results in an empirical study, this descriptive research allows us to learn from the actual and anticipated behaviors of our sample relevant to a hurricane evacuation warning. Therefore the scope and generalizability of this study is focused on housing development residents. Indeed, the concentration of this population within specific housing sites may assist the PHA and county emergency managers to engage residents in the development of targeted hazard preparedness plans. We compare this to lessons learned from assisted living and nursing home facilities, and dormitory style housing on college campuses where population specific evacuation and response plans are vital to minimize injury and loss of life.

13.5 Conclusion

Understanding the role of social connections in disaster preparedness planning and evacuation decision-making is important to help mitigate one's vulnerabilities to a natural hazard. Individuals residing in public housing developments, particularly those located within an emergency evacuation zone, are likely to experience known factors that place them at greater risk in a disaster. Among these are age, gender, race and physical abilities. Overall, findings from this study support the literature on social ties as having an intangible experience that in the midst of a hazard may factor into decision-making and evacuation behavior (Ersing et al. 2020; Hawkins and Maurer 2010). Both the integration and mobilization of social connections are important buffers to a hazard event, it may be that the dependability of resources from even a small network of relationships carries more importance. This might be in line with the idea that even weak ties may be sufficient depending on their utility when needed (Granovetter 1973).

The results from this study offer insight to a specific at-risk population and therefore encourage confidence in using the outcomes here to generate lessons learned that can inform the PHA and local emergency preparedness planners. With this in mind, we believe additional research is warranted to further assess how neighbors in public housing may be able to offer support that family and friends both nearby or a distance away cannot. Likewise, empirical testing with a control group may better assess the influence of individual social and economic factors. Finally, PHA managers and emergency planners may be able to draw on the role of social connections as they continue to advance strategies for more effective communication, information sharing, and preparedness planning to save lives.

References

Alexander D, Gaillard JC, Wisner B (2012) Disability and disaster. In: The Routledge handbook of hazards and disaster risk reduction, vol 1. Routledge, New York, pp 413–423

Al-rousan TM, Rubenstein LM, Wallace RB (2014) Preparedness for natural disasters among older U.S. adults: a nationwide survey. Am J Public Health 104(3):506–511. https://doi.org/10.2105/AJPH.2013.301559

Atilano Pena-Lopez J, Sanchez-Santos JM (2017) Individual social capital: accessibility and mobilization of resources embedded in social networks. Soc Networks 49:1–11

Bakkensen LA, Ma L (2020) Sorting over flood risk and implications for policy reform. J Environ Econ Manag 104:102362

Bergstrand K, Mayer B, Brumback B, Zhang Y (2015) Assessing the relationship between social vulnerability and community resilience to hazards. Soc Indic Res 122:391–409

Bethel JW, Foreman AN, Burke SC (2011) Disaster preparedness among medically vulnerable populations. Am J Prev Med 40(2):139–143

Bethel JW, Burke SC, Britt AF (2013) Disparity in disaster preparedness between racial/ethnic groups. Disaster Health 1(2):110–116. https://doi.org/10.4161/dish.27085

Bousquet S, Klas ME (2017) Millions of Floridians who fled Irma are eager to get home. Patience will be necessary. Miami Herald, 11 Sept 2017. http://www.miamiherald.com/news/weather/hurricane/article172603391.html

Brewer DD (2000) Forgetting in the recall-based elicitation of personal and social networks. Soc Networks 22:29–43. https://doi.org/10.1016/S0378-8733(99)00017-9

Collins JM, Ersing R, Polen A (2017) Evacuation decision-making during Hurricane Matthew: an assessment of the effects of social connections. Weather Clim Soc 9:769–776

Collins JM, Ersing R, Polen A, Saunders M, Senkbeil J (2018) The effects of social connections of evacuation decision-making during Hurricane Irma. Weather Clim Soc 10:459–469

Curley AM (2009) Draining or gaining? The social networks of public housing movers in Boston. J Soc Pers Relation 26(2–3):227–247

Cutter SL (2017) The forgotten casualties redux: women, children, and disaster risk. Glob Environ Chang 42:117–121

Cutter SL, Emrich CT (2006) Moral hazard, social catastrophe: the changing face of vulnerability along the hurricane coasts. Ann Am Acad Pol Soc Sci 604(1):102–112

Cutter SL, Barnes L, Berry M, Burton C, Evans E, Tate E, Webb J (2008) A place-based model for understanding community resilience to natural disasters. Glob Environ Chang 18:598–606

David E, Enarson E (eds) (2012) The women of Katrina: how gender, race and class matter in an American disaster. Vanderbilt University Press, Nashville

Davies IP, Haugo RD, Robertson JC, Levin PS (2018) The unequal vulnerability of communities of color to wildfire. PLoS One 13(11). https://doi.org/10.1371/journal.pone.0205825

Davlasheridze M, Miao Q (2021) Natural disasters, public housing, and the role of disaster aid. J Reg Sci 61:1113–1135

Drolet J, Ersing R, Dominelli L, Alston M, Mathbor G, Huang Y, Wu H (2018) Rebuilding lives and communities postdisaster: a case study on migrant workers and diversity in the USA. Aust Soc Work 71:1–13

Dyson ME (2007) Come hell or high water: Hurricane Katrina and the color of disaster. Civitas Books, New York

Ersing RL, Pearce C, Collins J, Saunders ME, Polen A (2020) Geophysical and social influences of evacuation decision-making: the case of Hurricane Irma. Atmosphere 11:851–870. https://doi.org/10.3390/atmos11080851

Etikan I, Musa SA, Alkassim RS (2016) Comparison of convenience sampling and purposive sampling. Am J Theor Appl Statist 5(1):1–4. https://doi.org/10.11648/j.ajtas.20160501.11

Fothergill A (2004) Heads above water: gender, class, and family in the Grand Forks flood. State University of New York Press, Albany

Fothergill A, Peek LA (2004) Poverty and disasters in the United States: a review of recent sociological findings. Nat Hazards 32:89–110
Granovetter MS (1973) The strength of weak ties. Am J Soc 78:1360–1380. https://doi.org/10.1086/225469
Hamideh S, Rongerude J (2018) Social vulnerability and participation in disaster recovery decisions: public housing in Galveston after Hurricane Ike. Natural Hazards 93(3):1629–1648
Hawkins RL, Maurer K (2010) Bonding, bridging and linking: how social capital operated in New Orleans following Hurricane Katrina. Br J Soc Work 40:1777–1793
Henrici JM, Helmuth AS, Ferenandes R (2010) Mounting losses: women and public housing after Hurricane Katrina (report no. D491). Institute for Women's Policy Research. https://iwpr.org/iwpr-general/mounting-losses-women-and-public-housing-after-hurricane-katrina/
Henrici J, Childers C, Shaw E (2015) Get to the bricks: The experiences of black women from New Orleans public housing after Hurricane Katrina (report no. D506). Institute for Women's Policy Research. http://citeseerx.ist.psu.edu/viewdoc/download?doi=10.1.1.696.6925&rep=rep1&type=pdf
Hillsborough County Board of County Commissioners (2019) Hillsborough County, Florida Evacuation Zones and Emergency Shelter Map. Retrieved from https://hillsborough.maps.arcgis.com/apps/webappviewer/index.html?id=04f1084467564dff88729f668caed40a
International Strategy for Disaster Reduction (ISDR) (2004) Living with risk: a global review of disaster reduction initiatives 11. United Nations, New York
Kleit RG (2001) Neighborhood relations in suburban scattered-site and clustered public housing. J Urban Aff 23(3–4):409–430
Laakso J (2013) Flawed policy assumptions and HOPE VI. J Poverty 17(1):29–46. https://doi.org/10.1080/10875549.2012.748000
Lindell C, Lu J, Prater CS (2005) Household decision making and evacuation in response to Hurricane Lili. Nat Hazards Rev 6:171–179. https://doi.org/10.1061/(ASCE)1527-6988(2005)6:4(171)
Morrow BH (1999) Identifying and mapping community vulnerability. Disasters 23(1):1–8
Neumayer E, Plümper T (2007) The gendered nature of natural disasters: the impact of catastrophic events on the gender gap in life expectancy, 1981–2002. Ann Assoc Am Geogr 97(3):551–566
Ngo EB (2001) When disasters and age collide: reviewing vulnerability of the elderly. Nat Hazards Rev 2(2):80–89
Norris FH (2006) Disaster research methods: past progress and future directions. J Trauma Stress 19:173–184. https://doi.org/10.1002/jts.20109
O'Donnell C (2017) Local housing authorities warn public housing residents to prepare for storm and heed evacuation orders. Tampa Bay Times. 7 Sept 2017. Retrieved from https://www.tampabay.com/news/tampa-housing-authority-issues-warnings-calls-on-public-housing-residents/2336613/
Peacock WG, Van Zandt S, Zhang Y, Highfield W (2014) Inequities in long-term housing recovery after disasters. J Am Plan Assoc 80(4):356–371
Peek LA, Fothergill A (2008) Displacement, gender, and the challenges of parenting after Hurricane Katrina. Natl Women's Studies Assoc J 20(3):69–105
Peek L, Fothergill A, Pardee JW, Weber L (2014) Studying displacement: new networks, lessons learned. Sociol Inq 84(3):354–359
Perry RW, Lindell MK (1991) The effects of ethnicity on evacuation decision-making. Int J Mass Emerg Disasters 9:47–68
Pickering CJ, Dancey M, Paik K, O'Sullivan T (2021) Informal caregiving and disaster risk reduction: a scoping review. Int J Disaster Risk Sci 12:169–187
ProPublica (2019) Tampa Housing Authority. Retrieved 10 Nov 2021 from https://projects.propublica.org/hud/owners/FL003
Sanders S, Bowie SL, Bowie YD (2004) Chapter 2: Lessons learned on forced relocation of older adults. J Gerontol Soc Work 40(4):23–35. https://doi.org/10.1300/J083v40n04_03

Simplican SC, Leader G, Kosciulek J, Leahy M (2015) Defining social inclusion of people with intellectual and developmental disabilities: an ecological model of social networks and community participation. Res Dev Disabil 38:18–29

Sorensen JH, Mileti D (1991) Risk communication for emergencies. In: Kasperson R, Stallen P (eds) Communicating risks to the public: international perspectives. Kluwer Academic, Boston, pp 369–394

Sorensen JH, Sorensen BV (2007) Community processes: warning and evacuation. In: Handbook of disaster research. Handbooks of sociology and social research. Springer, New York. https://doi.org/10.1007/978-0-387-32353-4_11

Squires G, Hartman C (2006) There is no such thing as a natural disaster: race, class and Hurricane Katrina. Routledge, New York

Stallings RA (ed) (2002) Methods of disaster research. Xlibris, New York

Stough LM, Kang D (2015) The Sendai framework for disaster risk reduction and persons with disabilities. Int J Disaster Risk Sci 6(2):140–149

Substance Abuse and Mental Health Services Administration (SAMHSA) (2017) Greater impact: how disasters affect people of low socioeconomic status (Disaster Technical Assistance Center Supplemental Research Bulletin, July 2017). U.S. Department of Health and Human Services. https://www.samhsa.gov/sites/default/files/dtac/srb-low-ses_2.pdf

Tampa Housing Authority (2021) Cultivating affordable housing while empowering people and communities. Tampa Housing Authority. Retrieved 10 Nov 2021 from https://www.thafl.com/Default.aspx

Twigg J, Kett M, Lovell E (2018) Disability inclusion and disaster risk reduction: Overcoming barriers to progress. (Briefing Note, July 2018). Overseas Development Institute (ODI). https://www.urban-response.org/system/files/content/resource/files/main/12324.pdf

Van der Gaag MJ, Snijders TB (2005) The resource generator: measurement of individual social capital with concrete items. Soc Netw 27:1–29

Vatsa K (2004) Risk, vulnerability, and asset-based approach to disaster risk management. Int J Sociol Soc Policy 24:1–48

Vigdor J (2008) The economic aftermath of Hurricane Katrina. J Econ Perspect 22(4):135–154. https://doi.org/10.1257/jep.22.4.135

Wisner B, Blaikie P, Cannon T, Davis I (2003) At risk: natural hazards, people's vulnerability and disasters, 2nd edn. Routledge, Abingdon-on-Thames

Zoraster RM (2010) Vulnerable populations: Hurricane Katrina as a case study. Prehosp Disaster Med 25(1):74–78

Chapter 14
Geohome: Resilient Housing for Climate Hazard Mitigation

George Elvin

Abstract The Geohome is a hurricane-resistant dwelling whose unique design draws from nature. Its wood frame emulates the structure of the coastal live oak; its water-, wind- and fire-resistant cement shell evokes the tubular shape of coastal seagrass; its protective window and door coverings embody the protective features of coastal bobcats; and it perches high on protective stilts like a great blue heron to avoid flood damage. It combines these nature-based resilient design features with the aim of improving residential building performance in hurricane conditions. Typical light-wood framing may be unsuited for coastal construction, as evidenced by the $24 billion in North Carolina property damage by Hurricane Florence. Having completed the design and construction of a partial prototype in 2020, our team is preparing to construct a computational model of the entire Geohome and test it using finite element analysisis. With nature as our teacher, we are confident that we can adapt our dwellings to our climate crisis so that people of all means can live safely and securely in the years ahead.

Keywords Hazard mitigation · Resilient design · Nature-based design · Biomimicry

14.1 Introduction

The Geohome project re-envisions light wood residential construction, maintaining its benefits of affordability, sustainability, and familiarity while increasing its resilience in the face of disaster. Its goal is to offer protection from hurricane- and tornado-force winds, elevation to avoid flooding, and an innovative enclosure system to reduce damage from wind forces. In addition, its unique framing geometry distributes loads more efficiently, reducing wood quantities and cost, and its monocoque cladding system protects the building frame, thereby reducing the amount of

G. Elvin (✉)
College of Design, North Carolina State University, Raleigh, NC, USA
e-mail: gelvin@ncsu.edu

wood required and its accompanying cost. In Summer 2021, a full-scale prototype of a portion of the Geohome was partially constructed at a facility near the campus of North Carolina State University. The facility was closed due to COVID-19 before construction could be completed. In Spring 2022, finite element analysis will be conducted on a computer-based model of the Geohome. Data gathered from this testing will help determine if the structure can provide enhanced climate hazard mitigation. While the construction of a physical prototype prior to the construction of a 3D computational model may seem counterintuitive, prototyping was, in this case, a significant part of the design process rather than an outcome of the computational model-testing phase. This reversal of conventional sequencing has helped to ensure that the next step in the project, computational model construction and testing, will be based on an improved design.

The Geohome is a novel approach to hurricane resilient design. The project is ongoing, so this chapter focuses on the design approach and initial findings, as well as evolving plans for the work. The results of the work will be discussed in forthcoming publications. Here, my aim is to describe the project in its early stage, to present the concept, and to motivate additional hurricane resilient design work across the broader community.

14.1.1 Project Objectives

- Develop a collection of nature-based design lessons for resilient, hurricane-resistant housing design (completed).
- Construct a full-scale partial prototype of the Geohome (completed).
- Conduct a finite element analysis on the computational model of the structure to measure its structural performance characteristics related to wind resistance.
- Analyze and disseminate the project results intended to contribute to the body of knowledge on climate hazard mitigation and resilient design.
- Consider the potential for commercial development of the Geohome (Fig. 14.1).

14.2 Architectural Responses to Hurricane Risk

In 2018, Hurricane Florence caused $24 billion in property damage in North Carolina alone. Because many of North Carolina's most vulnerable citizens occupy low-quality housing in areas prone to hurricanes and flooding, they were particularly hard-hit by this disaster. In 2020, a record 22 billion-dollar-plus climate disasters totaled over $95 billion in damage nationwide. Such losses are, however, not necessary. Instead, they suggest an extreme gap between current building design and the increasing power and frequency of climate disasters (US Global 2021). Buildings must not only adapt to this new, multi-hazard environment, but they must also be adaptive to the unpredictable effects of a changing climate in the future. The

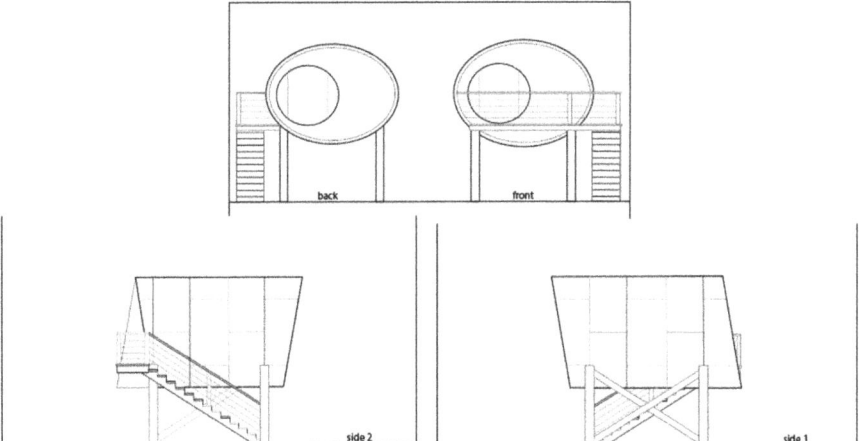

Fig. 14.1 The Geohome prototype

incremental change brought about by building code revisions is commendable but cannot keep pace with our changing climate and increasing disasters. As climate disasters intensify, it is all too likely that our building codes will fall further behind despite our best efforts to keep pace with an increasingly hazardous environment (Urbanek 2018).

With climate disasters in the United States alone costing over $100 billion per year, the need for resilient buildings capable of mitigating the effects of our climate crisis is urgent. Unfortunately, current buildings are not designed to withstand a 100-year flood every decade or a Category 5 hurricane every year. Moving forward, we face a choice. We can continue to build the way we have and face annual property damages in excess of $100 billion, we can pursue incremental change to building codes already lagging behind actual environmental conditions, or we can create a new kind of architecture adapted to the realities of life in the twenty-first Century (Urbanek 2018).

14.3 Learning from Nature

The Geohome research program is founded not on incremental change to building codes but on answering the fundamental question, "What would a disaster-proof building look like?" To answer this question, it looks to nature and how plants and animals adapt to extreme environments. It then applies nature's lessons to the design of buildings. Over the last decade, I have traveled to the world's hottest, wettest, windiest, and snowiest places trying to understand how plants and animals adapt to extreme environments. I have accumulated a collection of design principles based on my observations of these diverse environments.

14.4 Nature's Principles of Resilient Design

Resilience
Ability to adapt to change. Adaptability.

Regeneration
Renewal, restoration or replacement of components, relationships, and processes necessary to system health.

Efficiency
Minimal expenditure of energy for maximum achievement of or striving for goals.

Diversity
Rich variety of unique components, relationships and processes. Diversity of means to achieve goals.

Interdependence
Reliance of components, relationships and processes on each other for achievement of goals.

As a result, the Geohome draws design lessons from nature. Its framing emulates the structure of the coastal live oak; its water-, wind- and fire-resistant shell evokes the tubular shape of coastal seagrass and aquatic tusk shells, which can withstand water pressure even at 2000 m; its protective window and door coverings embody the eye structure of coastal bobcats; and it perches high on protective stilts like a great blue heron to avoid flood damage. The North Carolina coastal live oak, for example, entwines its roots and branches with others to create a storm-resistant shield where wildlife takes refuge from hurricanes. The Geohome also includes a unique root-like foundation system to distribute loads. These and other nature-based resilient design strategies may help the Geohome adapt to nature's forces rather than try to resist them. The result could be a new kind of building designed with nature-based solutions to withstand hurricanes, flooding, earthquakes, and wildfires (Table 14.1).

Table 14.1 Resilient plant and animal attributes and their architectural applications

Plant/animal	Attribute	Architectural application	Attribute
Coastal Live Oak	Entwines roots with other live oaks for increased wind and erosion resistance	Root-like foundation	Spreads out to resist hurricane-induced uplift
Seagrass	Hollow, tubular form with remarkable strength to weight and length ratio	Tubular building form	Reduces wind resistance
Bobcat	Third eyelid protects eye from dust storms	Pocket storm doors and windows	Protect glazing
Great Blue Heron	Long legs carry body over water	Raised pier foundations	Lift home above floodwaters
Tusk Shell	Tubular structure can withstand water pressure at 2000 m	Fiber-reinforced cement board cladding	Withstand hurricane-induced impacts

14.5 Theoretical Framework

The Geohome project combines these nature-based lessons in resilient design with the theoretical underpinnings of systems theory, human ecology, and environmental justice. Systems theory outlines a comprehensive method and cognitive framework for understanding complex systems (Bertalanffy 1969). It emphasizes the role of processes and relationships in understanding, analyzing, and modeling dynamic systems and synthesizing complex systems' environmental, social, and economic factors. It adopts an integrated human ecology approach to design to synthesize relevant aspects of human behavior, hazard analysis, and environmental design. Odum and Barrett (1994) have applied systems theory to the study of ecology, and their work heavily influences contemporary environmental studies. Human ecology targets the interrelationships between humans and their environment (Steiner 2002). However, a comprehensive systems approach has not been consistently applied to hazard mitigation research. The Geohome project is grounded in systems theory and human ecology to help ensure a comprehensive analysis of the social, environmental, and economic conditions at work in the coastal built environment.

Environmental justice – the principle that "all people and communities are entitled to equal protection of environmental and public health laws and regulations" – is a critical concern in the development of the Geohome as well (Bullard 1990). Socially vulnerable households often live in low-quality light wood frame housing that can sustain considerable damage during disasters. They also inhabit areas more prone to disasters. Environmental justice performance criteria are woven into the goals of the Geohome project and cross-referenced with other architectural and environmental performance criteria to help ensure that the project addresses the 17 Principles of Environmental Justice outlined by the First National People of Color Environmental Leadership Summit (Delegates 1991).

14.6 Project Research Methods

The project's research methods are grounded in environmental justice, data science, and computer simulation testing methods. Strategies for data collection, analysis, and synthesis are also grounded in whole building modeling techniques. Whole building modeling allows a comparison of a building's actual and expected performance in real time (Pang et al. 2012). Building performance simulation will be employed to gather data on computer-simulated building performance in the computer. This process of computational simulation seeks to create a realistic model of real-world building performance (Hensen and Lamberts 2012). Data will be collected through finite element analysis, then compared to standard building framing performance as documented by others. Experiemntal outcomes will test the primary research hypothesis that the structure can provide socially vulnerable households with safe, affordable housing less easily damaged during disasters.

Prior to the design and construction of the prototype, data on ecosystem resilience was collected onsite on the Outer Banks and Inner Banks of North Carolina. The purpose of this research was to learn how plants and animals adapt to extreme environments and then apply these nature-based lessons to the design of an innovative system for safer coastal housing. Once a satisfactory design was developed, a partial prototype structure was constructed at a facility in Raleigh. However, the facility was closed in August 2020 due to COVID-19, halting construction of the prototype. Nonetheless, valuable lessons were learned during the construction regarding the framing and structural shell of the building. Specifically, strengthening the framing was deemphasized in favor of developing a stronger, monocoque shell.

Once a computational model of the structure is constructed, finite element analysis will be conducted to guage system response to simulated hurricane forces. Finite element analysis employs engineering calculations to obtain information on the response of sytems to loads. Using finite element analysis to obtain such information will reduce the number of physical prototypes and experiments needed and produce test results in a measurable, mathematical format (Szabó and Babuška 1991).

14.6.1 Project Timeline

Fall 2019-present	Design prototype
Summer 2021	Construct prototype structural frame
Spring 2022	Construct digital 3D building model
Summer 2022	Collect data using finite element analysis
Fall 2022	Analyze data and synthesize results
Winter 2022	Document results
Spring 2023 and beyond	Disseminate results

14.7 Project Outcomes and Future Directions

Deliverables resulting from the project include a partial full-scale prototype for hurricane-resistant housing, a 3D computational model of the full Geohome, and test results from finite element analysis. Published results contributing to the body of knowledge on hazard mitigation and resilient design will also be produced. Testing will advance mitigation-related practice by determining the system's viability for further development and deployment in socially vulnerable communities often most affected by disasters.

Data collected from testing will help the research team refine the design, possibly leading to testing of a full-size, residential-scale building against the forces of wind, fire, and earthquakes. Such resilient structures could then play a role in improving building codes for a safer built environment and significantly reduced property damage. Looking forward, by partnering with industry and community leaders, we aim to create affordable, hurricane-resistant housing. Specifically, we anticipate that collaboration with the manufactured housing industry will facilitate market adoption of the resulting structures. With nature as our teacher, we are confident that we can adapt our dwellings to our climate so that people of all means can live safely and securely in North Carolina and beyond.

Acknowledgement The author would like to thank the Natural Hazards Center and North Carolina State University for partial funding of the project.

References

Bertalanffy L (1969) General system theory. George Brazilier, Inc, New York

Bullard RD (1990) Dumping in dixie: race, class, and environmental quality. Westview Press, Boulder

Delegates to the First National People of Color Environmental Leadership Summit (1991) 17 principles of environmental justice, Washington, DC

Hensen JLM, Lamberts R (eds) (2012) Building performance simulation for design and operation. Routledge, London

Odum EP, Barrett GW (1994) Fundamentals of ecology. Thompson Reuters, Egan

Pang X, Wetter M, Bhattacharya P, Haves P (2012) A framework for simulation-based real-time whole building performance assessment. Build Environ 54:100–108

Steiner F (2002) Human ecology. Island Press, Washington, DC

Szabó B, Babuška I (1991) Finite element analysis. Wiley, Hoboken

U.S. Global Change Research Program (2021) Fourth national climate assessment. Washington, DC. https://nca2018.globalchange.gov/chapter/2/

Urbanek L (2018) The climate is changing. So why aren't state building codes?, Natural Resources Defense Council, 4 Apr 2018. https://www.nrdc.org/experts/lauren-urbanek/climate-changing-why-arent-state-building-codes

GPSR Compliance
The European Union's (EU) General Product Safety Regulation (GPSR) is a set of rules that requires consumer products to be safe and our obligations to ensure this.

If you have any concerns about our products, you can contact us on

ProductSafety@springernature.com

In case Publisher is established outside the EU, the EU authorized representative is:

Springer Nature Customer Service Center GmbH
Europaplatz 3
69115 Heidelberg, Germany

www.ingramcontent.com/pod-product-compliance
Ingram Content Group UK Ltd.
Pitfield, Milton Keynes, MK11 3LW, UK
UKHW021250180426
11946UKWH00003B/51